Praise for
The New Climate War

"Fossil fuel companies have been, for decades longer than I have been alive, the largest contributors to the climate crisis that affects my generation today—all in pursuit of profits and growth. In *The New Climate War*, Michael Mann holds them to account, and shows us how we can take the bold steps we must all take together to win the battle to save this planet."

—GRETA THUNBERG, climate activist

"This book takes the reader behind the front lines into the decades long information war waged by the fossil fuel industry and those that share their interests. From his perspective as a leader in the battle for scientific reason, Michael Mann provides hope and a roadmap for all of us to address the systemic issues fueling climate change, and shows how we can come together to wage a new war in the fight for our future."

—LEONARDO DICAPRIO, actor and environmentalist

"Few people bear more scars from the climate wars than Michael Mann—and few have fought longer and harder for a basic, rational approach to dealing with this greatest of crises. Because of his persistence—and that of so many others—we are finally making progress!"

—BILL MCKIBBEN, author of *Falter: Has the Human Game Begun to Play Itself Out?*

"Michael Mann skillfully explains the complicated dynamics of global warming and vividly portrays the sophisticated and coordinated campaign by polluters to block the policies and solutions needed to solve the climate crisis. And most importantly, he proposes a path forward that is both realistic and optimistic, and should inspire readers to take action."

—former US vice president AL GORE

"With this book, Michael Mann details the challenges we face from enemies ('inactivists') both without and within while dropping critically important breadcrumbs for us to follow to lead us out of the forest of despair and set us on the path of victory in a battle we *must* win. We need an army of Michaels, stat!"

—DON CHEADLE, actor, activist,
and UN global goodwill ambassador

"Pulling no punches, Michael Mann lays out our predicament and tells the shocking story of persistent climate denial and corporate deception. We are in a war for the planet, but one we are now on the verge of winning. And he deftly cuts through the propaganda and shows us the path forward."

—JERRY BROWN, California governor, 1975–1983, 2011–2019

"Mann shows that corporations and lobbyists have been successful in convincing us that climate change will be fine, if we just recycle our bottles and turn out the lights. Instead, he says, global warming is a problem way too hot for any one person to handle. He's optimistic though, because he sees what we really can and will do. Read his book, and let's get to work."

—BILL NYE, science educator and CEO, the Planetary Society

"For over two decades, Michael Mann has been our Janus at the gates, defending climate science from corporate-funded insinuations of confusion and suspicion. We would not have progressed this far had it not been for his unflinching and brilliant rejoinders to the traffickers of doubt. This chronicle of ongoing climate injustice may make you mad, but hopefully it will make us act. This is the only civilization we have. Mann is its resolute champion once again."

—PAUL HAWKEN, founder, Project Drawdown

"*The New Climate War* is an insightful treatise on how the polluting fossil fuel industry and their right-wing allies have deflected the blame for the climate crisis. The book charts a common-sense course for collective actions to force government and corporations to make real solutions to the climate crisis—an existential threat to humanity and the planet."

—ROBERT D. BULLARD, professor of urban planning and environmental policy at Texas Southern University

The New Climate War is engaging, approachable, and ultimately deeply uplifting. Mann outlines a hopeful vision of the transformation we must undertake in order to create a better, brighter future on this planet. He makes the clear case that our species is capable of great change, laying out exactly why and how we can rise to overcome the grave challenges before us."

—SASHA SAGAN, author of *For Small Creatures Such As We*

"A fascinating journey through the minds and motivations of the champions of climate denialism as well as the more recent climate doomists. Along the way, we learn of the unequivocal scientific evidence and the rapid evolution of technological solutions. Most importantly, public opinion finally seems to be at a 'tipping point' to catalyze political will to leave the next generation a sustainable world—and not a moment too soon!"

—ROSINA BIERBAUM, professor, University of Michigan and University of Maryland, and former Acting Director of OSTP

"Blunt, lucid. . . . Consistently displaying his comprehensive command of climate science and the attendant politics. . . . An expert effectively debunks the false narrative of denialism and advocates communal resistance to fossil fuels."

—*Kirkus*

THE NEW CLIMATE WAR

THE NEW CLIMATE WAR

The Fight to Take Back Our Planet

MICHAEL E. MANN

PUBLICAFFAIRS

NEW YORK

PublicAffairs
Hachette Book Group
1290 Avenue of the Americas, New York, NY 10104
www.publicaffairsbooks.com
@Public_Affairs

Printed in the United States of America

Originally published in hardcover and ebook by PublicAffairs in January 2021
First Trade Paperback Edition: May 2022

Published by PublicAffairs, an imprint of Perseus Books, LLC, a subsidiary
of Hachette Book Group, Inc. The PublicAffairs name and logo is a trademark
of the Hachette Book Group.

The Hachette Speakers Bureau provides a wide range of authors for
speaking events. To find out more, go to www.hachettespeakersbureau.com
or call (866) 376-6591.

The publisher is not responsible for websites (or their content) that are not
owned by the publisher.

Print book interior design by Linda Mark

Library of Congress Cataloging-in-Publication Data
Names: Mann, Michael E., 1965– author.
Title: The new climate war : the fight to take back our planet / Michael E. Mann.
Description: First edition. | New York : PublicAffairs, 2021. | Includes
  bibliographical references and index.
Identifiers: LCCN 2020027822 | ISBN 9781541758230 (hardcover) |
  ISBN 9781541758223 (ebook)
Subjects: LCSH: Environmental policy—Citizen participation. | Climatic
  changes—Government policy—Citizen participation. | Green New Deal.
Classification: LCC GE170 .M365 2021 | DDC 363.738/74—dc23
LC record available at https://lccn.loc.gov/2020027822
ISBNs: 978-1-5417-5823-0 (hardcover), 978-1-5417-5822-3 (ebook),
  978-1-5417-0057-4 (int'l), 978-1-5417-5821-6 (paperback)

LSC-C

Printing 1, 2022

Michael Mann dedicates this book to his wife, Lorraine Santy, and daughter, Megan Dorothy Mann, and to the memory of his brother Jonathan Clifford Mann and mother, Paula Finesod Mann

Contents

THE NEW CLIMATE WAR

Introduction

"There is general scientific agreement that the most likely manner in which mankind is influencing the global climate is through carbon dioxide release from the burning of fossil fuels. . . . There are some potentially catastrophic events that must be considered. . . . Rainfall might get heavier in some regions, and other places might turn to desert. . . . [Some countries] would have their agricultural output reduced or destroyed. . . . Man has a time window of five to ten years before the need for hard decisions regarding changes in energy strategies might become critical. . . . Once the effects are measurable, they might not be reversible."

YOU MIGHT BE FORGIVEN FOR ASSUMING THOSE PROPHETIC words were spoken by Al Gore in the mid-1990s. No, they were the words of fossil fuel giant ExxonMobil senior scientist James F. Black in recently unearthed internal documents from the 1970s.[1] In the decades since, instead of heeding the warnings of its own scientists, ExxonMobil and other fossil fuel interests waged a public relations campaign contesting the scientific evidence and doing everything in their power to block policies aimed at curbing planet-warming carbon pollution.

As a result, our planet has now warmed into the danger zone, and we are not yet taking the measures necessary to avert the largest global crisis we have ever faced. We are in a war—but before we engage we must first understand the mind of the enemy. What evolving tactics are the forces of denial and delay employing today in their efforts to stymie climate action? How might we combat this shape-shifting Leviathan? Is it too late? Can we still avert catastrophic global climate change? These are all questions to which we deserve answers, and in the pages ahead, we'll find them.

Our story starts nearly a century ago, when the original denial and delay playbooks were first written. It turns out, the fossil fuel industry learned from the worst.[2] The gun lobby's motto—that "Guns Don't Kill People, People Kill People"—dates back to the 1920s. A textbook example of dangerous deflection, it diverts attention away from the problem of easy access to assault weapons and toward other purported contributors to mass shootings, such as mental illness or media depictions of violence.

The tobacco industry took a similar tack, seeking to discredit the linkage between cigarettes and lung cancer even as its own internal research, dating back to the 1950s, demonstrated the deadly and addictive nature of its product. "Doubt is our Product" read one of the Brown & Williamson tobacco company's internal memos.

Then there's the now iconic "Crying Indian" ad. Some readers may recall the commercial from the early 1970s. Featuring a tearful Indian named "Iron Eyes Cody," it alerted viewers to the accumulating bottle and can waste littering our countryside. The ad, however, wasn't quite what it appeared to be on the surface. A bit of sleuthing reveals that it was actually the centerpiece of a massive deflection campaign engineered by the beverage industry, which sought to point the finger at us, rather than corporations, emphasizing individual responsibility over collective action and governmental regulation. As a result, the global environmental threat of plastic pollution is still with us, a problem that has reached such crisis proportions that plastic waste has now penetrated to the deepest part of the world's oceans.

Finally, we get to the fossil fuel industry. Joined by billionaire plutocrats like the Koch brothers, the Mercers, and the Scaifes, companies such as ExxonMobil funneled billions of dollars into a disinformation campaign beginning in the late 1980s, working to discredit the science behind human-caused climate change and its linkage with fossil fuel burning. This science denial took precedence even as ExxonMobil's own team of scientists concluded that the impacts of continued fossil fuel use could lead to "devastating" climate-change impacts.

And the scientists were right. Decades later, thanks to that campaign, we are now witnessing the devastating effects of unchecked climate change. We see them playing out in the daily news cycle, on our television screens, in our newspaper headlines, and in our social media feeds. Coastal inundation, withering heat waves and droughts, devastating floods, raging wildfires: *this* is the face of dangerous climate change. It's a face that we increasingly recognize.

As a consequence, the forces of denial and delay—the fossil fuel companies, right-wing plutocrats, and oil-funded governments that continue to profit from our dependence on fossil fuels—can no longer insist, with a straight face, that nothing is happening. Outright denial of the physical evidence of climate change simply isn't credible anymore. So they have shifted to a softer form of denialism while keeping the oil flowing and fossil fuels burning, engaging in a multipronged offensive based on deception, distraction, and delay. This is the *new climate war*, and the planet is losing.

The enemy has masterfully executed a deflection campaign—inspired by those of the gun lobby, the tobacco industry, and beverage companies—aimed at shifting responsibility from corporations to individuals. Personal actions, from going vegan to avoiding flying, are increasingly touted as the primary solution to the climate crisis. Though these actions are worth taking, a fixation on voluntary action alone takes the pressure off of the push for governmental policies to hold corporate polluters accountable. In fact, one recent study suggests that the emphasis on small personal actions can actually undermine support for the substantive climate policies needed.[3] That's quite convenient for fossil fuel companies like ExxonMobil, Shell, and BP, which continue to make record profits every day that we remain, to quote former president George W. Bush, "addicted to fossil fuels."

The deflection campaign also provides an opportunity for the enemy to employ a "wedge" strategy dividing the climate advocacy community, exploiting a preexisting rift between climate advocates more focused on individual action and those emphasizing collective and policy action.

Using online bots and trolls, manipulating social media and Internet search engines, the enemy has deployed the sort of cyber-weaponry honed during the 2016 US presidential election. They are the same tactics that gave us a climate-change-denying US president in Donald Trump. Malice, hatred, jealousy, fear, rage, bigotry, all of the most base, reptilian brain impulses—corporate polluters and their allies have waged a campaign to tap into all of that, seeking to sow division within the climate movement while generating fear and outrage on the part of their "base"—the disaffected right.

Meanwhile, these forces of inaction have effectively opposed measures to regulate or price carbon emissions, attacked viable alternatives like renewable energy, and advocated instead false solutions, such as coal burning with carbon capture, or unproven and potentially dangerous "geoengineering" schemes that involve massive manipulation of our planetary environment. Hypothetical future "innovations," the argument goes, will somehow save us, so there's no need for any current policy intervention. We can just throw a few dollars at "managing" the risks while we continue to pollute.

With climate progress sidelined by the Trump administration's dismantling of climate-friendly Environmental Protection Agency (EPA) policies such as the Clean Power Plan, along with its rollbacks in regulations on pollutants, its greenlighting of oil and gas pipelines, its direct handouts to a struggling coal industry, and its cheap leases to drill on public lands, the fossil fuel industry has enjoyed free rein to expand its polluting enterprise.

The enemy is also employing PSYOP in its war on climate action. It has promoted the narrative that climate-change impacts will be mild, innocuous, and easily adapted to, undermining any sense of *urgency*, while at the same time promoting the inevitability of climate change to dampen any sense of *agency*. This effort has been aided and abetted by individuals who are ostensible climate champions but have portrayed catastrophe as a *fait accompli*, either by overstating the damage to which we are already committed, by dismissing the possibility of mobilizing the action necessary to avert disaster, or by setting the standard so high (say, the very overthrow of market economics itself, that old chestnut) that any action seems doomed to failure. The enemy has been more than happy to amplify such notions.

But all is not lost. In this book, I aim to debunk false narratives that have derailed attempts to curb climate change and arm readers with a real path forward to preserving our planet. Our civilization can be saved, but only if we learn to recognize the current tactics of the enemy—that is, the forces of inaction—and how to combat them.

My decades of experience on the front lines of the battle to communicate the science of climate change and its implications have provided me with some unique insights. The "hockey stick" is the name that was given to a curve my colleagues and I published in 1998 demonstrating the steep uptick in planetary temperatures over the past century.[4] The graph achieved iconic status in the climate-change debate because it told a simple story, namely, that we were causing unprecedented warming of the planet by burning fossil fuels and pumping greenhouse gases into the atmosphere. Decades later, the hockey-stick curve is still attacked despite the many studies that have not only reaffirmed but extended our findings. Why? Because it remains a threat to vested interests.

The attacks on the hockey stick in the late 1990s drew me—then a young scientist—into the fray. In the process of defending myself and my work from politically motivated attacks, I became a reluctant and involuntary combatant in the climate wars. I've seen the enemy up close, in battle, for two decades now. I know how it operates and what tactics it uses. And I've been monitoring the dramatic shifts in those tactics over the past few years in response to the changing nature of the battlefield. I have adapted to those shifting tactics, changing how I engage the public and policymakers in my own efforts to inform and impact the public discourse. It is my intent, in this book, to share with you what I've learned, and to engage you, too, as a willing soldier in this battle to save our planet from a climate crisis before it is too late.

Here's the four point battle plan, which we'll return to at the end of the book:

Disregard the Doomsayers: The misguided belief that "it's too late" to act has been co-opted by fossil fuel interests and those advocating for them. It's just another way of legitimizing business-as-usual and a continued reliance on fossil fuels. We must reject the overt doom and gloom that we increasingly encounter in today's climate discourse.

A Child Shall Lead Them: The youngest generation is fighting tooth and nail to save their planet, and there is a moral authority and clarity in their message that none but the most jaded ears can fail to hear. They are the game-changers that climate advocates have been waiting for. We should model our actions after theirs and learn from their methods and their idealism.

Educate, Educate, Educate: Most hard-core climate-change deniers are unmovable. They view climate change through the prism of right-wing ideology and are impervious to facts. Don't waste your time and effort trying to convince them. But there are many honest, confused folks out there who are caught in the crossfire, victims of the climate-change disinformation campaign. We must help them out. Then they will be in a position to join us in battle.

Changing the System Requires Systemic Change: The fossil fuel disinformation machine wants to make it about the car you choose to drive, the food you choose to eat, and the lifestyle you choose to live rather than about the larger system and incentives. We need policies that will incentivize the needed shift away from fossil fuel burning toward a clean, green global economy. So-called leaders who resist the call for action must be removed from office.

It is easy to become overwhelmed by the scale of the challenge ahead of us. Change is always hard, and we are being asked to make a journey into an unfamiliar future. It is understandable to feel paralyzed with fear at the prospect of our planet's degradation. It's not surprising that anxiety and fear abound when it comes to the climate crisis and our efforts to deal with it.

We must understand, though, that the forces of denial and delay are using our fear and anxiety against us so we remain like deer in the headlights. I have colleagues who have expressed discomfort in framing our predicament as a "war." But, as I tell them, the surest way to lose a war is to refuse to recognize you're in one in the first place.[5] Whether we like it or not, and though clearly not of our own choosing, that's precisely where we find ourselves when it comes to the industry-funded effort to block action on climate.

So we must be brave and find the strength to fight on, channeling that fear and anxiety into motivation and action. The stakes are simply too great.

As we continue to explore the cosmos, we are finding other planetary systems, some with planets that are even somewhat Earth-like in character. Some are similar in size to ours, and roughly the right distance from their star to reside in the so-called "habitable zone." Some may harbor liquid water, an ingredient that is likely essential for life. Yet we have still not encountered any evidence of life elsewhere in our solar system, our galaxy, or indeed the entire universe. Life appears to be very rare indeed, complex life even more so. And intelligent life? We may, at least for all intents and purposes, be alone. Just us drifting aboard this "Spaceship Earth." No other place to dock, no alternative ports at which to sojourn, with air to breathe, water to drink, or food to consume.

We are the custodians of an amazing gift. We have a Goldilocks planet, with just the right atmospheric composition, just the right distance from its star, yielding just the right temperature range for life, with liquid-water oceans and oxygen-rich air. Every person we will ever know, every animal or plant we will ever encounter, is reliant on conditions remaining just this way.

To continue to knowingly alter those conditions in a manner that threatens humanity and other life forms, simply so a few very large corporations can continue to make record profits, is not just unacceptable, or unethical—it would be the most immoral act in the history of human civilization: not just a crime against humanity, but a crime against our planet. We cannot be passive bystanders as polluters work toward making that eventuality come to pass. My intent with this book is to do everything within my power to make sure we aren't.

CHAPTER 1

The Architects of Misinformation and Misdirection

Doubt is our product, since it is the best means of competing with the "body of fact" that exists in the minds of the general public.
    —Unnamed tobacco executive, Brown and Williamson (1969)

THE ORIGINS OF THE ONGOING CLIMATE WARS LIE IN DISINFOR-mation campaigns waged decades ago, when the findings of science began to collide with the agendas of powerful vested interests. These campaigns were aimed at obscuring public understanding of the underlying science and discrediting the scientific message, often by attacking the messengers themselves—that is, the scientists whose work hinted that we might have a problem on our hands. Over the years, tactics were developed and refined by public relations agents employed to undermine facts and scientifically based warnings.

KILL THE MESSENGER

Our journey takes us all the way back to the late nineteenth century, to Thomas Stockmann, an amateur scientist in a small Norwegian town. The local economy was dependent on tourism tied to the town's medicinal hot springs. After discovering that the town's water supply was being polluted by chemicals from a local tannery, Stockmann was thwarted in his efforts to alert the townspeople of the threat, first when the local paper refused to publish an article he had written about his findings, then when he was shouted down as he attempted to announce his findings at a town meeting. He and his family were treated as outcasts. His daughter was expelled from school, and the townspeople stoned his home, breaking all the windows and terrifying his family. They considered leaving town but decided to stay, hoping—in vain—that the townspeople would ultimately come around to accepting, and indeed appreciating, his dire warnings.

That's the plot of the 1882 Henrik Ibsen play *An Enemy of the People* (made into a film in 1978 that starred Steve McQueen in one of his final and arguably finest performances). The story is fictional, but it depicts a conflict that would be familiar to audiences in the late nineteenth century. The eerie prescience of this tale today, when an anti-science president dismisses the media as an "enemy of the American people," and conservative politicians knowingly allow an entire city to be endangered by a lead-poisoned water supply, has not been lost on some observers.[1] *An Enemy of the People* is the canonical cautionary tale of the clash between science and industrial or corporate interests. And it serves as an apt metaphor for the climate wars that would take place a century later.

But before we get there, let us next flash-forward to the mid-twentieth century, where we encounter the granddaddy of modern industry disinformation campaigns. This campaign was orchestrated by tobacco industry leaders in their effort to hide evidence of the addictive and deadly nature of their product. "Doubt is our product," confessed a Brown and Williamson executive in 1969.[2] The memo containing the admission was eventually released as part of a massive legal settlement between the tobacco industry and the US government. This and other internal documents showed that the companies' own scientists had established the health threats of smoking as early as the 1950s. Nevertheless, the companies chose to engage in an elaborate campaign to hide those threats from the public.

Tobacco interests even hired experts to discredit the work of other researchers who had arrived at the very same conclusions. Chief among these attack dogs was Frederick Seitz, a solid-state physicist who was also the former head of the US National Academy of Sciences and a recipient of the prestigious Presidential Medal of Science. Those impressive credentials made him a valuable asset for the tobacco industry. Tobacco giant R.J. Reynolds would eventually hire Seitz and pay him half a million dollars to use his scientific standing and stature to attack any and all science (and scientists) linking tobacco to human health problems.[3] Seitz was the original science-denier-for-hire. There would be many more.

Pesticide manufacturers adopted the tobacco industry's playbook in the 1960s, after Rachel Carson warned the public of the danger that DDT (dichlorodiphenyltrichloroethane) posed to the environment. Her classic 1962 book *Silent Spring* ushered in the modern environmental movement.[4] Carson described how DDT was decimating populations of bald eagles and other birds by thinning their eggs and killing the embryos within. The pesticide was accumulating in food webs, soils, and rivers, creating an increasingly dire threat to wildlife—and ultimately, humans. Eventually, the United States banned DDT, but not until 1972.

Carson was awarded for her efforts with a full-on character assassination campaign by industry groups who denounced her as "radical," "communist," and "hysterical" (with all its misogynist connotations—misogyny, and racism as well, as we will see, have become inextricably linked to climate-change denialism). The president of Monsanto, the largest producer of DDT, denounced her as "a fanatic defender of the cult of the balance of nature."[5] Her critics even labeled her a mass murderer.[6] Even today, the industry front group known as the Competitive Enterprise Institute (CEI) continues to defame the long-deceased scientist by insisting that "millions of people around the world suffer the painful and often deadly effects of malaria because one person sounded a false alarm. That person is Rachel Carson."[7] What Carson's posthumous attackers don't want you to know is that Carson never called for a ban on DDT, just an end to its indiscriminate use. It was ultimately phased out not because of the environmental damages that Carson exposed but because it had steadily lost its effectiveness as mosquitoes grew resistant to it. That was something that Carson, ironically, had warned would happen as a result of overuse.[8] And here we are thus afforded an early example of how the short-sighted practices of greedy corporations looking to maximize near-term profits often prove self-defeating.

Credibility and integrity are a scientist's bread and butter and greatest asset. It is the currency that allows scientists to serve as trusted communicators to the public. That's why the forces of denial targeted Carson directly, accusing her of all manner of scientific misconduct. In response to the controversy, President John F. Kennedy convened a committee to review Carson's claims. The committee published its report in May 1963, exonerating her and her scientific findings.[9] Science denialists are never deterred by pesky things like "facts," however. And so the attacks continue today. Consider a 2012 commentary that appeared in conservative *Forbes* magazine entitled "Rachel Carson's Deadly Fantasies," by Henry I. Miller and Gregory Conko. Miller and Conko are Fellows at the aforementioned Competitive Enterprise Institute. Miller is also a scientific advisory board member of an industry front group known as the George C. Marshall Institute (GMI), and, unsurprisingly, a tobacco industry advocate.[10] In the piece, they accuse Carson of "gross misrepresentations," "atrocious" scholarship, and "egregious academic misconduct," despite the fact that her scientific findings have been overwhelmingly affirmed by decades of research.[11] Though bird populations continue to be imperiled by pesticides, more sonorous springs did largely return. And for that, we owe a great debt of gratitude to Rachel Carson.[12]

Due to the work of Carson and other scientists studying the effects of industrial toxins on humans and the environment, awareness of other threats emerged in the 1970s. Lead pollution generated by the gasoline and paint industries, for example, came under scrutiny. Enter Herbert Needleman, whose story is disturbingly reminiscent of Thomas Stockmann's from Ibsen's play. Needleman was a professor and researcher at the University of Pittsburgh School of Medicine. His research identified a link between environmental lead contamination and childhood brain development. Sounding a familiar note, lead industry advocates sought to discredit him and his research, engaging in a character assassination campaign that included unfounded accusations against him of scientific misconduct.[13] He was exonerated—*twice*. The first exoneration was the result of a thorough investigation by the National Institutes of Health. Then, in what might sound like the scientific equivalent of double jeopardy, there was a separate investigation by his university, during which he was locked out of his own files, with bars placed on his file cabinets. No evidence of impropriety ever emerged. His research on how to detect chronic lead exposure—validated by numerous independent studies in the intervening decades—likely has saved thousands of lives and prevented brain damage in thousands more.[14] "Enemy of the People" indeed.

DENIAL GOES GLOBAL

In the 1970s and 1980s we begin to see the emergence of truly *global* environmental threats, including acid rain and ozone depletion. Industry groups whose bottom line might be impacted by environmental regulations began to significantly step up their attacks on the science demonstrating these dangers, and of course on the scientists themselves.

Frederick Seitz—the granddaddy of denialism who was enlisted by the tobacco industry in its war on science—was provided lavish industry funding in the mid-1980s to create the George C. Marshall Institute.[15] Seitz recruited as partners astrophysicist Robert Jastrow (founder of the venerable NASA Goddard Institute for Space Studies) and oceanographer William Nierenberg (onetime director of the revered Scripps Institution for Oceanography in La Jolla, California). These three individuals, as Naomi Oreskes and Erik M. Conway noted in their 2010 book *Merchants of Doubt*, were what could be called *free-market fundamentalists*. None of them had training in environmental science. What they *did* possess was an ideological distrust of efforts to limit what they saw as the freedom of individuals or corporations. As such, they played willfully into the agenda of regulation-averse special interests.[16] Borrowing from the very same tactics Seitz had cut his teeth on as a tobacco industry attack dog a decade earlier, the GMI crew would sow doubt in the areas of science that proved threatening to the powerful vested interests they represented.

One of these scientific issues was acid rain, a phenomenon I'm intimately familiar with, having grown up in New England during the 1970s. At that time, lakes, rivers, streams, and forests throughout eastern North America were being destroyed by increasingly acidic rainfall. The scientist Gene Likens and others discovered the origins of the problem: midwestern coal-fired power plants that were producing sulfur dioxide pollution. Likens would later become the "environmental sustainability czar" for the University of Connecticut.

In April 2017, I gave a lecture at the University of Connecticut in which I revealed some of my own experiences in the crosshairs of the climate-change-denial machine. At the dinner following the lecture, Likens was seated next to me. He turned to me and said, "Your stories sound a lot like mine!" As we ate our salads, he regaled me with stories that were disturbingly familiar: nasty letters and complaints to his bosses; hostile reception by conservative politicians; attacks from industry-funded hatchet men and politicians seeking to discredit his scientific findings. As Likens said some years ago in an interview, "It was bad. It was really nasty. I had a *contract* put out on me."

Likens was referring to a coal industry trade group known as the Edison Electric Institute that had offered nearly *half a million dollars* to anybody willing to discredit him.[17] William Nierenberg, the aforementioned member of the GMI trio, in essence took up that challenge when Ronald Reagan appointed him to chair a panel investigating the acid rain issue. The facts, however, proved stubborn, and the panel's conclusions, published in a 1984 report, largely reaffirmed the findings of Likens and other scientific experts. But hidden away in an appendix written by a contrarian scientist, S. Fred Singer, was a passage suggesting that, as Oreskes and Conway put it, "we really *didn't* know enough to move forward with emissions controls." The passage was just dismissive enough to allow the Reagan administration to justify its policy of inaction.[18]

Fortunately, the forces of denial and inaction did not prevail. Americans recognized the problem and demanded action, and politicians ultimately responded. That's precisely how things are supposed to work in a representative democracy. In 1990, it was a Republican president, George H.W. Bush, who signed the Clean Air Act, which required coal-fired power plants to scrub sulfur emissions before they exited the smokestacks. He even introduced a vehicle known as "cap and trade," a market-based mechanism that allows polluters to buy and sell a limited allotment of pollution permits. Cap-and-trade policy is, ironically, now pilloried by most Republicans. It was the brainchild of Bush's EPA administrator, William K. Reilly, a modern environmental hero whom I'm proud to know and call my friend.

My family frequently goes on vacation to Big Moose Lake in the western Adirondacks. My wife's family has been going there for seventy years. Her parents remember back in the 1970s when the lake was so acidic you literally didn't need to take any showers. A jump in the lake would clean you right off. The waters were crystal clear, because they were lifeless. The wildlife has returned now—I see and hear it when we're there, from the bugs to the fish and frogs to the ducks and snapping turtles, along with the haunting sound of the loons. You sometimes see small teams of scientists out in boats collecting samples of the water in the various lakes, examining its chemistry and contents. The affected ecosystems still haven't recovered completely. Environmental pollution can disrupt food chains, forest ecosystems, and water and soil chemistry in a way that can persist for decades or centuries even after the pollutants themselves are gone. But we are on the road to recovery in the Adirondacks, thanks—dare I say it—to *market-based* mechanisms for solving an environmental problem.

In the 1980s, scientists recognized that chlorofluorocarbons (CFCs), used at the time in spray cans and refrigerators, were responsible for the growing hole in the ozone layer in the lower stratosphere that protects us from damaging, high-energy ultraviolet radiation from the Sun. The erosion of the ozone layer brought with it an increasing incidence of skin cancer and other adverse health impacts in the Southern Hemisphere. My friend Bill Brune, former head of the Department of Meteorology at Penn State, was one of the original scientists researching the relevant atmospheric chemistry. As he has written, "Some of the scientists who carried out this seminal research decided to become advocates for action to mitigate the likely harm from a depleted ozone layer. These scientist-advocates were subjected to intense criticism."[19] That criticism, as Bill noted, took several forms: "Manufacturers, users, and their government representatives initiated public relations campaigns designed not to illuminate but to obscure, to throw doubt on the hypothesis and the weight of scientific evidence, and to otherwise convince lawmakers and the public that the data were too uncertain to act upon." He added, "When results inevitably began to refute their views, or whenever their own work was proven wrong or rejected for publication, these contrarian scientists, government representatives, and industry spokesmen then changed tactics, to denigrate the entire peer-review process." Among those contrarian scientists was the very same S. Fred Singer we encountered in the context of acid rain denial. Get used to that name.

Disregarding the naysayers, in 1987 forty-six countries—including the United States under Reagan—signed the Montreal Protocol, banning the production of CFCs. Since then, the ozone hole has shrunk to its smallest extent in decades. Environmental policy *actually* works. But, with both acid rain and ozone depletion, policy solutions came only because of unrelenting pressure on policymakers by citizens combined with continued bipartisan good faith and support on the part of politicians for systemic solutions to environmental threats. That good faith all but disappeared with the advent of the Trump administration. Indeed, after his 2016 election, Trump appointed individuals to important positions who not only denied the reality and threat of climate change but had played critical roles decades ago in industry-led efforts to deny both ozone depletion and acid rain. Think of them as all-purpose deniers-for-hire.[20]

You might also call them spiritual successors of the George C. Marshall Institute, Frederick Seitz's science-denying think tank. By the late 1980s, the GMI was largely focused on environmental issues. But as it happens, it was not acid rain or ozone depletion that brought the institute into existence in the first place. It was instead the threat that the findings of science posed to an entirely different vested interest: the military-industrial complex. During the late cold war, leading defense contractors, such as Lockheed-Martin and Northrop Grumman, were profiting from the escalating arms race between the United States and the Soviet Union. They stood to benefit in particular from Reagan's proposed Strategic Defense Initiative, otherwise known as Star Wars, an antiballistic missile program designed to shoot down nuclear missiles in space. Standing in their way, however, was, quite literally, one lone scientist.

SCIENTIST AS WARRIOR

Carl Sagan was the David Duncan Professor of Astronomy and Space Sciences and director of the Laboratory for Planetary Studies at Cornell University. He was a respected, accomplished researcher with an impressive record of achievement in earth and planetary science. Sagan did seminal work on the "Faint Young Sun Paradox," the surprising fact that Earth was habitable more than three billion years ago despite the fact that the Sun was 30 percent dimmer then. The explanation, Sagan realized, must be a magnified greenhouse effect. This work is so fundamental that it constitutes the first chapter in the textbook I've used to teach first-year Penn State students about Earth history.[21]

Sagan, however, was far more than a scientist. He was cultural phenomenon. He had an unmatched ability to engage the public with science. Not only could he explain it to the person on the street, he could get people excited about it. I can speak to this matter on a personal level. It is Carl Sagan who inspired *me* to pursue a career in science.

I had always had an aptitude for math and science, but it had constituted a path of least resistance, not a passion. Then Sagan's popular PBS series *Cosmos* premiered at the start of my freshman year in high school. Sagan showed me the magic of scientific inquiry. He revealed a cosmos that was more wondrous than I could have imagined, and the preciousness of our place in it as simple inhabitants of a tiny blue dot just barely discernible from the outer reaches of our solar system. And the questions! How did life form? Is there more of it out there? Are there other intelligent civilizations? Why haven't they contacted us? I pondered these questions and so many more that Sagan raised in the epic thirteen-part series. Sagan made me realize it was possible to spend a lifetime satisfying one's scientific curiosity by posing and answering such fundamental existential questions.

Sadly, I never got a chance to meet my hero. I finished my PhD in geology and geophysics in 1996, the very same year Sagan passed away. Being in the same field as Sagan, I almost certainly would have met him at meetings or conferences had I entered the profession just a few years earlier. But I have had the pleasure of getting to know him through his writings, and to make the acquaintance of some who knew him well. That includes his daughter, Sasha, a writer who is continuing her father's legacy of inspiring us about the cosmos and our place in it.[22]

Sagan was so compelling and charismatic a personality that he quickly became the voice of science for the nation. On Johnny Carson's *The Tonight Show*, he would mesmerize national audiences with his observations, insights, and often amusing anecdotes. In so doing, he literally knocked Carson's previous go-to science guy out of the lineup for good.[23] That was none other than astrophysicist Robert Jastrow, the aforementioned GMI cofounder. Which brings us back to the main thrust of our story.

Carl Sagan became increasingly political in the 1980s as he recognized the mounting threat of a nuclear arms race. He used his public prominence, media savvy, and unrivaled communication skills to raise awareness about the existential threat posed by a global thermonuclear war. Sagan explained to the public that the threat went well beyond the immediate death and destruction and the resulting nuclear radiation. The massive detonation of nuclear warheads during a thermonuclear war, Sagan and his colleagues argued in the scientific literature, might produce enough dust and debris to block out a sufficient amount of sunlight to induce a state of perpetual winter, or, as they termed it, "nuclear winter."[24]

Humanity, in short, might suffer the same fate the dinosaurs encountered following a massive asteroid impact: a sunlight-blocking dust storm that ended their reign sixty-five million years ago. Sagan helped bring about public understanding of that scenario through his various media interviews and in an article for the widely read Sunday newspaper insert *Parade* magazine.

Sagan feared that Reagan's Strategic Defense Initiative, which many cold war hawks and military contractors supported, would lead to an escalation of tensions between the United States and the Soviet Union and a dangerous buildup in nuclear arms, portending the very nuclear winter scenario he so feared. But, as Oreskes and Conway noted in *Merchants of Doubt*, the cold war–era physicists at GMI saw these legitimate concerns about SDI as scare tactics employed by Soviet-sympathizing peaceniks.[25] In their eyes, the very concept of nuclear winter was a threat to our security. Working with conservative politicians and industry special interests, the GMI trio sought to discredit the case for concern by going directly after the underlying science—first by discrediting the scientist, Carl Sagan, personally. The attacks took place in congressional briefings and in the pages of mainstream newspapers, where they solicited and wrote articles and op-eds to debunk the findings of Sagan and his colleagues. This campaign even included intimidating public television stations that considered running a program on nuclear winter.[26]

Here's why Sagan's anti-SDI campaign is germane to the central topic of this book: The nuclear winter simulations that Sagan and his colleagues conducted were based on early-generation global *climate models*. So if you didn't like the science of nuclear winter, you *really* weren't going to like the science of climate change, which revealed the culpability of the same powerful polluting interests that groups like GMI were defending. With the collapse of the cold war in the late 1980s, the GMI crew, as Oreskes and Conway noted, needed another issue to focus on. Acid rain and ozone depletion would keep them busy through the early 1990s. But as these matters faded from view (in substantial part because even *Republicans*—as noted earlier—ultimately supported action), GMI and like-minded critics needed another scientific boogeyman to justify their existence. Climate change surely fit the bill.

CHAPTER 2

The Climate Wars

There's no war that will end all wars.
　　　　　—HARUKI MURAKAMI

When the rich wage war, it is the poor who die.
　　　　　—JEAN-PAUL SARTRE

AND SO, IT BEGINS

In the early 1990s I was a graduate student working on my PhD in the field of climate science within the Department of Geology and Geophysics at Yale University. I had been lured away from the Physics Department, where I had been studying the behavior of matter at the quantum scale. Instead, I would now study the behavior of our climate system at the global scale. For an ambitious young physicist, climate science was the great western frontier. There were still big, wide-open questions where a young scientist with math and physics skills could make substantial contributions at the forefront of the science. This was my opportunity to realize the vision that Carl Sagan had instilled in me as a youth—a vision of science as a quest to understand our place in the larger planetary and cosmic environment.

My PhD adviser was a scientist named Barry Saltzman, who played a key role in the discovery of the phenomenon of "chaos"— one of the great scientific developments of the twentieth century.

Chaos is responsible, among other things, for the fact that one cannot predict the precise details of the weather beyond a week or so out. Barry was a skeptic—in the true and honest sense of the word. He was unconvinced in the early 1990s that we could establish the human impact on our climate. This was a tenable position then, given that the climate models being used were still quite crude and that the warming signal in the roughly one century of global temperature data was only perhaps just beginning to peek out from the background noise of natural variability.

There were other scientists, such as James Hansen, the prominent director of the NASA Goddard Institute for Space Studies (yes—the same institute that had previously been directed by none other than Robert Jastrow), who had a different view. Hansen felt that we could already demonstrate that human activity—specifically, the generation of greenhouse gases such as carbon dioxide from the burning of fossil fuels like oil, coal, and natural gas—was warming the planet. On a record hot June day in Washington, DC, in 1988, Hansen had testified to Congress, saying, "It is time to stop waffling. . . . [T]he evidence is pretty strong." The Reagan administration had become increasingly unhappy with Hansen's public statements even before that June day. As a NASA civil servant, he was subject to having his written congressional testimonies vetted by the administration, and starting in 1986, the White House's Office of Management and Budget had repeatedly edited them in such a way as to downplay their impact. Exasperated, Hansen finally announced in bombshell 1989 testimony that his words were being altered by the White House.[1]

As I began to study climate science in the early 1990s, my own position was closer to Barry Saltzman's than to Hansen's. My research involved the study of natural climate variability based on the use of theoretical climate models, observational data, and long-term paleoclimatic records, including tree rings and ice cores. This work suggested that there were important mechanisms that led to natural climate fluctuations with time scales of fifty to seventy years, almost as long as the instrumental temperature record itself. Such natural

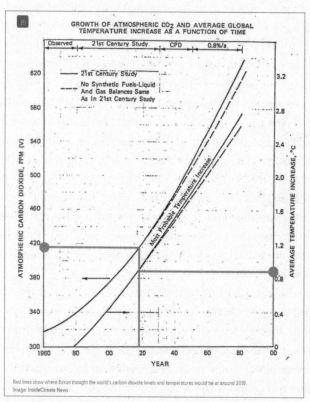

Prediction of future CO_2 rise and temperature increase from an internal 1982 ExxonMobil document. The current observed CO_2 level and global temperature increase are indicated by the thick horizontal and vertical lines. The actual values are 415 parts per million (ppm) CO_2 and a temperature increase of 0.8°C (1.44°F) since 1960, both within the range of the predictions. Figure 3 in Exxon report of November 12, 1982, subject line "CO_2 'Greenhouse' Effect," 82EAP 266, under Exxon letterhead of M. B. Glaser, manager, Environmental Affairs Program, posted by *Inside Climate News* at http://insideclimatenews.org/sites/default/files/documents/1982%20Exxon %20Primer%20on%20CO2%20Greenhouse%20Effect.pdf, p. 7.

long-term climate fluctuations, at the very least, obscured the impacts of human-caused climate change.[2]

It's important to keep some perspective here. Although scientists were still debating whether we had yet detected a human impact on the climate, there was a broad consensus on the basics—i.e., that burning fossil fuels and increasing the concentration of greenhouse gases in the atmosphere would substantially warm the planet, something that had been established by the great Swedish scientist Svante Arrhenius in the late 1800s. And it is worth recalling, from the introduction, the words of ExxonMobil's own experts in the 1970s: *"There is general scientific agreement that . . . mankind is influencing the global climate . . . through carbon dioxide release from the burning of fossil fuels"* (emphasis added.)[3] The famous Danish physicist Niels Bohr is reported to have once said, "Predictions are hard. Especially about the future." Well, Exxon's own scientists made an impressive one back in 1982, more or less predicting spot-on the increase in CO_2 concentrations and the resulting warming we would now see given business-as-usual burning of fossil fuels.[4] The coal industry also knew, as far back as the 1960s, that their carbon emissions were warming the planet.[5]

Nonetheless, the fact that there was still some real division within the climate research community on a matter as seemingly fundamental as whether we had yet firmly detected a human influence on climate meant there was a preexisting cleavage into which the forces of denial could attempt to drive a wedge and generate uncertainty and controversy about the science. For the fossil fuel industry, time was of the essence, because policy action aimed at addressing the problem appeared imminent.

During the presidential election of 1988, George H.W. Bush had pledged to meet the "greenhouse effect with the White House effect." He appointed as his science adviser a physicist named David Allan Bromley. Bromley was a professor from the Yale Physics Department, where I was doing my degree at the time, and I still remember him returning to New Haven to give a special departmental seminar on climate change and climate modeling. Bromley was no left-leaning environmentalist. But he understood the irrefutable physics behind climate change. Meanwhile, Bush's EPA administrator, the aforementioned William K. Reilly, *was* an environmentalist, and he strongly supported action on climate. By 1991, Bush had signaled that he would sign the United Nations Framework Convention on Climate Change (UNFCCC).

But there was some dissent within the administration. Bush's chief of staff, a Massachusetts Institute of Technology (MIT)–trained engineer named John Sununu, was—and remarkably enough, remains today—a climate-change denier. He drew heavily from an unpublished 1989 white paper by the GMI trio of Jastrow, Seitz, and Nierenberg (published the following year as a book, *Global Warming: What Does the Science Tell Us?*) that blamed global warming on solar activity. In his capacity as a representative of GMI, Nierenberg secured a meeting with White House staff, where he presented their dismissive view of climate change. At the very least, this helped create a schism within the Bush administration and blunted the momentum behind climate action.[6]

With the advent in 1988 of the United Nations Intergovernmental Panel on Climate Change (IPCC), the task of refuting the scientific evidence for human-caused global warming became too great for a single small organization like GMI. The cavalry would soon arrive, however. A consortium of fossil fuel interests known as the Global Climate Coalition, which included ExxonMobil, Shell, British Petroleum (BP), Chevron, the American Petroleum Institute, and others, came together in 1989, joining forces with other industry think tanks and front groups, including the genteel-sounding "Heartland Institute" and "Competitive Enterprise Institute." Collectively they constituted what Oreskes and Conway analogized in *Merchants of Doubt* as a "Potemkin Village," a facade of impressive-sounding organizations, institutions, and individuals who would challenge—through newspaper op-eds, public debates, fake scientific articles, and any other means available—the basic science of climate change. They would seek to carry the argument that the science was too uncertain, the models too unreliable, the data too short and too error-ridden, the role of natural variability too unknown to establish any clear human role in global warming and climate change.

David and Charles Koch, otherwise known as the "Koch brothers"—the owners of the largest privately held fossil fuel interest (Koch Industries)—are best known for their highly visible role in funding climate-change denialism in recent years. But they played a key early role here as well, something that has only recently come to light.[7] Under the auspices of the Cato Institute, the libertarian think tank that they founded and funded, they held the very first known climate-change-denial conference back in June 1991. Titled "Global Environmental Crisis: Science or Politics?," it was a sort of Council of Elrond of climate-change denialism. It featured two scientists in particular who would join the ranks of Seitz, Jastrow, and Nierenberg, leveraging their scientific and academic credentials to grant an air of legitimacy to broadsides aimed at discrediting mainstream climate science.

Among the invited speakers was Richard S. Lindzen of MIT, who was quoted in the brochure advertising the conference as saying there was "very little evidence at all" that climate change was a threat. His credentials, like Seitz's, were impressive. He was a chaired professor of meteorology at MIT and a member of the National Academy of Sciences. Like Seitz, he has also received money from fossil fuel interests for his advocacy on their behalf.[8] Scientifically speaking, Lindzen is best known for his controversial insistence that climate models overestimate the warming effect of increasing greenhouse gas concentrations because of processes—related to clouds or atmospheric moisture—that he continues to claim are either missing or poorly represented in the models. Such processes in principle can either tend to increase warming (in a "positive feedback") or decrease warming (in a "negative feedback"). Lindzen, however, has remained focused only on the latter. It seems, indeed, that he has never met a negative feedback he didn't like. He has spent much of his professional career arguing for supposedly missing negative feedbacks, only to have other scientists continually shoot them down.[9] Lindzen has even been so bold as to argue that a doubling of CO_2 concentrations (which we will reach in a matter of decades, given business-as-usual burning of fossil fuels) would raise global temperatures only a very minimal 1°C (1.8°F). The claim strains credulity, given that the planet has now *already* warmed up more than that after only a roughly 50 percent increase in CO_2 concentrations. Indeed, a vast array of evidence, including the response of the climate to volcanic eruptions, the coming and going of the ice ages, and past warm periods, such as the early Cretaceous, when dinosaurs roamed the planet, all point toward warming that is roughly three times (3°C, or 5.4°F) as large as Lindzen predicted.

Also among the speakers at this influential early conference was S. Fred Singer, whom we can now begin to recognize as a sort of all-purpose denier-for-hire. Like Seitz's, Singer's origins were as an academic and a scientist, and like Seitz, he would leave the academic world in the early 1990s to advocate against what he called the "junk science" of acid rain, ozone depletion, tobacco health threats, and, of course, climate change, receiving substantial industry funding for his efforts.[10]

Singer's most significant role relates to the legacy of the revered atmospheric scientist Roger Revelle. Revelle made fundamental contributions to our current understanding of human-caused climate change, providing key evidence in the 1950s that the burning of fossil fuels was increasing greenhouse gas concentrations. He made some of the early projections of future warming. Revelle is also credited with having inspired Al Gore's concern about climate change when Gore was a student at Harvard.

Shortly before Revelle passed away in 1991, Singer added him as a coauthor to a paper he had written for the journal *Cosmos*, published by the Cosmos Club, a Washington, DC, intellectual society. The paper was nearly identical to an earlier dismissive article by Singer. It disputed the evidence that climate change is human caused. Both Revelle's secretary and his former graduate student Justin Lancaster have suggested that Revelle was uncomfortable with the manuscript and that the dismissive framing was added after Revelle, who was gravely ill (and died just months after the paper's publication), had an opportunity to see the final version. Lancaster has stated that Singer hoodwinked Revelle into adding his name to the article and that Revelle was "intensely embarrassed that his name was associated" with it. Lancaster characterized Singer's behavior as unethical and, furthermore, said he had a strong suspicion of Singer's ultimate objective: to discredit Al Gore and his campaign in the early 1990s to raise public awareness about the threat of climate change. Lancaster stands by these charges despite legal threats against him by Singer.[11]

THE BATTLEFIELD TAKES SHAPE

We now fast-forward a few years, to late 1995, when it would all come to a head. The scientific evidence for human-caused climate change had grown ever more compelling. The observations, the model simulations, all seemed to be coming into clear alignment. My once skeptical PhD adviser Barry Saltzman and I were coauthors on an article making this very case.[12] Industry-funded resistance to the science, however, had grown proportionately. Dozens of front groups and scientist deniers-for-hire now occupied an increasingly fortified Potemkin Village of industry-funded climate-change denial. The battlefield had taken shape, the forces were mobilized. Climate change was the defining political issue of the time.

By late November 1995, the Intergovernmental Panel on Climate Change would hold its final plenary in Madrid for its Second Assessment Report. The purpose of the report was to summarize the current consensus among the world's scientists on climate change. As remarked earlier, that consensus was rapidly converging toward acceptance of the reality and threat of climate change. Nonetheless, a fierce dispute arose between the scientists authoring the report and government delegates representing a small subset of countries— Saudi Arabia and Kuwait, in particular, two major oil-exporting nations that profited greatly from the continued extraction and sale of fossil fuels. As science journalist William K. Stevens put it, these nations "made common cause with American industry lobbyists to try to weaken the conclusions of the report."[13]

The question was whether one could state with confidence that human-caused climate change was now detectable. The scientist with primary responsibility for the relevant section of the report was Ben Santer, a climate researcher at the US Department of Energy's Lawrence Livermore National Laboratory in California who had published a series of important articles on the topic. The recipient of a MacArthur "genius award" in recognition of his fundamental contributions to our understanding of climate change, Santer and his IPCC coauthors concluded, based on the existing climate literature, that "the balance of evidence suggests an *appreciable* human influence on climate."[14]

The Saudi delegate complained that the word "appreciable" was too strong. For two whole days, the scientists haggled with the Saudi delegate over this single word in the "Summary for Policy Makers" of the report—the part of the report most likely to be read by politicians and most likely to be reported upon by journalists. They purportedly debated nearly thirty different alternatives before IPCC chair Bert Bolin found a mutually acceptable word: "discernible." The term acknowledged that human activity played at least some role in observed climate change, as the scientists had argued, while making it sound like one almost had to squint to see it, conceding a level of uncertainty that no doubt pleased the oil-rich Saudis.

The fact that two entire days at the final plenary were devoted to debating a single word in the report's summary gives you some idea of how politically charged the debate over climate change had become by late 1995. Ben Santer was the scientist most directly connected to the emerging scientific consensus. In the tradition of Rachel Carson, Herbert Needleman, and Gene Likens, he would be savaged by industry groups and their now familiar attack dogs in an effort to undermine his credibility.

Just a few months after the IPCC plenary, in February 1996, S. Fred Singer published a letter in the journal *Science* attacking Santer. He disputed the key IPCC finding that model predictions matched the observed warming, claiming that the observations instead showed cooling. This was, of course, wrong. They showed nothing of the sort.

They demonstrated clear evidence of warming. But climate-change deniers would cling to one curious dataset—a satellite-derived estimate of atmospheric temperatures produced by two contrarian scientists from the University of Alabama at Huntsville, John Christy and Roy Spencer—that appeared to contradict all the other evidence of warming. The cooling claimed by Christy and Spencer would later be shown to be an artifact of serial errors on their part. But not before climate-change deniers would milk it for all it was worth.[15]

Singer went on to claim that inclusion of Santer's work in the report somehow violated IPCC rules because the work hadn't yet been published. In fact, most of the work had been published, and in any case, the IPCC rules did not require a work cited to be published at the time of the report, but simply that the work be available to reviewers upon request.

Meanwhile, the aforementioned Global Climate Coalition circulated a report to Washington, DC, insiders repeating these false allegations and accusing Santer of "political tampering" and "scientific cleansing." The latter charge, echoing language of the Third Reich, was especially odious given that Santer had lost relatives in Nazi Germany. The claims were of course false. At the request of the IPCC leadership, Santer had simply removed a redundant summary to ensure that the structure of the chapter on which he was lead author would conform to that of the other chapters. A few months later, Frederick Seitz published an op-ed in the *Wall Street Journal* repeating the same false allegations against Santer.[16]

Climate-change deniers were able to spread false charges about Santer faster (and in more prominent venues) than he—or the rest of the scientific community—could possibly hope to refute them. Santer's integrity was impugned, and his job and his life were threatened. It's an example of what I later termed the "Serengeti Strategy," in which industry-funded attackers go after individual scientists just as predators on the Serengeti plain of Africa hunt their prey: attempting to pick off vulnerable individuals by isolating them from the rest of the herd. When my work was prominently featured in the next IPCC report, Ben Santer commented, "There are people who believe that if they bring down Mike Mann, they can bring down the IPCC."[17] They thought I was easy prey.

THE SEITZ DECEPTION

Fast-forward a couple more years, to 1997. The Kyoto Protocol, an addition to the United Nations Framework Convention on Climate Change, had just been adopted. It would commit the countries of the world to substantial reductions in carbon emissions with the aim of avoiding "dangerous anthropogenic interference with the climate system."[18] The pressure on policymakers was mounting. The forces of denial and delay would need to marshal additional forces if they were to forestall action on climate.

In so doing, they would find common cause with some increasingly odd characters. Consider Arthur B. Robinson, a chemist with admittedly impressive credentials. A onetime protégé of Nobel Prize–winning chemist Linus Pauling, Robinson heads up a family-run outfit in Cave Junction, Oregon, that calls itself the Oregon Institute of Science and Medicine. Robinson has advanced some very odd scientific hypotheses over the years, including the discredited claim that vitamin C causes cancer. He has also shown an interest in collecting and analyzing other people's urine. And yes—I know you're wondering—Robinson is also a climate-change denier, a position that has more recently ingratiated him with the right-wing climate-change-denying Mercer family as well as the Trump administration.[19]

In 1998, one year after Kyoto, Robinson joined forces with our old friend Frederick Seitz to undermine support for the protocol. The two organized a petition drive opposing the international agreement. To this day, the "Oregon Petition," with thirty-one thousand nominal "scientist" signatories, is touted as evidence of widespread scientific opposition to the research underlying models of human-caused climate change. This is in spite of the fact that few of the supposed signatories were actually *scientists* (the list included the names Geri Halliwell, one of the Spice Girls; and B. J. Hunnicutt, a character from the TV series *M*A*S*H*). Not to mention that a majority of signatories who actually *were* scientists indicated they no longer supported the petition or couldn't remember signing the petition, or were deceased, or failed to respond when they were contacted by *Scientific American*.[20]

The petition was mailed out to an extensive list of scientists, journalists, and politicos along with a cover letter and an "article" attacking the scientific evidence for climate change. The article, titled "Environmental Effects of Increased Atmospheric Carbon Dioxide," was coauthored by Robinson, his son Noah, and climate-change contrarian Willie Soon. It was formatted to appear as if it had been published in the prestigious *Proceedings of the National Academy of Sciences* (PNAS), the official journal of the hallowed National Academy of Sciences (NAS). Seitz even signed the enclosed letter using his past affiliation as NAS president. The NAS, in response, took the extraordinary step of publicly denouncing Seitz's efforts as a deliberate deception, noting that its position on the issue—that there was now a consensus that climate change is real and human caused—was very much the opposite of what Seitz was saying.

The entire episode, coincidentally enough, played out just days before the publication of our "hockey-stick" article, which appeared in the journal *Nature* on April 22 (Earth Day), 1998.[21] The curve demonstrated the unprecedented nature of modern global warming. It would become a symbol in the climate-change debate. It—and I— would soon become a major target of attack.

THE HOCKEY FIGHT

Let us skip ahead a few more years, to 2002, where we encounter the now infamous "Luntz Memo." Frank Luntz is a professional pollster who has long advised the GOP on matters of policy based on insights derived from polling and focus groups. In a 2002 memo that was leaked by an organization known as the Environmental Working Group, Luntz warned his fossil-fuel-industry-coddling Republican clients that "Should the public come to believe that the scientific issues are settled, their views about global warming will change accordingly."[22] He advised using less threatening language in characterizing the phenomenon, favoring "climate change" over "global warming." Ironically, the very same scientific community that climate-change deniers accuse of being alarmist would increasingly favor the use of that term as well, simply because it's a more comprehensive description of the problem. Climate change involves not only the warming of Earth's surface, but the melting of ice, sea-level rise, the shifting of rainfall and desert belts, altered ocean currents, and so on. Luntz also suggested that Republicans "reposition global warming as theory [rather than fact]." This, too, is ironic, for a *theory* is the most powerful of scientific entities. Gravity is just "a theory." That hardly makes it safe to jump off a cliff.

Luntz warned that "the scientific debate is closing [against Republicans] but not yet closed. There is still a window of opportunity to challenge the science," by which he meant to insert doubt into the public mindset. Following Luntz's prescription, fossil fuel interests and the politicians and attack dogs doing their bidding doubled down in their assault on the science, engaging in a "shoot the messenger" strategy designed to discredit the science underpinning concern over human-caused climate change. I found myself at the center of the attack because of the hockey-stick curve, which soon took on an iconic status in the climate debate. It would be featured in the "Summary for Policy Makers" of the 2001 Third Assessment Report of the IPCC as *the* key new piece of climate-change evidence, supporting the conclusion that recent warmth was unprecedented over at least the past one thousand years.[23]

In reality, it was only one of many independent pillars of evidence that now existed. Human influence on the climate had already been established, as readers will recall, with the publication of the IPCC's Second Assessment Report in 1995. But the hockey stick was far more compelling to the layperson than the rather abstract statistical work behind the key findings of the previous report. One didn't need to understand the physics, mathematics, or statistics underlying climate research to understand what the striking visual was telling us.

The long, gentle cooling trend that characterizes the descent from the relatively warm conditions of the eleventh century into the so-called Little Ice Age of the seventeenth to nineteenth centuries resembles the downturned "handle" of a hockey stick, and the abrupt warming spike of the past century is the upturned "blade." The fact that this dramatic recent warming accompanies the rapid increase in atmospheric carbon dioxide concentrations from the industrial revolution conveys an easily understood, unmistakable conclusion: the warming we are experiencing is unprecedented in modern history. Fossil fuel burning and other human activities are the cause.

That the hockey stick rose to prominence at precisely the same time that climate-change deniers were planning renewed and heightened attacks on the science was a coincidence of timing that had profound implications for my own career. In *The Hockey Stick and the Climate Wars*, I describe the efforts by fossil fuel interests and their hired guns to discredit the hockey stick and me personally.[24] Those efforts included attacks against me and my work by right-wing media outlets like Fox News and the *Wall Street Journal* as well as hostile congressional hearings and investigations by climate-change-denying politicians such as Oklahoma senator James Inhofe, former Texas congressman Joe Barton, and former Virginia attorney general Ken Cuccinelli. All were Republicans. All were recipients of substantial fossil fuel largesse. I was subject to legal assaults by fossil-fuel-industry front groups seeking to abuse open records laws to obtain my personal emails—in the hope of finding something embarrassing with which to discredit me, or something that could be taken out of context and misrepresented to cast doubt on my research. Most of the people and groups behind the effort—the Koch brothers, the Heartland Institute, the George C. Marshall Institute, Fred Singer—are familiar by now.

The good thing about science is that it possesses what the great Carl Sagan described as "self-correcting machinery." The processes of peer review, replication, and consensus, mixed with a healthy dose of skepticism—real skepticism, not the fake kind that is passed off as such by climate-change deniers—keeps science on a path toward truth. If a scientific claim is wrong, other scientists will demonstrate it to be so. If it's right, other scientists will reaffirm it, perhaps improve it and extend it. Climate-change deniers like to claim that scientists simply seek to reaffirm the prevailing paradigm, because that's how you secure funding and get published in the leading journals. As with most things climate-change deniers assert, the opposite is in fact true. *Disproving* the conventional wisdom, refuting a landmark study—that's the path to fame and glory in the world of science.

Accordingly, challenges to the hockey-stick curve in leading scientific journals, such as *Nature* and *Science*, have helped launch the careers of ambitious young scientists. Yet the hockey stick has withstood those and other challenges. Two decades of research by dozens of independent teams, using different data and methods, has time and again reaffirmed our findings. There is now a veritable hockey league of studies that not only confirm our original conclusion—that the recent warming is unprecedented over the past millennium—but in fact extend it to at least the past two millennia and, more tentatively, at least the past twenty thousand years.[25] Our basic finding has stood the test of time and the scrutiny of skeptical scientists. Accordingly, it has now been incorporated into the scientific consensus, and the scientific investigations have moved on, extending our findings and providing additional context. That's how science works.

That is not at all to say that efforts to discredit the hockey stick have ceased. And here we must distinguish between the world of science and the world of politics. The former is driven by the self-correcting machinery that Sagan so eloquently spoke of, in that scientific findings are always subject to appropriate scrutiny and (largely) good-faith challenges. The latter obeys no such rules. The hockey stick continues to be attacked in the conservative media based on the most cynical and disingenuous misrepresentations of the facts.[26] In the world of politics today, almost anything—it seems—goes; reality and logic have gone out the window, replaced by ideologically and agenda-driven "alternative facts."

Nearly two decades ago, in his book *The Demon-Haunted World*, Sagan presaged with some trepidation the world we now live in:

> I have a foreboding of an America in my children's or grandchildren's time—when the United States is a service and information economy; when nearly all the manufacturing industries have slipped away to other countries; when awesome technological powers are in the hands of a very few, and no one representing the public interest can even grasp the issues; when the people have lost the ability to set their own agendas or knowledgeably question those in authority; when, clutching our crystals and nervously consulting our horoscopes, our critical faculties in decline, unable to distinguish between what feels good and what's true, we slide, almost without noticing, back into superstition and darkness.[27]

Sagan's fears have no doubt been realized when it comes to the climate wars, if not our societal discourse writ large. And there is no better example of this pathology than the pseudo-scandal manufactured by the fossil fuel industry that came to be branded as "Climategate," a last gasp, if you will, of hard-core climate-change denial.

CLIMATEGATE—A LAST GASP?

In a more recent counterpart to the infamous 1972 Watergate affair that brought down the presidency of Richard M. Nixon, hackers with links to Russia and WikiLeaks broke into an email server and released stolen emails in a massive, carefully orchestrated disinformation campaign designed to impact the course of American politics.[28]

You could be forgiven for thinking that I'm talking about the now well-established conspiracy between Russia and the campaign of Donald Trump to steal the US presidential election of 2016, a scandal that has since been branded *Russiagate*. But no. I'm talking about the affair in November 2009 that would come to be known as *Climategate*.

Advocates for climate action anticipated an opportunity for meaningful action on climate heading into the United Nations Climate Change Conference in Copenhagen in December 2009. A successor to the Rio and Kyoto conferences, Copenhagen was a source of great hope to climate campaigners; indeed, many referred to it as *Hopenhagen*. With the growing public recognition of the climate threat, thanks to ever greater clarity about the impacts of climate change (the unprecedented disaster of Hurricane Katrina was still fresh in the memory of Americans) and Al Gore's wildly successful documentary *An Inconvenient Truth*, it seemed we were turning a corner. Perhaps, finally, the world was ready to act on climate.

The forces of denial and delay, however, would intercede once again, manufacturing a fake "scandal" in the weeks leading up to the summit. Even the name they successfully attached to the affair— "Climategate"—was the product of a carefully crafted narrative foisted on the public and policymakers in a collaborative effort by fossil-fuel-industry front groups, paid attack dogs, and conservative media outlets. Thousands of emails between climate scientists (including me) around the world were stolen from a university computer server in Great Britain late that summer. Bits and pieces of the emails were disingenuously rearranged and taken out of context by climate-change deniers to misrepresent both the science and the scientists.[29]

Before long, climate deniers had combed through the emails and organized them into a searchable archive. Taking individual words and phrases out of context to distort the original meanings, they claimed to have found the "smoking gun" that revealed climate change to be an elaborate hoax. Terms that were entirely innocent in context—for example, the word *trick*, which mathematicians and scientists use to denote a clever shortcut to solving a problem—were extracted and deliberately misinterpreted.

Climate-change deniers used these misrepresentations to claim that scientists were cooking the books, engaged in an elaborate scheme to *trick* the public! Front groups connected with the Koch brothers and industry-funded critics wanted the public to distrust the climate science by suspecting the climate scientists. Right-wing media outlets—especially the Murdoch media empire (e.g., Fox News and the *Wall Street Journal*) and conspiracy-theory-promoting bottom-feeders such as the *Drudge Report*, Breitbart "News," and Rush Limbaugh—served as a megaphone for outrageous untruths, filling the airwaves, television screens, and Internet with false allegations, smears, and innuendo.

Right-wing politicians joined the fray. James Inhofe, who had famously dismissed the overwhelming scientific evidence of climate change as "a hoax," embraced, with no sense of irony, the *true* hoax that was Climategate. Based on the untruthful Climategate allegations, he called for the criminal investigation of seventeen climate scientists, including Presidential Medal of Science recipient Susan Solomon of MIT, Michael Oppenheimer of Princeton, and Kevin Trenberth of the National Center for Atmospheric Research (NCAR). And yes, I was honored to be on that list as well.

Two years later, roughly a dozen (depending on how you count) different investigations in the United States and the United Kingdom had exonerated the scientists. There had been no data fudging, no attempt to mislead the public about climate change. The only wrongdoing that was established was the criminal theft of the emails in the first place—another cruel case of irony, given that the Watergate scandal, the origin of the "-gate" suffix, was about the theft of documents, not their content.[30]

In the meantime, however, climate-change deniers milked the fake scandal for all it was worth. Readers might recall Saudi Arabia's efforts to dilute the conclusions of the IPCC's Second Assessment Report back in 1995. Here, fifteen years later, the Saudis were still up to their usual mischief, attempting to sabotage the already delicate negotiations at the Copenhagen Summit. The lead Saudi climate-change negotiator, Mohammad al-Sabban, insisted that the pilfered emails would have a "huge impact" on the negotiations. Fifteen years after the IPCC had concluded that there was a discernible human influence on climate, Sabban, remarkably, asserted that "it appears from the details of the scandal that there is no relationship whatsoever between human activities and climate change." Somehow, a few misrepresented emails managed to negate more than a century of physics and chemistry and the overwhelming consensus of the world's scientists.

In understanding the role that both Saudi Arabia and the Murdoch media empire played in promoting Climategate smears and lies, it is worth noting that there is a curious connection between the two. Prince Alwaleed bin Talal of the Saudi royal family and Rupert Murdoch are close allies, and the two have financial ties. Until recently, Prince Alwaleed owned, via his company (Kingdom Holding), 7 percent of News Corporation's shares, making him the second-largest shareholder after Rupert Murdoch and his family (Alwaleed sold off his shares when he was arrested for corruption in 2017, knowing his assets would likely be frozen). Murdoch, News Corp, and the Saudi royal family all share a motive for opposing climate action.[31]

The Climategate thieves were never caught. What we do know is that Russia and Saudi Arabia both played roles in hosting and helping to distribute the stolen emails. Saudi Arabia made direct use of the false Climategate allegations in its efforts to halt progress toward a meaningful global climate treaty in Copenhagen. Recent evidence suggests that the "hacker" who broke into the server did so from Russia.[32] In light of Russia's tampering in the 2016 US presidential election, it seems relevant that Climategate used the same modus operandi and involved some of the same actors (WikiLeaks and Julian Assange) who were part of that campaign. Indeed, it could be argued that it was the very same *motive*.[33]

Vladimir Putin had an interest in Hillary Clinton's defeat in the 2016 election not just for geopolitical reasons, but because fossil fuels are Russia's primary asset, with much of the Russian economy dependent on fossil fuel exports. A prospective Trump presidency was of mutual benefit to both Russia and the world's largest fossil fuel company, ExxonMobil, offering the prospect of a collaborative venture between ExxonMobil and the Russian state oil company Rosneft to develop the largest currently untapped oil reserves in the world—Arctic, Siberian, and Black Sea petroleum reserves worth an estimated $500 billion.

The two companies signed a partnership in 2012 that was stymied when the Obama administration placed economic sanctions on Russia in 2014 for its annexation of part of the Ukraine (Crimea). It is almost certain that Hillary Clinton would have kept those sanctions in place. But not Donald Trump. At the July 2016 Republican National Convention, with Trump the presumptive Republican presidential nominee, his campaign, led by Paul Manafort—an individual who had worked for more than a decade as a lobbyist for Viktor Yanukovych, the Russian-backed former president of Ukraine—altered the official Republican platform to remove language supporting the sanctions.

Once in office, Trump appointed ExxonMobil CEO Rex Tillerson as his secretary of state. His administration attempted (unsuccessfully, thanks to some vestigial backbone among Senate Republicans) to lift the sanctions that stood in the way of the ExxonMobil–Russia oil partnership. We now know, thanks to the special counsel investigation led by former FBI director Robert Mueller, that Russia attempted to influence the election in favor of Donald Trump. It is plausible, if not probable, that a half-trillion-dollar oil deal was the primary impetus. Quid meet quo.

That brings us back to Climategate, which involved the use of stolen emails to influence the Copenhagen Summit of December 2009. It, too, advanced the agenda of fossil fuel interests, including ExxonMobil and Rosneft, by attempting to undermine the single greatest argument against continued fossil fuel exploitation—the threat of human-caused climate change. With regard to Russia's motivations, it is also worth noting that Vladimir Putin is on record dismissing the notion of any human causality to climate change, arguing that the solution is to simply adapt to the changes anyway, and asserting that global warming would actually be a good thing for Russia.[34]

Climategate was, in fact, an early test run for the larger assault on climate action by a small coalition of petrostates that is underway today. "US and Russia Ally with Saudi Arabia to Water Down Climate Pledge," read the headline in *The Guardian* on December 9,

2018, roughly eight years after Copenhagen.[35] Those three countries (and Kuwait for good measure) formed a small coalition opposing a UN motion to welcome the conclusions of a recent IPCC special report warning of the dangers of planetary warming in excess of 1.5°C (2.7°F).[36] And while we're at it, what about Brexit? Or the "Yellow Vest" carbon tax protests in France? Or similar revolts in Australia, Canada, and the state of Washington? Might these episodes, too, be tied to the efforts of rogue state actors to block international climate policy progress? We will return to that question later.

YOU CAN'T FOOL MOTHER NATURE

Climategate could in fact be viewed as the opening skirmish in the new climate war. It marked the critical juncture wherein the forces of denial and inaction all but conceded that they could no longer make a credible, good-faith case against the basic scientific evidence. So they would instead deploy new, more nefarious strategies in their effort to block action on climate.

One of the strategies is simply *lying*. That's what Climategate was all about. Prevarication has become so normalized in the era of Trump (who lies so often that journalists have a hard time keeping up with the count[37]) that climate-change deniers have felt emboldened to dissemble with abandon. With a majority of the public now accepting the reality of climate change, their efforts are targeted at a shrinking minority of people who are motivated by ideology and tribal political identity over fact—a subset of the "conservative base." Polling from 2019 suggests that the percentage of these so-called dismissives in American society now number only in the single digits.[38] But their apparent prominence in the public sphere appears far greater thanks to the megaphone provided by the fossil-fuel-funded climate-change denial machine. The megaphone includes Fox News and the rest of the Murdoch media empire as well as bot armies that are deployed online to flood our social media with misinformation and disinformation. The collective effect is to make extreme positions appear more popular than they actually are. The problem also encompasses fake reports and public debates sponsored by fossil-fuel-industry front groups intended to lend a veneer of credibility to climate-change denial.[39] These efforts provide right-wing politicians with talking points and political cover as they continue to do the bidding of the fossil fuel interests who fund their campaigns instead of the people's business.

It is important to combat this rear-guard assault on the basic facts, not because we are likely to convince the diminishing and increasingly irrelevant denialist fringe—we're not. But they still threaten to infect the larger public discourse. As a result of the denialist echo chamber, people tend to *perceive* that a far greater proportion of the public denies climate change than actually does.[40] That flawed perception, in turn, inhibits people from engaging their friends, neighbors, and acquaintances on climate. If we perceive a topic as contentious and likely to raise conflict with our prospective interlocutors, we often shy away from it entirely. The less we talk about the issue, the less prominent it is in our larger public discourse, and the less pressure that is brought to bear on policymakers to act.

To the extent climate denial persists, it tends to be more in the form of downplaying the *impacts* rather than outright denial of the basic physical evidence. To be specific, much of the residual promoted denialism involves dismissal not of climate change itself, but of the negative impacts that it is having now and will have in the near future. One of the best examples involves the extensive wildfires that have recently afflicted California. Contrarians sought to divert attention from the clear role that climate change—in the form of unprecedented heat and drought—was playing in these record wildfires.[41] Denier-in-chief Donald Trump infamously disparaged state officials by blaming them for "gross mismanagement" of the forests, attributing the problem specifically to an absence of "raking" of forests.[42] In an ironic twist, given the false Climategate accusations that climate scientists had been subject to a decade earlier, released emails in 2020 actually *did* indicate data manipulation—by the Trump administration—to downplay the linkage between climate change and the devastating California wildfires.[43]

Other denialist heads of state have followed suit. President Jair Bolsonaro of Brazil tried to blame environmentalists, rather than his pro-deforestation policies (and climate change), for the widespread Amazon wildfires in 2019. But perhaps an even better illustration lies in the events I witnessed during my sabbatical in Australia during late 2019 and early 2020. As I wrote at the time, "Take record heat, combine it with unprecedented drought in already dry regions, and you get unprecedented bushfires like the ones . . . spreading across the continent. It's not complicated."[44]

The conservative prime minister of Australia, Scott Morrison, is dismissive of climate change. He has promoted Australian coal interests, helped sabotage the 25th Conference of the Parties (COP25) of the United Nations Framework Convention on Climate Change, in Madrid in December 2019, and vacationed in Hawaii while Australians were suffering the impacts of unprecedented heat and wildfires.[45] He and other conservative politicians and pundits sought to deflect attention from the true underlying cause, instead blaming greens for supposedly preventing the government from thinning out forests. The Murdoch media machine, including *The Australian* (described by the independent media watchdog SourceWatch as a paper that "promotes climate change denial in a way that is sometimes . . . so astonishing as to be entertaining"[46]), the *Herald Sun*, and Sky News television, meanwhile, promoted the myth that the massive bushfires engulfing Australia were a result of arson. Rupert Murdoch's own son James chose to speak out, publicly stating that he was "particularly disappointed with the ongoing denial" by his father's media empire.[47]

The impacts of climate change have become too obvious for a reasonable, honest person to deny. They are lapping at our feet— quite literally, when it comes to flooding and coastal inundation by sea-level rise and supercharged hurricanes, and figuratively when it comes to unprecedented droughts, heat waves, and wildfires. Climate change has touched my own life numerous times in recent years. The record flooding in the summer of 2016 where I live in central Pennsylvania was one. Watching my alma mater, the University of

California, Berkeley, shut down in late October 2019 by a historic wildfire in the East Bay hills was another. But my sabbatical during the Australian summer of 2019/2020 was when I truly came face to face with the climate crisis.

Climate change now threatens our economy to the tune of more than a trillion dollars a year.[48] A recent study commissioned by the Pentagon warns of a scenario in which electricity, water, and food systems might collapse by midcentury as a result of the effects of climate change.[49] What was once largely perceived as an *environmental* threat is now viewed as an *economic* and *national security* threat. That reality is bringing increasing numbers of political conservatives to the table—people like Bob Inglis, former Republican congressman from South Carolina, who now heads up an organization called republicEn that promotes free-market climate solutions.

There is also a growing bipartisan Climate Solutions Caucus in the US House of Representatives. Thanks largely to the efforts of the Citizens' Climate Lobby, an international grassroots movement that trains volunteers to engage their representatives on climate issues, there are now twenty-three Republican members of the caucus who support taking action to mitigate climate risk. Even some of the most conservative Republicans in the House of Representatives— including Matt Gaetz of Florida, often regarded as Donald Trump's pit bull in Congress—recognize that the people of their states don't have the luxury of debating the science of climate change, because they are suffering its consequences *now*. Indeed, Gaetz has chided fellow Republicans who still deny the science.[50]

There are indications that some of the leaders of the conservative movement are moderating their stance on climate. There is antitax crusader Grover Norquist, for example, who has at least alluded to the possibility of support for a revenue-neutral carbon tax.[51] I met with Norquist in the fall of 2019 and found him to be informed and thoughtful about the climate issue. And then there is Charles Koch, the remaining "Koch brother," his brother David having passed away in August 2019. In a November 2019 interview, Charles Koch was quoted as saying, "What we want them to do is to find policies that will actually work, actually do something about reducing CO_2 emissions, manmade CO_2 emissions, and at the same time not make people's lives worse."[52] Those words sound encouraging, but until the sole remaining Koch brother calls off his attack dogs—the front groups and dark-money outfits that continue to attack the science and scientists—and demonstrates a good-faith willingness to entertain real climate solutions, it is appropriate to remain skeptical.

Indeed, the "solutions" being advanced by conservatives are often not real solutions. Consider, for example, Marco Rubio's suggestion that the people of Florida can simply "adapt" to the impacts of sea-level rise (What does that mean? Growing gills and fins?).[53] But it is a welcome sea change (forgive the pun) that Republicans seem to be moving on from outright science denial to a more worthy debate over climate policy.

The forces of inaction—that is, fossil fuel interests and those doing their bidding—have a single goal—inaction. We might henceforth call them *inactivists*. They come in various forms. The most hard-core contingent—the *deniers*—are, as we have seen, in the process of going extinct (though there is still a remnant population of them). They are being replaced by other breeds of *deceivers* and *dissemblers*, namely, *downplayers, deflectors, dividers, delayers,* and *doomers*—willing participants in a multipronged strategy seeking to deflect blame, divide the public, delay action by promoting "alternative" solutions that don't actually solve the problem, or insist we simply accept our fate—it's too late to do anything about it anyway, so we might as well keep the oil flowing. The climate wars have thus not ended. They have simply evolved into a new climate war. The various fronts on which this war is being waged constitute the subject of subsequent chapters.

CHAPTER 3

The "Crying Indian" and the
Birth of the Deflection Campaign

Good actions give strength to ourselves and inspire good actions in others.
—PLATO

But our energy woes are in many ways the result of classic market failures that can only be addressed through collective action, and government is the vehicle for collective action in a democracy.
—SHERWOOD BOEHLERT (R-NY), former chair of the
House of Representatives Science Committee

VESTED INTERESTS HAVE OFTEN EMPLOYED WHAT'S KNOWN AS a *deflection* campaign in their efforts to defeat policies they perceive as disadvantageous to their cause. Deflection campaigns seek to divert attention from—and dampen enthusiasm for—calls for regulatory reforms to rein in bad industry behavior posing threats to consumers and the environment. The onus is instead placed on personal behavior and individual action. There are numerous examples from recent US history involving the tobacco industry and the gun lobby, but the archetypal deflection campaign is undoubtedly the Crying Indian public service announcement (PSA) of the early 1970s.

Past deflection campaigns set the stage for understanding the current debate over the relative roles of individual and collective action in addressing the climate crisis. Deflection is a critical component of the multipronged strategy now being employed by the fossil fuel industry in its battle against efforts to regulate their activities, and an important front in the new climate war.

DEFLECTION CAMPAIGNS

The slogan "Guns Don't Kill People, People Kill People," used by the National Rifle Association (NRA), provides a textbook example of deflection. Its intent is to divert attention away from the problem of easy access to assault weapons toward other purported contributors to mass shootings, such as mental illness or media depictions of violence. The campaign based on this sloganeering has been remarkably effective in forestalling commonsense gun-law reform. A recent poll indicated that 57 percent of the public believes that mass shootings reflect "problems identifying and treating people with mental health problems," while only 28 percent attribute the phenomenon to overly lax gun laws. A whopping 77 percent believe that the tragic 2018 Parkland High School shooting in Florida could have been prevented by more effective mental health screening.[1]

As gun violence expert Dennis A. Henigan has explained, "the gun lobby's political power will never be overcome until these myths are destroyed. . . . The source of the NRA's disproportionate political power is not simply its money and the intensity of its supporters' beliefs; it is also its effective communication of several simple themes that resonate with ordinary Americans and function to convince them that gun control has little to do with improving the quality of their lives." He noted that the "Guns Don't Kill People" slogan "has been remarkably effective in diverting attention from the issue of gun regulation to the endless, and often fruitless, search for more 'fundamental' causes of criminal violence."[2] Or, as journalist Joseph Dolman put it, it is "about the power of an interest group to impede what looks to most of us like genuine public progress." Thanks in substantial part to the gun lobby's successful deflection campaign, roughly forty thousand Americans die from gun violence every year.[3]

Equally craven, if less well known, is the tobacco industry's effort to divert attention from the dangers of cigarette-initiated house fires by pointing the finger instead at *flammable furniture*. "It wasn't that they argued that cigarettes don't cause fires, they just argued that the better way to address the problem was to have flame retardant furniture," according to Patricia Callahan and Sam Roe, a pair of *Chicago Tribune* journalists. They wrote a series of articles about this classic deflection campaign by the tobacco industry (in concert with the chemical industry).[4]

Burn victim and firefighter groups had been campaigning for laws requiring the tobacco industry to develop fire-safe cigarettes, which would stop burning when not being smoked. Tobacco executives insisted this would be an onerous requirement that would diminish the quality of the smoking experience and the appeal of their product. So they instead sought to neutralize the efforts of firefighting organizations, and, even more audaciously, to co-opt them. Charles Powers, a top executive at the Tobacco Institute—a tobacco-industry front group—boasted that "many of our former adversaries in the fire service defend us, support us and carry forth our federal legislation as their own."[5]

How did the tobacco industry accomplish such a seemingly Herculean task? Through the classic tool of all successful deflection campaigns—subterfuge. Industry supporters infiltrated fire safety organizations to influence their messaging, buying off many of the people working for genuine reform. Tobacco Institute vice president Peter Sparber initiated the effort in the mid-1980s. By the late 1980s, he had left the institute to run his own lobbying firm while still representing the Tobacco Institute, which became a major client. In this capacity, he continued to advance the institute's interests while maintaining plausible deniability with regard to any ostensible *direct* ties to Big Tobacco.

Sparber's crowning achievement was organizing (and ultimately weaponizing) a group of governor-appointed state fire marshals into the National Association of State Fire Marshals (NASFM). He volunteered to serve as the group's legislative consultant and serve on its executive board (it is noteworthy that a somewhat similarly themed group, the Association of State Climatologists, was later weaponized by the forces of climate-change denial[6]). Sparber was even listed on the NASFM's official letterhead and shared its Washington, DC, office.

Among its first actions under Sparber's leadership, the NASFM endorsed an industry-backed federal bill calling for further study of fire-safe cigarettes—in place of a competing bill that would have actually required them. Sparber sought to redirect attention to the ostensible need for flame-proof furniture. Enter *flame retardants*.

Flame retardants are chemicals—to be specific, they are polybrominated diphenyl ethers (PBDEs) that are added to products such as televisions, computers, infant car seats, strollers, textiles, and, yes, furniture, to inhibit flammability. They are also toxic and accumulate in the human body over time. Studies show that PBDEs may inhibit brain development in children and impair sperm development and thyroid function. They have been banned in several states.[7] Here we had a marriage made in heaven (or, rather in hell) between Big Tobacco and the chemical industry, whose interests were suddenly aligned. The tobacco industry needed a scapegoat—flammable furniture. And the chemical industry provided a putative solution—flame retardants.

Another classic tool of deflection campaigns is the use of front groups masquerading as grassroots efforts. Americans for Prosperity, for example, is a Koch brothers front group that advances the agenda of the fossil fuel industry by attacking climate science and blocking action on climate (it also advocates for the tobacco industry[8]). Citizens for Fire Safety was a front group for the chemical industry. It actively opposed legislation seeking to ban the use of hazardous fire retardants in furniture. Its executive director, Grant Gillham, came out of the tobacco industry. The group's mission, according to tax records, is to "promote common business interests of members involved with the chemical manufacturing industry," and most of its money goes toward lobbying efforts in state legislatures where bans on flame retardants are being considered.

Another player in this effort was the scientific-sounding "Bromine Science and Environmental Forum," funded by chemical manufacturers for the purpose of "generating science in support of brominated flame retardants."[9] The Madison Avenue advertising and public relations firm of Burson-Marsteller represented the group. Burson-Marsteller also represented a chemical-industry front group calling itself the Alliance for Consumer Fire Safety in Europe, which, among other things, preyed on instinctive fears of fire by touting an "interactive burn test tool," with which visitors could envision, with horror, their sofas catching on fire. Remember the name Burson-Marsteller. It's not the last time we'll hear it.

Seemingly compelling but inauthentic storytelling is a common theme in deflection campaigns. And here, as detailed by Callahan and Roe, we have one of the very best examples. Dr. David Heimbach was a retired Seattle doctor and burn surgeon; he was also a former president of the American Burn Association. And he, too, preyed on fear. He repeatedly testified, often to the gasps of audiences, about the horrific child burn victims of flame-retardant-free furniture. Among these victims, he said, was a nine-week-old patient who died in a candle fire in 2009. In Alaska, he told lawmakers about a six-week-old patient who was fatally burned in her crib in 2010. Then there was a seven-week-old baby girl who was burned in a candle-ignited fire as she lay on a flame-retardant-free pillow. "Half of her body was severely burned," Heimbach told California lawmakers. He went on to describe how "she ultimately died after about three weeks of pain and misery in the hospital." "Heimbach's passionate testimony about the baby's death made the long-term health concerns about flame retardants voiced by doctors, environmentalists and even firefighters sound abstract and petty," as Callahan and Roe put it.[10]

"But there was a problem," Callahan and Roe noted, "with his testimony." The stories weren't true. There was no evidence of any dangerous pillow or candle fires like the ones he described. None of the victims existed—not the nine-week-old, the six-week-old, or the seven-week-old. The only thing that seemed to be true was that

Heimbach was indeed a burn doctor. He was also, it turns out, a shill for industry, helping bolster the dubious claim that chemical retardants save lives.

It was the Citizens for Fire Safety that sponsored Heimbach and his lurid testimonies about burned babies. Its website featured a photo of smiling children standing in front of a red brick fire station, brandishing a handmade banner reading "Fire Safety" with a heart dotting the "i." Heimbach referred to the image when claiming to lawmakers that Citizens for Fire Safety was "made up of many people like me who have no particular interest in the chemical companies: numerous fire departments, numerous firefighters and many, many burn docs."[11] One person's Astroturf, after all, is another person's grassroots. Who's to say?

As a result of the joint efforts of the tobacco and chemical industries, fire retardants have proliferated broadly throughout the environment—indeed, so much so that these dangerous chemicals can now be detected in North American kestrels and barn owls, in bird eggs in Spain, in fish in Canada, and even in Antarctic penguins and Arctic killer whales. They have been found in honey, in peanut butter—and in human breast milk.[12] And it's all the result of an industry-promoted deflection campaign.

THE CRYING INDIAN

My most vivid early memories date back to the early 1970s, when I was five or six years old. I would like to tell you they are fond remembrances of meaningful moments from my youth: a summer family vacation at the Maine seashore, a holiday gathering with my grandparents and cousins, my first sleep-away camp. But no, my most lucid early memories are of television. Commercials, to be precise.

"I'd like to buy the world a Coke." I can still hear the jingle promising that world peace could be achieved if only everyone would just choose to drink Coca-Cola. It's as if it played on the radio just this morning. Imprinted, too, in my memory is Smokey Bear's stern admonition: "Only you can prevent forest fires." This PSA helped instill in me an early appreciation for nature and the importance of preserving it.

But *one* of those ultra-sticky ads became embedded in my very soul. If you're close to my age, and you grew up in the United States, you know the ad. You can perhaps still picture it: A chiseled, traditionally clothed Native American is paddling his canoe down a river. Slightly ominous music plays in the background, accompanied by a steady drumbeat. As he proceeds down the river, he encounters an increasing volume of flotsam and jetsam. Behind him factories belch smog into the air. The music grows louder and more foreboding. He finally lands his canoe along the river's edge, where it is inundated with more strewn litter.

The man (named "Iron Eyes Cody") makes his way onto land, trampling through yet more discarded refuse, and approaches the edge of a highway. A passenger in a passing car throws a bag of trash out the window, the contents of which splatter at his feet and onto his clothing. He looks down at the mess as we hear a voiceover. It's authoritative and stern, reminiscent of the *Twilight Zone*'s Rod Serling. "*Some* people have a deep abiding respect for the natural beauty that was once this country," it tells us in an almost admonishing tone. "Some people don't." It goes on: "*People start pollution. People can stop it.*" As the camera closes in on the man's sullen face, a tear drips from the corner of his eye and down his cheek. He looks sadly at the camera, a polluted American landscape in the background.

His tears were our tears. His pain was our pain. Our great land— and the legacy of indigenous peoples—was now imperiled by our own profligate behavior. Could we save our rivers, fields, and forests before it was too late? *Were we willing to change?*

You *bet* we were. As the newly commissioned stewards of the environment we now were, we knew our mission. No scrap of litter would escape our seizure. Thus was born a new generation of *litter* scoopers. To this day, I have difficulty letting a piece of strewn trash lie there on the roadside. I instinctively look for a nearby receptacle where I can dispose of it. That commercial changed me—and so many people of my generation—for the better, it would seem reasonable to argue.

The ad harnessed the power of an incipient environmental movement. Part of the larger "Keep America Beautiful" campaign, the PSA premiered on the first anniversary of the inaugural Earth Day, April 22, 1971.[13] The famous Cuyahoga river fire in Ohio was a recent memory, ingrained in the collective American psyche. That event arguably triggered a tipping point in public consciousness, spurring a new era of environmental awareness and a flurry of new environmental policies: the creation of the Environmental Protection Agency, the Clean Water Act, the Clean Air Act. This was the dawning of the Age of Aquarius—a new age of environmentalism.

The "Crying Indian" ad, as it came to be known, was the right message at the right time, simple and empowering. You and I can solve this problem. We just need to put our minds to it, get our act together, roll up our sleeves. It's been rated one of the most effective ads in history. Environmental organizations such as the Sierra Club and the Audubon Society embraced the ad, and even served on the advisory council for the campaign. Cody soon made his way onto highway billboards. He became an icon of the modern environmental movement.

But if you're looking for a feel-good story here, you'd better get used to disappointment. Some things aren't quite what they seem. Scratch beneath the surface of the Crying Indian, as historian Finis Dunaway did in a 2017 commentary in the *Chicago Tribune*, and a different picture—indeed, a *dramatically* different picture—begins to emerge.[14]

FALSUS IN UNO, FALSUS IN OMNIBUS

Dunaway emphasized the role that Native Americans played in the early 1970s counterculture. The Crying Indian tapped into that ethos, drawing upon the peace movement in much the same way that the "Buy the World a Coke" jingle did, airing around the same time. (We will see later that the connection with Coca-Cola is not just incidental.) Two of the most memorable films from my childhood, both released in 1971, drew upon these same themes. *Billy Jack* tells the story of a man who was half American Navajo. His pacifist aspirations are at odds with his temper and his penchant for vigilante justice, and he defends a counterculture school of peaceniks and hippies, many of whom are from indigenous tribes, from hostile, bigoted townspeople. *Bless the Beasts and Children* tells the story of a group of teen misfits who find meaning, validation, and empowerment by freeing a herd of buffaloes that are being hunted by ruthless men for sport. It isn't just the buffaloes that are being destroyed; it is the spirit of the Native Americans, for whom the American buffalo is an enduring symbol. The underlying themes in both films were unmistakable: the struggle between peace and empowerment, and *the central symbolic role* in that struggle played by indigenous peoples and their plight. In playing on that very same symbolism, the Crying Indian captured the zeitgeist of the early 1970s. It harnessed all that power.

And it is in the Native American symbolism of the Crying Indian PSA where we encounter our first betrayal, if a seemingly minor and superficial one. For, as it turns out, "Iron Eyes Cody" was not a member of a Native American tribe. He wasn't even Native American at all. Born Espera Oscar de Corti, he was an Italian American who often portrayed Indians in Hollywood films, including "Chief Iron Eyes" in the 1948 Bob Hope film *The Paleface*. This was not, as we shall see, the only sleight of hand in the Crying Indian ad. Nor was it the most significant.

The Crying Indian PSA must be viewed through the prism of the growing problem of highway litter, from bottles and cans in particular, that followed the construction of the interstate highway system in the 1950s. The problem had reached crisis proportions by the late 1960s. It had become noticeable and disturbing. Who wouldn't want to "Keep America Beautiful," after all? It was clear we had a problem. But *how* best to fix it? And *who* pays the cost?

Consumer advocate Ralph Nader founded Public Interest Research Groups (PIRGs), a network of groups throughout the United States focused on consumer and environmental advocacy, in 1971, the same year the Crying Indian first aired. The PIRGs would play a critical role in advancing one particular vision of how to fix the problem and who should pay, namely, *the bottle bill*. The bottle bill was legislation that placed a deposit on bottles and cans (typically five or ten cents) that would be refunded to consumers upon their return, promoting returnable and refillable bottles and encouraging consumers to recycle rather than toss. The legislation placed an additional burden on supermarkets, grocery stores, and package stores, but the *beverage industry*—i.e., Coca-Cola, Anheuser-Busch, PepsiCo, and so on—would bear the brunt of the responsibility and costs, as they would be required to process the returned bottles and cans. That would add to their expenses and decrease their profits.

Now, fast-forward to the summer of 1984, thirteen years after the PIRGs were founded and the Crying Indian first aired. I had just graduated from high school and I needed a summer job. MassPIRG, which had a base in my hometown of Amherst, Massachusetts, seemed like a perfect opportunity to make some money; I could learn about modern environmental and consumer history and help the environment all at the same time.

MassPIRG—the Massachusetts PIRG affiliate—was best known for its efforts to help pass a bottle bill in the state of Massachusetts. During my training, a veteran canvasser accompanied me as I went door to door soliciting contributions from the residents of a small western Massachusetts town. Among other things, my trainer worked with me as I refined my "rap"—the short statement a canvasser quickly recites during that awkward but pivotal moment between when someone has answered the door and when they've had a chance to say anything other than "hello." My trainer expressly discouraged me from mentioning, in my rap, the *bottle bill*—MassPIRG's signature achievement. At least not when canvassing blue-collar and conservative neighborhoods. Instead, I was encouraged to talk about the *lemon law*—a less prominent piece of legislation MassPIRG had sponsored that protects car buyers from defective car purchases. How odd, I thought at the time, that MassPIRG wouldn't take every opportunity to tout its crowning achievement.

The story of the Massachusetts bottle bill dates back to 1973, when I was eight years old. Working together with the Massachusetts

Audubon Society and other environmental organizations, MassPIRG helped lobby the state legislature to place a five-cent deposit on bottles and cans. After several failed attempts to pass the bill, they were eventually able to get it on the ballot as a referendum. They were opposed, however, by a $2 million ad campaign funded by the beverage industry, funneled through two front groups: the Committee to Protect Jobs, and Use of Convenient Containers. The beverage industry succeeded in defeating the bill, albeit by a very narrow margin (less than 1 percent). In 1977, bottle-bill legislation garnered majority support in the state House of Representatives but failed in the state Senate. In 1979, the bill cleared both House and Senate, but was immediately vetoed by Governor Edward King (who was a Democrat at the time, but would later become a Republican).

Perhaps sensing the growing support behind the bottle bill, King proceeded to help push an alternative to the bill, promoted by the beverage industry, through a front group calling itself the Corporation for a Cleaner Commonwealth. It involved hiring kids to pick up bottle and can litter. The solution, you see, isn't regulation of industry. It's individual action. You've just been introduced to the *deflection campaign*. You'll get to know it much better soon enough.

The bottle bill once again passed both the House and the Senate in 1981. King exercised his veto a second time, calling the bill an embodiment of "everything that is wrong with big government." He claimed it imposed an undue financial burden on individuals and would have adverse impacts on the state's economy. This type of argument—that regulatory solutions to environmental problems are supposedly bad for the economy—will also become all too familiar by the time our story is over.

Subject to an intense lobbying campaign from MassPIRG and others, the state legislature, both the House and the Senate, voted to override King's veto. The bill officially became law on November 16, 1981. But the beverage industry wasn't simply going to roll over. It funded a campaign to repeal the bottle bill, getting a referendum onto the ballot. The referendum failed, but 40 percent of voters were in favor of repeal. The bottle bill was implemented on January 17,

1983, with divided public support, after a bruising battle with millions of dollars of negative advertising spent against it. A little more than a year later, during the summer of 1984, I would be discouraged from talking about the bottle bill in all but the most progressive neighborhoods as I canvassed for MassPIRG.

Similar dramas played out in other states. Oregon was actually the first state to enact a bottle bill, in 1971. Next was the very green New England state of Vermont in 1973. The relatively progressive states of Connecticut, Delaware, Iowa, Massachusetts, Maine, Michigan, and New York all followed suit by the early 1980s. Bottle bills failed in numerous other states, however, once again as a result of intensive lobbying and campaigning by the beverage industry. One ad even depicted a group of sad kids in scout uniforms searching in vain for bottles and cans (to the chagrin of the Boy Scouts and Campfire Girls of America, who complained about their brands being appropriated for dubious political motives without their approval). The bottle bill, you see, would put a damper on their money-making recycling efforts. *Deflection* again.

The beverage industry, in other words, employed a crafty, multi-pronged strategy in opposing the passage of bottle bills. It fought them by lobbying state legislatures and through advertising campaigns aimed at voters depicting such bills as costly for consumers and bad for the business community.

Through the advertising campaigns, the industry did its best to make sure the bottle bills were bruised and battered into marginal popularity, if not toxicity, in the states where they did pass—enough so that any attempts at a national bottle bill, such as the ones proposed by Edward Markey (D-MA) in the US House of Representatives in 2007 and 2009, would be dead in the water. But there was one more critical task—the industry needed to dampen the enthusiasm of the "base," that is, environmentalists intent on action. The week of the very first Earth Day, in April 1970, environmental activists dumped a huge pile of nonreturnable Coke bottles in front of the Coca-Cola headquarters in Atlanta, Georgia, as a means of pressuring the company into supporting a bottle bill.[15] The beverage industry knew it would be difficult to convince those folks and their growing number of followers that a bottle bill wasn't a good thing for the environment. But you *might* be able to convince them that a bill wasn't necessary, or that it wouldn't be effective. That's where the Crying Indian came in.

Putting aside the Native American imagery for now, consider the larger message of the Crying Indian ad: Those bottles and cans that were littering our countryside? They were the result of our bad personal behavior. That's a convenient message to promote if you're an industry whose practices generate massive metal and plastic pollution and you're trying to fight regulations aimed at requiring you to package and process that waste.

Enter Coke and Madison Avenue. The consortium of American corporations behind the "Keep America Beautiful" campaign included Coca-Cola, Pepsi, Anheuser-Busch, and tobacco giant Philip Morris. They had been collaborating with the Ad Council—a nonprofit organization that produces and promotes PSAs on behalf of various sponsors, including environmental groups. In 1971, they worked with New York advertising giant Marsteller (whose public relations firm is Burson-Marsteller, whom you may recall from our earlier discussion of tobacco flame retardants) to create the Crying Indian PSA.

Environmental groups like the Sierra Club and the Audubon Society were initially partners in the campaign, believing it would be a powerful way to raise awareness around littering. But they ultimately distanced themselves from the effort when they realized they'd been had. That's because they realized the "Crying Indian" was, in fact, a public relations ploy hatched by beverage industry groups.

That campaign, part of the multipronged effort to defeat the bottle bill, achieved its primary objective. As we know, only a limited number of blue states passed bottle bills, and a national bottle bill is nowhere on the horizon even decades later. Meanwhile, the growing mass of discarded plastic bottles has given rise to another great environmental crisis of our time—global plastic pollution. It's no longer just littering the countryside: today plastic pollution is so extensive that it has been found in the Mariana Trench, thirty-six

thousand feet underwater. Indeed, it's even in the air: *Wired* magazine ran a story in June 2020 proclaiming, "Plastic Rain Is the New
Acid Rain." *Wired* cited research finding that many tons of micro-
plastics are falling onto wilderness areas every year.[16]

In 2006, ironically, the Ad Council would run another PSA, this
time about ocean plastic pollution. Featuring the title character from
Disney's *The Little Mermaid*, the message was all about how individuals can act by properly disposing of trash. There was no mention of
the beverage industry's role in plastic pollution. Environmental organizations—like the Environmental Defense Fund—were cosponsors. Fool me once . . .

Reflecting on the true story behind the Crying Indian, it's hard
for me not to feel *personally* betrayed, as if the innocence of our
youth was an illusion, as if I—and everyone else of my generation—
was led astray by a false prophet, for the motive of corporate profit.

When it comes to the wider legacy of the Crying Indian, Finis
Dunaway offered this assessment:

> The answer to pollution, as Keep America Beautiful would have it,
> had nothing to do with power, politics or production decisions; it was
> simply a matter of how individuals acted in their daily lives. Ever since
> the first Earth Day, the mainstream media have repeatedly turned
> big systemic problems into questions of individual responsibility. Too
> often, individual actions like recycling and green consumerism have
> provided Americans with a therapeutic dose of environmental hope
> that fails to address our underlying issues.[17]

Which leads us back to the issue of climate change.

WHAT TO DO?

The dual epigraphs that begin this chapter embrace the duality of
the lever arms of action in a functioning democracy. Progress requires individual action—what is a collective, after all, but a group
of individuals? Doing the right thing sets an example for others to

follow and creates a more favorable environment for change. But we also need systemic change, which requires collective action aimed at pressuring policymakers who are in a position to make decisions about societal priorities and government investment.

There are plenty of lifestyle changes that should be encouraged, many of which make us happier and healthier, save us money, and decrease our environmental footprint. Demand-side pressure by consumers can certainly influence the market (indeed, millennials have been accused of killing off various traditional products and services, including landline phones, men's suits, and fast-food chains, with their purchasing decisions[18]). But consumer choice doesn't build high-speed railways, fund research and development in renewable energy, or place a price on carbon emissions. Any real solution must involve both individual action and systemic change.

We must beware of efforts to make it seem as if the former is a viable alternative to the latter. Studies suggest that a solitary focus on voluntary action may actually undermine support for governmental policies to hold carbon polluters accountable.[19] So there is a delicate middle ground—which we must seek out—that encourages personal responsibility and individual action while continuing to use all of the lever arms of democracy (including voting!) to pressure politicians to support climate-friendly governmental policies.

Those who mount deflection campaigns are not truly interested in solving problems—if they were, they'd advocate for multipronged approaches that benefit society at large. Instead, their intent is to sabotage systemic solutions that might be disadvantageous to moneyed interests through subterfuge and misdirection. We can see this in the gun lobby's efforts to deflect the focus away from gun control reform, by shifting attention from the large number of poorly regulated weapons in the country to the mental health of individual perpetrators of gun violence. Big Tobacco, with help from the chemical industry, denied us safe-burning cigarettes but gave us toxic peanut butter and breast milk. The beverage industry largely defeated efforts to pass bottle bills and gave us the problem of global plastic pollution.

And today, fossil fuel interests and the climate inactivists work-ing on their behalf are seeking to block policies aimed at regulating carbon emissions by employing a Crying Indian–like deflection cam-paign. It is instructive to note some of the striking parallels between the current campaign and deflection campaigns in the past. The silly contention by right-wing figures that climate advocates "want to take away your burgers," for example, sounds a lot like the NRA's admo-nition that gun-law reform advocates "want to take away your guns." Both reflect an attempt to prey upon fears of big government and limitations on liberty that are prevalent among political conservatives.

The climate action deflectors, moreover, as we shall see, are also attempting to drive a wedge into preexisting rifts within the climate activist community. That includes rifts arising from the ongoing de-bate over the role of personal behavior versus systemic change. (It also includes rifts involving the politics of identity, and matters of gender, age, and race.) When the climate discourse devolves into a shouting match over diet and travel choices, and becomes about personal purity, behavior-shaming, and virtue-signaling, we get a di-vided community unable to speak with a united voice. We lose. Fos-sil fuel interests win.

In June 2019, I made these points in a *USA Today* op-ed I coau-thored with my Penn State colleague Jonathan Brockopp.[20] In it, we noted the similarity between the Crying Indian deflection campaign of the 1970s and current efforts to equate climate action almost ex-clusively with personal responsibility. A prominent state politician from Vermont who campaigns against air travel responded angrily to the comparison. "You do realize that ten states have bottle bills?" he asked, seeming to think he was disproving my point. In debating, this is sometimes referred to as an "own goal." The fact that fewer than a dozen (all *deep blue*) states have a bottle bill—and the fact that a national bottle bill is nowhere on the horizon—speaks to the success of that past deflection campaign. It serves as a cautionary note when it comes to the current deflection campaign on climate, which is the focus of the next chapter.

CHAPTER 4

It's YOUR Fault

A house divided against itself cannot stand.
—ABRAHAM LINCOLN

DEFLECTION IS A PARTICULARLY DEVIOUS STRATEGY FOR INAC-
tivists. In addition to directing attention away from the need for collective action—such as pricing or regulating carbon, removing fossil
fuel subsidies, or providing incentives for clean energy alternatives—
it divides the community of climate advocates by generating conflict
and promoting finger-pointing, behavior-shaming, virtue-signaling,
and purity tests. It also provides a means for tarring leading climate
advocates as hypocrites and firing up political conservatives by emphasizing the purported personal sacrifice and loss of personal liberty
that climate action demands. It all starts with a simple deflection . . .

DEFLECTION COMES TO CLIMATE CHANGE

The current preoccupation with personal behavior didn't arise in a
vacuum. Much like the Crying Indian's focus on the role of individuals in cleaning up bottle and can litter, the focus on the individual's
role in solving climate change was carefully nurtured by industry.

The concept of a "personal carbon footprint" was something that
the oil company BP promoted in the mid-2000s. Indeed, BP launched
one of the first personal carbon footprint calculators, arguably as part

of a larger public relations effort to establish the company as *the* environmentally conscious oil company.[1] I still recall the inspiring ads BP ran at the time billing itself as "Beyond Petroleum." Whether this reflected a genuine embrace of green energy or a cynical "greenwash" gambit, we'll never know. The Deep Horizon oil spill in the Gulf of Mexico in April 2010 would seem to have ended any hope of BP carving a niche out for itself as the green oil company, though it has recently been trying to talk a good game on climate again anyway.[2]

As environmental author Sami Grover opined, "contrary to popular belief, fossil fuel companies are actually all too happy to talk about the environment. They just want to keep the conversation around individual responsibility, not systemic change or corporate culpability."[3] And as Malcolm Harris wrote in *New York Magazine*, "these companies aren't planning for a future without oil and gas, at least not anytime soon, but they want the public to think of them as part of a climate solution. In reality, they're a problem trying to avoid being solved."[4]

It's thoroughly unsurprising, given the emphasis on deflection, that an oil industry site, for example, promoted an article titled "Is Eating Meat Worse Than Burning Oil?"[5] A bit more surprising is the fact that none other than the *New York Times* has been trafficking in messaging that deflects responsibility from fossil fuel interests and their abettors.

In August 2018, the *Times* heavily promoted an article by Nathaniel Rich titled "Losing Earth: The Decade We Almost Stopped Climate Change."[6] It was the cover story for the *New York Times Magazine*, and the entire issue was devoted to just this one story. The *Times* even produced a video trailer to go with it. In the piece, which focuses on the period from 1979 to 1989, Rich dismisses the fossil fuel industry as a mere "boogeyman" when it comes to climate policy inaction. He absolves the Republican Party of blame as well. And he does so through the transparently tenuous argument that the fossil fuel disinformation campaign didn't really ramp up until the 1990s, so the failure of the climate movement, which began in the 1980s, cannot be tied to it. Robinson Meyer of *The Atlantic* argued that Rich's thesis was wrong—and irrelevant anyway. As we saw in Chapters 1 and 2, the basic infrastructure behind the climate disinformation campaign was already in place in the mid-1980s. The lengthy subtitle of Meyer's piece provides a cogent summary: "By portraying the early years of climate politics as a tragedy, the [*New York Times Magazine*] lets Republicans and the fossil-fuel industry off the hook."[7] The penultimate line in Rich's piece is classic deflection: "Human nature has brought us to this place; perhaps human nature will one day bring us through."

Since the publication of Rich's story, the *Times* has published dozens of pieces emphasizing the role of individual behavior in combatting climate change. That includes what we eat ("The Facts About Food and Climate Change"), how we move around ("One Thing We Can Do: Drive Less"), whether we should vacation ("If Seeing the World Helps Ruin It, Should We Stay Home?"), and how we view our overall role in tackling the climate crisis ("I Am Part of the Climate-Change Problem. That's Why I Wrote About It").[8]

During the heightened meat safety concerns of the coronavirus pandemic in the spring of 2020, the *Times* promoted one particular commentary ("The End of Meat Is Here") with the tagline, "If you care about the working poor, about racial justice, and about climate change, you have to stop eating animals."[9] For those in the back, what the *Times* was trying to tell us, in not-so-subtle terms, is that we can't really claim to care about climate change (or racial justice, or really, just about anything) if we choose to eat meat. That's some weapons-grade psychological meat-shaming manipulation there, *New York Times* editors!

In an unusual move during the lead-up to the 2020 Democratic primaries, the *New York Times* endorsed *two* candidates, Minnesota senator Amy Klobuchar and Massachusetts senator Elizabeth Warren. Both candidates supported action on climate (great!), but the *Times* editorial board couldn't resist making this (not so great) critique of one of them: "Ms. Warren often casts the net far too wide, placing the blame for a host of maladies from climate change to gun violence at the feet of the business community when the onus is on society as a whole."[10] If it looks like deflection, sounds like deflection, and smells like deflection, it's probably deflection.

All of this is not to say that the *Times* has been acting as a willing coconspirator in spreading fossil-fuel-industry propaganda; rather, it has been, at the very least, an unintentional enabler of deflection by buying into framing that overwhelmingly emphasizes individual responsibility over systemic change. Perhaps the *Times* and other mainstream media outlets are victims of "seepage"—the infiltration of contrarian framing into the mainstream climate discourse—arguably a consequence of the fossil fuel industry's constant barrage of dis-informative propaganda.[11]

Sami Grover noted how successful the deflection campaign has proven to be: "Ask your average citizen what they can do to stop global warming, and they will say 'go vegetarian,' or 'turn off the lights,' long before they talk about lobbying their elected officials."[12] My graduate institution, Yale University, has bought into this logic. In defending its decision not to divest of fossil fuel holdings, the Yale Corporation issued a statement arguing that "ignoring the damage caused by *consumers*, is misdirected," blaming individuals over corporate behavior.[13]

In an essay titled "The Futility of Guilt-Based Advocacy," ethicist Steven D. Hales argued that the entire discipline of philosophical ethics has been hijacked by such misleading framing. Hales reported that he "recently attended an international ethics conference, and the overwhelming take-away was the realization that philosophical ethics remains obsessed with individuals . . . what you should do, how you should act, who you should become. It is not appreciated that all the really serious moral issues of our time are collective action problems. . . . The talks that did address collective action issues were keen on making them ultimately a matter of individual responsibility or blame."[14]

It is possible that behavior-shaming—about dietary preference, for example—is fed by a sense of powerlessness, despair, and doom—a pervasive sense of inevitability that is promoted by a form of inactivism we will call *doomism*. After all, pointing fingers is something we can all do, even if we perceive climate mitigation as a lost cause. Climate scientist Daniel Swain wondered aloud on Twitter "why so many people apparently think [personal shaming and advocacy of extreme austerity] is somehow helpful." *New York Times* climate reporter John Schwartz replied, "For a lot of people, shaming is its own reward: that feeling of standing on the moral high ground is a hell of a drug."[15] Is behavior-shaming the modern opiate of the climate-anxiety-stricken masses? And are the inactivists the pushers?

Climate-messaging expert Max Boycoff of the University of Colorado agrees that the focus on individual behavior is a product, at least in part, of the seemingly "overwhelming" nature of the challenge and of our failure thus far to have engaged in the kind of collective response that is needed. He argues, however, that framing the problem solely as one of individual responsibility does little to address it, and that "flight-shaming," for example, "is one of the more unproductive ways to have a conversation." He noted that all shaming does is make people feel bad; it's "blaming other people while not actually talking about the structures that give rise to the need or desire to take those trips."[16]

The messaging from some climate pundits, however, has actually played *into* the individual-behavior deflection campaign. Consider David Victor, a policy researcher at the University of California, San Diego. Victor has been criticized for having taken funding from fossil fuel interests (BP and the electric-utility-backed Electric Power Research Institute, to be specific), for serving as a witness for the Trump administration in opposing the Our Children's Trust lawsuit against fossil-fuel-industry polluters, and for his opposition to the use of actionable warming targets in guiding climate policy.[17] In December 2019, he wrote an op-ed in the *New York Times* downplaying the responsibility of leading emitters for the failure of international climate summits, including COP25, to achieve substantial emissions reductions.[18] Dr. Genevieve Guenther, founder and director of the media watchdog organization End Climate Silence, characterized Victor's op-ed as "a master class" in inactivist "tropes," from "downplaying the responsibility of the US, to exaggerating the costs of action, to focusing on the hard-to-decarbonize sectors, to calling climate policy . . . 'a religion.'"[19]

Victor has also raised eyebrows by insisting that climate advocates should avoid the rhetoric of "corporate guilt" and accept the fact that "we're all guilty." This language echoes that of the fossil fuel companies. In 2018, for example, Chevron argued in court that "it's the way people are living their lives" that's driving climate change.[20] I cannot speak to Victor's intent or motives. What I *can* speak to is how this sort of framing—deflecting responsibility from corporate polluters to individual behavior—is seized upon and exploited by the inactivists. Victor's comments precipitated a particularly heated argument between climate advocates on social media, which leads us to an important related discussion: how deflection plays into the inactivists' efforts to sow division within the climate community.[21]

ANTISOCIAL MEDIA AND BOTS OF WAR

At the center of the acrimonious debate over individual action versus systemic change is a false dilemma. *Both* are important and necessary. But this debate is increasingly being used to drive a wedge within the community of climate advocates. This is what's known as a wedge campaign, and it's nothing new. It has its origins in decades-old disinformation campaigns.

This style of information warfare was first called "the wedge strategy" more than two decades ago by the Discovery Institute, a religious organization devoted to undermining public acceptance of the theory of evolution.[22] The Discovery Institute sought to introduce the teaching of creationism in schools, driving a wedge into the scientific community by appropriating scientific terminology (including the use of the scientific-sounding term "intelligent design" to describe what is actually thinly veiled young Earth creationism) and articulating an agenda that reasonable science educators might buy into.

As with any broad and diverse community, there are natural divisions within the climate movement that fall along lines, for example, of age, gender, ethnicity, political identity, and, of course, *lifestyle* choices—we'll come back to that. Taking advantage of these preexisting fault lines, a particular subclass of the inactivists, whom we shall call the *dividers*, employ the wedge strategy to sow division and discord within the climate community. The principle is simple: divide climate advocates so they cannot speak with one voice, and use this internal division to distract, disable, preoccupy, and nullify.

Today we know that state actors—Russia, in particular—manipulated social media by promoting fake news articles and deploying trolls and bot armies during the 2016 US presidential election.[23] Their mission was to divide Democratic voters by siphoning off potential supporters of Democratic candidate Hillary Clinton and convincing them to support alternative candidates like Bernie Sanders (in the primaries) or Jill Stein (in the general election), or else discouraging voters so they simply stayed home and didn't vote at all. All they needed to do was depress Democratic turnout enough to tip the election to their preferred candidate, Donald Trump. They were successful.

Some of the fiercest online attacks against Hillary Clinton in 2016 appeared to come from the environmental left, criticizing her climate policies (for example, her position on fracking). We now know that many of those attacks were actually Russian trolls and bots seeking to convince younger, greener progressives that there was no difference between the two candidates (so they might as well stay home). As an adviser to the Clinton campaign on energy and climate, I can attest that there was a *world* of difference between Trump and Clinton when it came to climate.[24] But the fact that younger voters remained under the false impression that Clinton was no better than Trump on climate almost certainly played a role in keeping them from voting, effectively handing the election to Trump—and Russia.[25]

We have seen that Russia's efforts were likely motivated—in part, if not entirely—by an agenda of fossil fuel extraction—in particular, a half-trillion-dollar partnership between Russian state oil company Rosneft and American oil giant ExxonMobil. That partnership was stymied by US sanctions against Russia over its 2014 invasion of Ukraine. A President Clinton would have continued to support the sanctions. A President Trump would not. The scandal that has come to be known as Russiagate might actually be so simple it can be summarized in two words: *fossil fuels.*

The fact that Russia is continuing to use social media to manipulate the American public is well known. What has gotten less attention is *why.* And again, the answer may be the same two words. The Russian economy is dependent on the continued extraction and monetization of Russia's primary economic asset—oil reserves. And we know that Vladimir Putin has made dismissive comments about climate change, misguidedly arguing that Russia might actually benefit from it.

Seen in this light, it would be surprising if Russia did *not* use the tactics it has honed over the past decade to continue to manipulate public opinion on climate change to its advantage in the United States and beyond. As a *Chicago Tribune* reporter summarized it, a 2018 congressional report found that "Russian trolls used Facebook, Instagram and Twitter to inflame U.S. political debate over energy policy and climate change, a finding that underscores how the Russian campaign of social media manipulation went beyond the 2016 president election."[26] Their efforts continue to bear fruit.

The deeply flawed notion that there is no difference between Trump and the Democrats on climate remains pervasive among green progressives.[27] In December 2019, when I posted a link to a *New York Times* op-ed emphasizing the role the Republican Party has played in enabling climate-change denial and inaction, I received a number of angry responses insisting that both parties were equally to blame.[28] One user, whose Twitter home page was emblazoned with an image of Bernie Sanders, insisted, "Climate denial is absolutely bipartisan. Frankly, establishment Democrats are worse because they say it's real and still pursue policies that will kill us all." A pile-on ensued: "This establishment Democratic partisan absolutely proves your point," and, "What difference does it make if Democrats believe the science, but then still promote fracking, drilling." A defender did enter the fray (my climate scientist colleague Eric Steig of the University of Washington), tweeting, "There's so much wrong with your way of thinking about this, I don't even know where to start. I'll just make one point: Democrats were in the [2015] Paris agreement. Trump took us out. Yes, this matters." But have no fear, nihilism always finds a way. Responded one final interlocutor, "But the Paris agreement isn't doing anything and isn't strong enough to halt warming even if everyone fully participated, so . . ." Wrong and misguided. And just what the *dividers* ordered.[29]

Of course, it's not just bad state actors who are involved. Fossil fuel interests and their front groups are known to be using similar methods to manipulate public opinion, and the collective effect is a poisoning and weaponization of social media to advance the cause of denial, deflection, doomism, and delay. The basic strategy is as follows: Use professional trolls to amplify a particular meme on social media, and send in an army of bots to amplify it further, baiting genuine individuals to join the fray. The idea is to create a massive food fight (the term is apropos, given that many of the online tussles, as we will see, are actually *about* individual food preferences) and thereby generate polarization and conflict. A favored approach is to seed a prospective online discussion with trolls or bots aggressively advocating opposite positions and acrimoniously attacking each other. Pretty soon a melee unfolds. This has, of course, been done specifically in the climate arena.[30]

Christopher Bouzy is a software developer and founder of Bot Sentinel, a platform that estimates how likely a specified Twitter account is a "bot" based on an evaluation of its pattern of tweeting. *Inside Climate News* reported an episode that Bouzy observed in the wake of CNN's climate forum in early September 2019. An unusually high number of mentions of the term "climate change" (seven hundred) were made over a twenty-four-hour period by the roughly one hundred thousand Twitter accounts Bot Sentinel was tracking as "trollbots" (bots that have been programmed to engage in divisive trolling behavior). According to Bouzy, whenever a particular phrase like "climate change" is trending among trollbots, there is likely some amount of coordination involved: "What we are noticing is these phrases are more than likely being pushed by accounts that have an agenda," he said, adding, "It's fascinating to see this stuff happen in real time. Sometimes we can see literally 5 or 10 accounts able to manipulate a hashtag because they have so many people following them. It doesn't take that many accounts to get something going."[31] *The Guardian* reported that "the social media conversation over the climate crisis is being reshaped by an army of automated Twitter bots." It estimated that "a quarter of all tweets about climate on an average day are produced by bots," with the effect of "distorting the online discourse to include far more climate science denialism than it would otherwise."[32]

DIVIDE AND CONQUER

When it comes to polarizing rhetoric, there is no greater opportunity to divide people than when it comes to lifestyle choices, for they are tied directly to one's sense of identity. Since individual action is tied directly to lifestyle choices, it provides a perfect opportunity for in-activists. They are more than happy to weaponize individual action and deflection in an effort to generate a wedge within the climate movement.

Social media, as we have already seen, provides a perfect tool for dividers to generate a "personal behavior" wedge. Some of our more combative inclinations—finger-pointing, behavior-shaming, virtue-signaling, and grievance-airing—are on full display on social media, and the bots and trolls exploit them to turn minor fissures into major rifts. Aggressive guilt-based engagement is known to be counterac-tive to progress, as a result of something known as the "boomerang effect." One study concluded, for example, that when it comes to reducing household energy use, "it is particularly important to focus on positive norms (desirable behaviors) rather than negative norms (undesirable behaviors)." Behavior-shaming of individuals may even be counteractive to climate action. Think of this as a "bonus" for the inactivists.[33]

Behavior-shaming isn't *always* a bad thing. When it is directed at the inactivist politicians, industry shills, and climate-change deniers seeking to block action on climate, it is wholly appropriate. Which brings us to Greta Thunberg, the seventeen-year-old Swedish girl who has become the leader of a global youth climate movement. Thunberg has chided politicians and opinion leaders who fail to support climate-friendly policies. But interestingly—and relevant to our larger thesis here—much of the media and social media emphasis has been on her *personal* behavior—in particular, her refusal to fly (owing to the carbon footprint of commercial flight), which resulted in her much-publicized transatlantic crossing by sailboat to participate in the September 2019 United Nations Climate Change Summit in New York City.

Thunberg seems to be a favorite target for those looking to create polarization. Consider the former Fox News executive Ken LaCorte, who operates a website called Liberal Edition News. According to the *New York Times*, it uses "Russian tactics" by feeding readers "a steady diet of content guaranteed to drive liberal voters further left or to wring a visceral response from moderates." One example of LaCorte's divisive content, according to the *Times*, was a story that "singled out an Italian youth soccer coach who called Greta Thunberg, the teenage climate activist, a 'whore.'" It is thoroughly unsurprising that inactivists have employed Thunberg in their wedge-generation efforts.[34]

I have even found myself at the center of those efforts. I have repeatedly emphasized in my public outreach that *both* individual and collective action are important when it comes to climate-change mitigation.[35] I've been outspoken in my support of Thunberg's efforts, for she exemplifies this view in her own actions. But facts be damned if there's a wedge that can be created.[36]

In November 2019, two academics wrote a click-bait commentary on the *Forbes* website with the leading title "Does Greta Thunberg's Lifestyle Equal Climate Denial? One Climate Scientist Seems to Suggest So."[37] Yes, that "climate scientist" was purportedly me. The authors sought to manufacture a fake conflict between Thunberg and me by misrepresenting statements I had made in an interview with *The Guardian*.[38] In the interview, I had criticized inactivists for deflecting attention from systemic solutions to personal action. The authors of the *Forbes* piece claimed, without any evidence, that my criticism was directed at Greta Thunberg, brazenly insisting that "Thunberg v. Mann is now the debate to watch!" The accusation, as others noted, was absurd.[39]

Ironically, in the very same *Guardian* interview, I had warned, "We should also be aware how the forces of denial are exploiting the lifestyle change movement to get their supporters to argue with each other." Mission accomplished! While there's no evidence that these authors were working directly on behalf of inactivists, their efforts were indeed exploited by them.

If you're thinking such disingenuous claims are a one-off, a fluke, an isolated incident, think again. Just a few months earlier, Anthony Watts, a climate-change denier affiliated with the Koch brothers–funded Heartland Institute, leveled a very similar attack against Katharine Hayhoe, a leading climate science communicator. Watts posted a commentary falsely claiming that Hayhoe had attacked Greta Thunberg's "Climate 'Shaming' Crusade."[40] Hayhoe responded immediately on Twitter, saying, "[I] don't normally bother to call out liars on Twitter but I will here, as they're trying to invent a disagreement to drive a wedge between us. @GretaThunberg is not personally shaming anyone: she is acting according to her principles."[41]

The above examples show how the inactivists are seeking to pit climate scientists against climate advocates. They are indisputably also drawing upon generational and cultural divides, seeking, for example, to pit the ivory-towered Baby Boomer and Gen X academics against the down-in-the-trenches, upstart millennial, Gen Z youngsters. Such divisive efforts appear to be catching on ("Okay, boomer," anyone?).[42]

The clashes have spilled out into the streets. Consider this example involving former California governor Jerry Brown. Brown made climate change the signature issue of his final term. He spearheaded the most ambitious climate target in history with an executive order committing California to total, economy-wide carbon neutrality by 2045.[43] He led the opposition to the Trump administration's threat to withdraw from the 2015 Paris Agreement, leading the "we're still in" coalition of states, municipalities, corporations, and businesses that vowed to uphold their commitments. When rumors emerged shortly after the 2016 election that Trump was planning to defund climate data-collecting satellites, Brown famously said, "California will launch its own damn satellite." I was there and heard him say it.[44]

While I will concede that I'm a bit biased—I am both an adviser to, and a friend of, former governor Brown—I was genuinely taken aback to see him vilified as the *enemy* by youth and social justice protesters at the Global Climate Action Summit that Brown held in San Francisco in September 2018. One journalist described the protesters as "part of the environmental justice movement, many people of color from less well-off, more polluted places. . . . They want the fight for climate change to result in cleaner air for them right now," adding, "The people they are protesting (Governor Brown and his allies) are part of the mainstream environmental movement, which is often wealthier and whiter."[45] The protesters attacked Brown for failing to ban oil and natural gas drilling (sometimes called fracking) in California. One banner read, "Climate Leaders Don't Frack." If ever there was a case of the perfect being the enemy of the good, this was it. As Brown pointed out, "We're trying to do more . . . but we have a legislature; we have courts; we have a federal Congress and federal courts. . . . We've got a lot of elements in the political landscape that in a free society we have to deal with."[46]

You might argue that this was a clash between youthful idealism on one side and jaded pragmatism and realpolitik on the other, but there were elements of race and culture and perhaps gender undergirding the confrontation. One former Green Party candidate for the US Senate and Congress, Arn Menconi, posed the rhetorical question, "Have you noticed the other hit pieces coming from our own side that when you don't agree with them you're a sexist and racist?"[47]

Though inactivists are constantly looking for wedges they can use to divide the climate movement, sometimes the climate movement makes it easy for them. Consider, for example, the recent remarks of Roger Hallam, cofounder of the climate activist group Extinction Rebellion, who downplayed the Holocaust. Combined with his controversial past statements on sexism, racism, and democracy, his statements have led to the German branch of Extinction Rebellion distancing itself from the cofounder of the very organization with which they're affiliated.[48]

Let us return, however, to the matter of individual action versus systemic change, for this is at the very center of the wedge campaign. Here we have seen even the most well-meaning of environmental organizations contribute to the growing schisms within the climate movement. The Sierra Club has played a key role in promoting climate action in recent years under the leadership of its executive director, Michael Brune. But readers may also recall that the Sierra Club was one of the (presumably unwitting) initial supporters of the infamous beverage-industry-hatched "Crying Indian" deflection campaign. It appears that the Sierra Club has more recently, at least at times, also fallen victim to the broader climate-change deflection campaign.

In an editorial in the organization's magazine, *Sierra*, titled "Yes, Actually, Individual Responsibility Is Essential to Solving the Climate Crisis," the magazine's editor, Jason Mark, commented on a *USA Today* op-ed I wrote in June 2019 with my Penn State colleague Jonathan Brockopp.[49] In the piece, called "You Can't Save the Climate by Going Vegan. Corporate Polluters Must Be Held Accountable," we opined that *both* individual action and systemic change are important, and that the former cannot be a substitute for the latter. Jason Mark implied that we had presented a "half-truth"—that we had "traded lifestyle scolding for now being scolded if you mention the importance of personal lifestyle." What we had *actually* said was that an *exclusive* focus on individual action, to the point of neglecting systemic change, is problematic: "Though many of these actions are worth taking, and colleagues and friends of ours are focused on them in good faith, a fixation on voluntary action *alone* takes the pressure off of the push for governmental policies to hold corporate polluters accountable" (emphasis added). To be fair, Mark did subsequently revise the online version of his commentary to clarify that we had not in fact argued against the role of lifestyle change.

The Sierra Club has been and remains an important ally in the climate wars. This example serves as a reminder that even allies can buy into unhelpful framing, especially when inactivists are so intent on nurturing that framing.

FOOD FIGHT

Down in the trenches of social media, the wedge campaign is clearly bearing fruit. The dividers have successfully generated a veritable "food fight"—in fact, a literal one, getting people to argue about their dietary preferences, as well as their preferred means of transportation, how many children they have, and other matters of lifestyle and personal choice. "If nobody is without carbon sin, who gets to cast the first lump of coal?" I asked in a commentary for *Time* magazine. "Who is truly walking the climate walk? The carnivore who doesn't fly? The vegan who travels to see family abroad?" The opportunities for finger-pointing seem endless.[50]

Meat, however, is central to the food fight. Especially the meat that so many seem to have a beef with—beef. Though beef consumption is responsible for only 6 percent of total carbon emissions, it often seems to fill close to 100 percent of my Twitter feed.[51] The meat melee was fed by a highly successful and influential 2014 documentary, *Cowspiracy*, which promoted the false notion that meat-eating is the primary contributor to human-caused climate change. *Cowspiracy* diverted—you might even say *deflected*—attention from the real conspiracy on the part of fossil fuel interests to confuse the public about the role of fossil fuel burning. Billed as "The Film That Environmental Organizations Don't Want You to See," *Cowspiracy* maintained that the livestock industry has conspired with leading environmental organizations to hide the fact that meat consumption, not the burning of fossil fuels, is responsible for the lion's share of warming.

But that thesis isn't the product of incisive research and investigative journalism: it's a product of dubious non-peer-reviewed claims built on bad math, amplified by a polemic film.[52]

As a result, there is now a seemingly limitless and very animated community of vegan activists who are convinced that meat-shaming is the solution to climate change. I myself have been attacked on social media for my ostensible lust for meat—despite being up front about the fact that I don't eat it.[53] Animal rights activists often attack climate scientists simply for pointing out that the contribution from meat-eating is small compared to that from the burning of fossil fuels. Such hostility toward fact-based discourse from the ostensible left finds much in common, ironically, with the attacks on climate scientists from the climate-change-denying right. Unwittingly or not, a sizable number of vegan activists have been weaponized by the dividers and deflectors for their cause.

As I frequently point out, I get my electricity from renewables, have one child, drive a hybrid vehicle, and don't eat meat.[54] I have chosen a meat-free lifestyle for several reasons—solidarity with my daughter (who has chosen a meat-free diet), ethical considerations, and a desire to minimize my environmental footprint. I think it's important to set a good example. But I do not attempt to police others' lifestyles. We have already seen that this can be counteractive to climate progress. Gentle encouragement, and, most importantly, incentives for lifestyle change, are a better path.

There are many worthwhile societal problems to confront—animal rights, a cleaner environment, social justice, income inequality, the list goes on. I respect and appreciate those who advocate for tackling these pressing challenges. But I am wary of individuals who seem to be trying to hijack the climate discourse for the cause of worldwide veganism (or, for that matter, the abandonment of market economics—something we'll discuss later). It represents another form of deflection and plays right into the agenda of the inactivists.

Of course, personal behavior isn't limited to food choices. Methods of travel, particularly flight, are another source of great contention. Unlike overland transportation (automobiles, buses, trains), which can be electrified and powered by renewable energy, flight requires moving great amounts of mass over large distances, and at present it is very difficult to decarbonize. There is promising research into high-density biofuels, and with better battery technology, electrification of aviation should eventually become possible. In the meantime, those who rely on air travel have no choice right now but to keep emitting carbon—and, for those of us, like myself, who have the luxury of being able to pay for it, purchasing verifiable carbon offsets.

Still, flying receives a disproportionate amount of shaming, considering its carbon contribution. Air travel only accounts for about 3 percent of global carbon emissions. It is dwarfed by emissions from the rest of the transportation sector, as well as the power sector and industry. In fact, alternatives can sometimes incur a greater carbon footprint than flying—one study has found that when a full life-cycle analysis is performed (i.e., accounting for construction and maintenance of vehicles and required infrastructure), train travel can sometimes come with a higher carbon footprint than air travel.[55] It might also be noted that taking three days off work to travel to a destination by train, or three weeks to travel by boat, is a luxury most working people can't afford. Insisting they do it anyway could be argued to be laden in implicit privilege.

So why does flying, like meat-eating, elicit such passion and so much attention when it comes to our individual carbon footprints? Perhaps deflectors want us to focus on the purported hypocrisy of elites—jet-setting celebrities and conference-attending scientists—who advocate for climate action. We'll return to this point later. Perhaps class warfare is at work. Like age, gender, and ethnicity, class is something that divides us, and it provides a natural wedge for those who want people to fight with each other rather than working together to improve things. Part of the opposition to flight seems to be that it is mostly well-off people who do it. Business-related air travel is a matter of routine for businesspeople, and their companies and corporations pay for it. The greater someone's disposable income, the more likely that person is to fly to distant locations for vacations. Flight-shaming is a good fit with those who see capitalism itself as the enemy. We'll return to this point later, too.

Meanwhile, let's recognize one other matter of personal choice that has substantial implications for our individual carbon footprints: having children. All other things being equal, doubling your family size doubles your collective carbon footprint. It is, of course, possible to raise a child in a carbon-friendly manner—in fact, a friend of mine wrote a book called *The Zero-Footprint Baby* on how to do just that.[56] But doing so requires even more diligence than managing your own carbon footprint—if you're not very good at the latter, it's exceedingly unlikely you'll be any good at the former.

Responding to the criticism she gets from behavior-shamers about her own personal behavior—why she flies to conferences or why she had children—climate scientist Katharine Hayhoe responded that "flying and eating and having children is often framed as a purity test. It's like, 'So you say you care about climate change. But if you have a child or eat meat or, heaven forbid, have ever stepped in an airplane, then you are not one of our allies. You are one of our enemies.'" Hayhoe noted that the sorts of attacks on climate scientists that used to come from climate-change deniers now come from "people who are not only concerned but are part of the fight."[57] Jonathan Foley, a climate scientist who is now executive director of Project Drawdown, an organization devoted to societal decarbonization, calls flight-shaming "the climate movement eating its own."[58]

I must confess that I share the concern and bewilderment of my climate scientist peers. We've chosen to devote our lives to raising awareness about the climate crisis, and yet many ostensible climate advocates would gladly throw us under the bus because we're not living our lives as off-the-grid vegan hermits. Who, today, can truly afford to live without carbon "sin"? The flightless vegan with four children? The childless vegetarian whose job requires transcontinental air travel? The childless, flightless celiac sufferer who relies upon meat for protein? We all face real-world challenges and tough choices that complicate the effort to completely decarbonize our lives in a system that is still reliant on fossil fuel infrastructure. We must change that system. Individual efforts to reduce one's carbon footprint are laudable. But without systemic change, we will not achieve the massive decarbonization of our economy that is necessary to avert catastrophic climate change.

Would Hayhoe and other leading climate scientist-advocates be as effective as they are if they chose to operate entirely outside the system that exists? While some have argued so, I think a compelling argument can be made that advocates for change have the greatest reach, as measured by media accessibility, public speaking opportunities, and engagement with policymakers and stakeholders, working *within* the system that exists.

Nonetheless, many prominent climate advocates continue to advance the "individual action" primacy frame. *Guardian* columnist George Monbiot, known for his advocacy for climate action and his efforts to call out bad corporate actors in the climate space, has insisted that "we are all killers" (that is, those of us who continue to fly are killing the planet).[59] Meteorologist Eric Holthaus, the former editor of the *Wall Street Journal*'s weather blog, decided back in 2013 to swear off air travel for good, insisting "there is no other way that makes sense" when it comes to reducing carbon emissions. He reportedly also considered taking the extraordinary measure of having a vasectomy to guard against his own contribution to population growth.[60]

Holthaus, who garnered a fair amount of media coverage at the time for his no-flying pledge (some of it rather mixed[61]) went further, however. He demanded that others, too, refuse to fly, scolding science writer Andrew Freedman when he said that "those of us that report on weather & climate . . . should be the ones leading by example. There is no neutrality anymore."[62] He also excoriated Jason Rabinowitz, a travel aficionado and host of an aviation podcast, when Rabinowitz shared his plans to fly to Madrid, despite having no particular reason to go, because he had found cheap tickets. Holthaus equated Rabinowitz's pleasure trip to "taking a gun and just firing blindly into the air towards a crowd just because you think it's fun to shoot a gun." Holthaus didn't stop there, going on to call Rabinowitz a "selfish, entitled a**hole," adding, "You don't care who you're hurting. You just care about yourself." He capped off this diatribe with the statement that "unexamined privilege like this is literally causing the biggest existential threat we've ever faced as a species." Note the insertion of yet another wedge issue—socioeconomic class—into the conversation. Conservative media outlets seemed to delight in reporting on the fracas.[63]

Holthaus's critiques afford us a window into how wedge-generation seems to feed on itself, beginning with one form (e.g., behavior-shaming), and expanding to others (e.g., identity politics). In June 2019, I found myself at the center of Holthaus's criticism over my association with the climate-change-themed documentary *Ice on Fire*. The basis of the critique? I—as a white male scientist—had appeared in a trailer for the film (produced by HBO) that he felt didn't have enough gender and racial diversity. It is worth noting that I had no editorial role in either the trailer or the film, that the director was a respected woman filmmaker (Leila Connors), and that the film itself featured a diversity of voices and near equal balance with respect to gender.

Indeed, I have frequently found myself immersed in the toxic online environment that prevails in the climate space today, and my own experiences are representative of a much broader pattern in today's online discourse. The dividers, as we have seen, will feed on any organic disagreement, amplifying divisive messages with bot armies and professional trolls. The friendly fire from ostensible climate advocates provides them with more than ample fodder.

YOU HYPOCRITE!

Dividers have sought to target influential experts and public figures in the climate arena as "hypocrites" by accusing them of hedonistic lifestyles entailing huge carbon footprints. It's a brilliant strategy, because it associates concern about climate with social elites, creating a class/culture wedge and discrediting critical thought leaders, which in turn limits the effectiveness of their messaging efforts. All the while, it once again places the emphasis on the _behavior of individuals_, deflecting attention away from the need for systemic change and policy action. It's a perfect storm of divide, discredit, and deflect.

In many cases, the accusations are false and misleading. But what if they were accurate? What if these thought leaders really did have outsized carbon footprints? Would the "hypocrite" accusation be fair? As cogently articulated by environmental writer David Roberts, "there's a hidden premise here . . . that personal emission reductions are an important part of the fight against climate change—if you take climate seriously, you take on an obligation to reduce your own emissions." But, as we have already seen, personal action means little without systemic change. As Roberts noted, individuals targeted as hypocrites typically "are not advocating sacrifice or asceticism" but instead articulating the case for systemic change. "If they advocate for, and are willing to abide by, taxes and regulations designed to reduce emissions, then such folks are being true to their beliefs. You might think they are wrong . . . but they are not hypocrites."[64]

Take Al Gore, the public figure most closely associated with the climate crisis. Gore has for years been pilloried by the right-wing press, including Fox News and the rest of the Murdoch media empire, for the size of his home, his electricity bills, and even his weight. It's all part of an effort to portray him as a gluttonous hypocrite who doesn't practice what he preaches when it comes to his personal carbon footprint.

Now consider what happened upon the release of Gore's breakthrough documentary _An Inconvenient Truth_ in 2006. Credited for raising international public awareness about climate change, the film pulled no punches, pointing the finger directly at polluting interests for society's collective failure to act on climate. Shortly after the film was released, a Koch brothers dark-money interest front group, the Tennessee Center for Policy Research (now the Beacon Center of Tennessee), issued a "report" claiming that the former vice president's home used twenty times more energy than the average American home.[65]

Never mind that the "twenty times" number was inflated (it was probably closer to twelve), or that, as Gore's spokespeople pointed out, his residence was also his office, which was staffed with employees, and that Gore paid a premium to get his electricity from green energy sources.[66] There was a smear to be had! The story was promoted by the right-wing media; indeed, the "hypocrite" narrative ultimately proved irresistible even to mainstream media outlets like ABC News, which ran the headline, "Al Gore's Inconvenient Truth: a $30,000 Energy Bill."[67]

Betraying the true agenda behind the smear campaign—to discredit a key climate messenger at a critical moment *and* to deflect attention from systemic change toward individual behavior—the Tennessee Center's twenty-seven-year-old "president," Drew Johnson, gleefully quipped that "as the spokesman of choice for the global warming movement, Al Gore has to be willing to walk the walk, not just talk the talk, when it comes to home energy use."[68]

As the most prominent celebrity climate activist, Leonardo Di-Caprio has been similarly targeted for attack, being portrayed as a jet-setting hedonist by the conservative media. "Hollywood Hypocrite's Global Warming Sermon," read one headline in the Murdoch-owned *Herald Sun* of Australia.[69] Another Murdoch-owned paper, the *New York Post*, ran an article titled "Leo DiCaprio Isn't the Only Climate Change Hypocrite," which also name-checked actor-turned-climate-activist Mark Ruffalo, President Barack Obama, and . . . Pope Francis.[70] Yes, not even the pope is above being smeared by Rupert Murdoch when fossil fuels are on the line. In case those headlines seem too subtle, Britain's right-wing *Daily Mail*, which has attacked DiCaprio as a supposed hypocrite numerous times, gives us "Eco-Warrior or Hypocrite? Leonardo DiCaprio Jets Around the World Partying . . . While Preaching to Us All on Global Warming."[71] You get it? DiCaprio is a *hypocrite*, because he has a *large carbon footprint*, and so we shouldn't be concerned about *global warming*.

Absurd, you say? And yet it's an incredibly effective message, and not just with conservatives, but with the privilege-averse environmental left as well. There is something about perceived hypocrisy—and the sense that someone else is getting more than their fair share—that seems to tap directly into the reptilian part of our brains, bypassing the logic circuits and eliciting instinctive outrage and anger. The intent is to focus that outrage and anger on climate champions—and, ideally, the entire climate movement.

So are the charges fair? As the journalist David Roberts pointed out in a 2016 article for *Vox*, there is not "any evidence that DiCaprio has advocated personal emission reductions or told anyone they ought to forgo planes or boats." Roberts noted that "his focus is on [the need for] political leadership. . . . So the 'hypocrisy' charge fails. You're not a hypocrite for not doing things you haven't said anyone else should do either."[72]

But what about the claim that figures like DiCaprio should simply do better, and that, as an opinion leader devoted to raising awareness about climate change, he should "walk the walk," signaling to others the behavioral changes that are necessary? Roberts answers that charge as well: "If signaling is the issue, well, DiCaprio is supporting electric cars and pushing for clean energy in the film industry and building eco-resorts and supporting clean energy campaigns and starting a friggin' climate charity. Oh, and making heartfelt appeals in front of 9 million people at the Academy Awards. That's a lot of signaling! . . . DiCaprio has a long history of serious work on this issue. By any measure, he's doing better on signaling than the vast majority of wealthy, influential people."[73]

In spite of her extraordinary actions to reduce her carbon footprint, even Greta Thunberg has become a favorite target of these types of attacks. Inactivists have been doing their best to dismiss her and her fellow youth climate activists as hypocrites. After Thunberg famously crossed the Atlantic in a boat to participate in events surrounding the September 2019 UN Climate Change Summit in New York City, Anthony Watts posted an article on his website pouncing on her for, among other things, traveling in a boat that was made partly of "non-recyclable plastic"—really![74]

On September 20, 2019, Thunberg marched with thousands through the streets of New York City. Millions more in other cities around the world joined into this Global Youth Climate Strike, an event seeking to draw attention to the climate crisis. In an effort to undermine the message of the strike, a hoax photo was promoted on Facebook by a dubious group calling itself the Australian Youth Coal Coalition. It claimed to show trash strewn by climate strikers in Sydney's Hyde Park. The caption read, "Look at the mess today's climate protesters left behind in beautiful Hyde Park. So much plastic. So much landfill. So sad." The post was shared nineteen thousand times in twelve hours, and thousands more times by copycats, both on Facebook and on other social media platforms. Unsurprisingly, the image wasn't from the climate strike; it wasn't even from Australia. It was taken in April 2019 at London's Hyde Park following a marijuana festival. In fact, it had already been falsely used before to smear a previous climate protest by the climate activist group Extinction Rebellion—ironically, the Extinction Rebellion folks had actually tried to *clean up* the mess left behind from the festival.[75]

Alleged "hypocrisy" is often the weapon of choice, and it has been wielded against numerous leading public figures in the climate arena. Murdoch's *Herald Sun*—and, of course, Anthony Watts—attacked Bill McKibben, a leading climate activist and founder of the international activist organization 350.org, for his use of air travel.[76] In fact, a Republican Party opposition group had him trailed, trying to get photographs of him—god forbid—using plastic grocery bags.[77] Murdoch's *New York Post* found fault with Green New Deal advocate and freshman congresswoman Alexandria Ocasio-Cortez (AOC) of New York for having the temerity to use an automobile.[78] The Murdoch press criticized the mayor of Sydney, Australia, Clover Moore, for flying. But she was wise to what was going down, tweeting, "This isn't about flights . . . The underlying issue is what climate scientist @MichaelEMann refers to as 'deflection.' While those against action on climate change used to flatly deny climate science, their tactics have matured. Now they don't deny; they deflect."[79]

Sometimes we witness the buckshot approach. Katie Pavlich of the Young America's Foundation—a Koch brothers–supported outfit—managed to get *The Hill*, an ostensibly mainstream media outlet, to publish an op-ed hit piece ("The Frauds of the Climate Change Movement") invoking all the classic hypocrisy tropes, going after the youth climate protesters, Al Gore, and Barack and Michelle Obama, too, for good measure. Pavlich argued that the Obamas' interest in purchasing a seaside home contradicted their belief in climate change and sea-level rise. I kid you not: "If the former president is truly concerned about sea levels rising as a result of climate change . . . his latest real estate purchase places doubts on his sincerity."[80]

Climate scientists, too, as we have already seen, have been deemed fair game for such assaults. I am frequently subject to specious criticisms about my individual lifestyle choices by those looking to discredit my messaging on climate action. Back in November 2018, in response to a widely misunderstood and misrepresented statement by AOC about how much time is left to avert dangerous warming, I tweeted, "We have ZERO years left to take action. Ten years is (an overly conservative estimate of) when we cross into dangerous territory in the absence of immediate action on climate."[81] One critic replied, "Is @MichaelEMann vegan or vegetarian? Does he drive an EV and have solar at home?"—critiques that, as we've already seen, are misguided in my case.[82]

During my sabbatical in Sydney, Australia, in early 2020, the Australian Broadcasting Corporation invited me onto a news program to discuss the unprecedented heat and bushfire outbreak that Australia was experiencing at the time. Tweeted one critic, "So Michael, you flew over on a jet plane to criticise the democratically elected Australian government on its Climate Policy?"[83] It was a deft use of deflection, behavior-shaming, and hypocrisy-leveling all at the same time by an anonymous user who had just joined Twitter, followed nobody, was followed by nobody, and had only ever tweeted a single thing—that reply to me. Dark money meet dark Twitter.

Why would dividers go after *scientists*? The public perceives scientists to possess both authority and integrity; they rank as the most trusted messengers in society.[84] Undermine the public's trust in scientists, and you undermine their message. That premise underlies the "kill the messenger" strategy by climate-change deniers that we encountered in Chapter 1. The problem the deniers face is that the evidence is so overwhelming that attacks on the science itself aren't credible. Exposing scientists as supposed "hypocrites," however, is a relatively effective way of undermining their perceived integrity along with their credibility as climate messengers.

One might think that scientists would push back fiercely against these cynical and exploitive charges of hypocrisy. Yet the response of many climate scientists and climate communicators has been to internalize the criticism. It's an example of the aforementioned phenomenon of "seepage," a kind of academic twist on the Stockholm syndrome. In short, scientists and communicators absorb the bad-faith criticisms of hypocrisy leveled against them by their adversaries and make dramatic changes in their lifestyle, unwittingly *affirming* the flawed and misleading premise that it's all about personal action rather than policy. Leonardo DiCaprio—a hero in my book—has nonetheless spearheaded a global publicity campaign pointing the finger at the eating of meat.[85] A whole bevy of climate scientists and advocates now advertise the fact that they no longer fly, have turned to vegan diets, or have chosen not to have children. These individuals are trying to do what they believe to be the right thing, and attempting to lead by example. But they seem surprisingly unaware that when they seem to make it all about personal choices and the need for sacrifice, they are in fact unwittingly playing into the inactivist agenda. The Crying Indian PSA redux.

This framing was furthermore encouraged by a peer-reviewed study in the journal *Climatic Change* purporting to demonstrate that scientists who didn't "walk the walk" in limiting their personal carbon emissions were less likely to be believed or trusted than those who did. More specifically, the authors claimed that "the communicators' carbon footprint massively affect their credibility and intentions of their audience to conserve energy," and, "their carbon footprint also affects audience support for public policies advocated by the communicator." Or, as the study's lead author, Shahzeen Attari, put it, "It's like having an overweight doctor giving you dieting advice."[86]

Here's the problem: The study didn't actually demonstrate *any* of these things. The protocol of the study was akin to what is known in political polling as a "push poll," that is, a poll that poses a question in such a way as to elicit a particular response. Specifically, at the beginning of the study, the researchers presented respondents with information about a hypothetical communicator's carbon footprint. That, of course, planted the seed in the respondent's mind that personal carbon footprints are the important thing to be focusing on when it comes to climate solutions. I have done thousands of media interviews about climate change during the past few years and I cannot recall a single instance when the interviewer asked me about my personal carbon footprint (other than the one or two cases when the article itself was actually *about* individual carbon footprints). Rarely, if ever, would viewers or readers have any information about my individual carbon footprint. So how could it possibly influence their appraisal of what I have to say? Only in the very artificial laboratory created by this study's protocol could a communicator's personal behavior so directly influence the impact of their message.

Furthermore, the advice the study offered was completely impractical. If climate advocates had to live off the grid, eat only what they could grow themselves, and wear only the clothes they'd knitted from scratch, there wouldn't be much of a climate movement. That level of sacrifice is unacceptable to most. Climate communicators must operate within the system that exists to be effective while articulating the case for changing that system.

Finally, and perhaps most importantly, the study neglected a growing body of research demonstrating that an inordinate focus on individual action can erode support for systemic solutions to the climate-change problem—that is, governmental climate policy.[87] Given that effective policy is far more critical than individual behavior in actually achieving the necessary carbon emissions reductions to stave off catastrophic climate change, the case could easily—I would even say *convincingly*—be made that attempts to redirect focus to communicators' individual carbon footprints are antithetical to action on climate.

The impact of the *Climatic Change* article and its framing has nonetheless reached well beyond the scientific and academic communities. There is now a popular myth, for example, that it is youth climate activist Greta Thunberg's lifestyle decisions (for example, her unwillingness to use air travel and her vegan diet) that are responsible for her outsized impact on the climate conversation. One commentator in *Forbes* argued that the *Climatic Change* study "may explain why Greta Thunberg has succeeded more than others at communicating the climate crisis and galvanizing social action."[88] Science historian Cormac O'Rafferty took issue with the assertion, however, writing, "The authors . . . assert that Greta's lifestyle may be one reason for her great impact. No evidence in the paper is presented to support this hypothesis. It doesn't hurt of course, but I suspect there are many reasons for the impact of Greta in particular."[89] Indeed, I think it's much more defensible to argue that it is Greta Thunberg's fearlessness in speaking truth to power and her steadfast demand that policymakers support the needed systemic changes that make her so powerful and compelling a figure.

RED MEAT FOR THE BASE

There is another adverse consequence of the "individual responsibility" deflection campaign that I have thus far only alluded to.[90] Requiring climate activists to live ascetic lives gives the appearance that they expect everyone else to give up meat, or travel, or other pleasures. And that is politically dangerous: it plays right into the hands of the inactivists who want to portray climate champions as freedom-hating totalitarians. In other words, there is the danger that efforts appearing to place constraints on individual behavior unduly antagonize and energize the conservative opposition to climate action.

Anger can be stirred up among both progressives and conservatives. With progressives, it's typically about issues of perceived injustice. With conservatives, it often involves the perceived loss of personal liberty. Making climate action all about personal sacrifice re-inforces the dominant conservative framing: "They'll take away your hamburgers and plastic straws . . . then your guns, too." Predictably, this very notion has become a tribal rallying call for conservatives.

Observe the messaging by trolls and trollbots promoting this framing. I encounter it constantly in my own Twitter feed. One example, which occurred in the lead-up to the September 2019 UN Climate Change Summit in New York City, was initiated by denier-for-hire Patrick Moore.[91] (Moore is perhaps most famous for saying that the Monsanto-produced weed killer glysophate was safe enough that "you can drink a whole quart of it and it won't hurt you." When presented with a cup of it during a live interview, he refused to drink it and stormed off the set.[92]) Moore, in a character-istically distasteful ad hominem attack, referred to me as a "death-wish cultist" for noting that stabilizing warming below dangerous levels will require us to eventually reach zero net carbon emissions (which can be achieved through both decarbonization of our econ-omy and use of technologies to capture and sequester carbon). Re-plied one account with a "problematic" (65 percent) trollbot rating from Bot Sentinel, "With the anti red meat brigade you wont even be allowed to wear animal skins when you return to the stone age."[93]

Marianne Lavelle of *Inside Climate News* documented eerily sim-ilar messaging patterns by trollbots ahead of the 2019 UN Climate Change Summit. "Democrat #Socialists want to ban:—everything made from plastic—red meat—nuclear power . . . I don't think even Venezuelans have ever been this brainwashed!" tweeted an account with an 84 percent trollbot rating. "The Dems position is ban straws, portion our meat, take our guns, take our cars, abort on a massive scale for population control, and all so we can die from climate change in 11 years!!" tweeted another account with a suspicious rating.[94]

Here's why this is important: Climate change is no longer a wedge issue in and of itself. Climate dismissives poll in the single digits, and recent polling shows that most conservatives are on board with clean energy solutions.[95] So the inactivists need something else to mobilize conservatives for their cause. Framing action on climate as a big-government "they'll take away your burgers" power grab that threatens personal freedom is tailor made, ideologically speaking, for this purpose.

Unfortunately, climate messengers often play right into this agenda when they characterize action on climate in terms of "sacrifice." Energy and climate author Ramez Naam put it this way: "I think a focus on personal sacrifice . . . turns off many in the middle. It makes getting political action on climate harder rather than easier."[96] The reason that conservative media outlets like *Twitchy* and *Conservative Edition News* were so delighted to report on Eric Holthaus's flight-shaming episode was that it fed the narrative of environmental extremists trying to dictate how other people should lead their lives.[97] It was a gift from deflectionist heaven.

The premise that climate action demands sacrifice is itself deeply flawed. If anything, the opposite is actually the case. The cost of _inaction_ on climate, as measured in the damage done by devastating wildfires, heatwaves, wildfires, floods, and superstorms, is far greater than the cost of taking action. The real sacrifice would be if we *fail* to act, and subject ourselves to ever more dangerous and damaging climate-change impacts. *That's* the appropriate frame to be using here.

Instead we see climate advocacy organizations often reinforcing the notion of a need for "personal sacrifice" in their messaging. Consider again the Sierra Club article by Jason Mark about the importance of individual action. The subtitle for the article? "Personal Sacrifices Are Necessary, and Greens Should Be Honest About That."[98]

Climate scientists, too, can inadvertently play into this unhelpful framing when they emphasize the sacrifices they have chosen to make in an effort to decrease their personal carbon footprints. In an Associated Press article, journalist Seth Borenstein observed, "Some climate scientists and activists are limiting their flying, their consumption of meat and their overall carbon footprints to avoid adding to the global warming they study." Borenstein focused primarily on one particular climate scientist, Kim Cobb, explaining that, "she is about to ground herself. . . . Cobb will fly just once next year, to attend a massive international science meeting in Chile. . . . This year she passed on 11 flights, including Paris, Beijing and Sydney."[99]

Borenstein emphasized that "there's a cost," noting that "Cobb was invited to be the plenary speaker wrapping up a major ocean sciences conference next year in San Diego. It's a plum role. Cobb asked organizers if she could do it remotely," but "they said no. . . . Conference organizers withdrew the offer." Cobb is a colleague and a friend of mine, and there's no doubt in my mind that she's earnest in her efforts to be a positive force for change and to set a good example for other scientists. But do we want to be sending the message that scientists (and by implication, others) must make career sacrifices for the sake of their individual carbon footprints?

Shahzeen Attari, author of the "scientists must walk the walk" study, seems to think so. Attari said that while she doesn't "want to clip [Cobb's] wings," she will be "judged" by her audience on how much energy she uses.[100] We've already discussed why that is unlikely to be true. And to the extent that "sacrifice" is emphasized in scientists' messaging, it might have the unintended effect of alienating moderate and conservative audiences, playing into the notion that climate action is just an excuse to impose a Spartan lifestyle upon the populace.

There is a sizable contingent within the conservative base that frankly seems predisposed to aversion when it comes to a strong, smart, bold, young, powerful Latina. New York City congresswoman Alexandria Ocasio-Cortez fits the bill—and thus served as the perfect foil when she became the principal proponent of the Green New Deal (GND). But the GND has been used to erode conservative support for climate action in more substantive ways, too.

The GND is a nod to President Franklin Roosevelt's New Deal, a major government initiative of the 1930s that used massive government stimulus spending in an effort to lift the United States out of the Great Depression. The New Deal did this primarily by boosting employment through major construction projects, including highways, dams, national park infrastructure, and so on. As originally conceived in the Obama era, the GND also embraced market mechanisms, such as carbon taxes and subsidies for green energy, to tackle environmental challenges.[101]

As reinvented by AOC (and Senator Ed Markey), however, the GND has taken on a considerably larger agenda with an additional focus on diverse social programs. The formal resolution, introduced by AOC and cosponsors on February 7, 2019, supports a "10-year national mobilization over the next 10 years," which includes among its planks the following:

- "Guaranteeing a job with a family-sustaining wage, adequate family and medical leave, paid vacations, and retirement security to all people of the United States."
- "Providing all people of the United States with—(i) high-quality health care; (ii) affordable, safe, and adequate housing; (iii) economic security; and (iv) access to clean water, clean air, healthy and affordable food, and nature."
- "Providing resources, training, and high-quality education, including higher education, to all people of the United States."[102]

Though I broadly support the GND's goals, I have some concerns about the ambitious scope of this specific proposal, as I expressed in a commentary in *Nature* magazine: "Saddling a climate movement with a laundry list of other worthy social programmes risks alienating needed supporters (say, independents and moderate conservatives) who are apprehensive about a broader agenda of progressive social change."[103]

I'm hardly alone in this view. In a January 2020 op-ed in *The Guardian*, Harvard economist Jeffrey Frankel expressed very similar concerns, writing, "In the US, the 'green new deal' signals commitment to the climate cause. But I fear that the legislative proposal that its congressional supporters have introduced will do more harm than good. It includes extraneous measures such as a federal jobs guarantee. This proposal creates a factual basis for a lie that US climate-change deniers have long been telling: that global warming is a hoax promoted as an excuse to expand the size of government."[104]

In my *Nature* commentary, I also questioned the conspicuous absence in the proposal of support for market mechanisms such as carbon pricing, noting that, too, might alienate political centrists who could otherwise be brought on board. We'll have much more to say on this topic in the next chapter.

Some advocates of climate-change action have gone further, including author and activist Naomi Klein. As I stated in the *Nature* commentary, "[Klein's] thesis is that neoliberalism—the prevailing global policy model, predicated on privatization and free-market capitalism—must be overthrown through mass resistance [and that] climate change can't be separated from other pressing social problems, each a symptom of neoliberalism: income inequality, corporate surveillance, misogyny and white supremacy."[105] Such framing fans the flames of the conservative fever swamps, reinforcing the right-wing trope that environmentalists are "watermelons" (green on the outside, red on the inside) who secretly want to use environmental sustainability as an excuse for overthrowing capitalism and ending economic growth.[106] Consider, for example, the admonition by one conservative commentator who said that the climate movement aims to cause "the decline of growth in world economies."[107] That's irresistible bait for many conservatives.

I want to draw attention to one additional plank in the GND resolution, which suggests "working collaboratively with farmers and ranchers in the United States to eliminate pollution and greenhouse gas emissions from the agricultural sector as much as is technologically feasible." Lest we underestimate the creativity of Murdoch's Fox News presenters when it comes to misrepresentation and bad-faith arguments, consider their translation of this rather reasonable proposal. "No more steak. I guess government-forced veganism is in order," said Fox News firebrand Sean Hannity.[108] "They want to take your pickup truck, they want to rebuild your home, they want to take away your hamburgers. . . . This is what Stalin dreamt about but never achieved," said paid Fox News contributor and (extremely) amateur Sovietologist Seb Gorka at a conservative conference.[109]

Fox News waged war against AOC and the GND, with headlines like "AOC Accused of Soviet-Style Propaganda with Green New Deal Art Series" and "Stealth AOC 'Green New Deal' Now the Law in New Mexico, Voters Be Damned," among others.[110] Unsurprisingly, studies show that Fox News weaponized the GND through such framing, dramatically eroding support for the proposal among conservative Republicans from 57 percent to 32 percent in just a matter of months.[111]

The inactivists, of course, have been more than happy to play both sides here. A leaked document that emerged in June 2020 revealed that fossil fuel companies, including Chevron, were behind a PR campaign aimed at exploiting the spring 2020 Black Lives Matter protests to sow racial division within the climate movement.[112] CRC Advisors, the conservative PR firm enlisted to conduct the campaign, had been circulating an email to journalists encouraging them to focus on how environmental groups supporting the Black Lives Matter movement supposedly advocated "policies which would hurt minority communities," and specifically, on how the GND would supposedly hurt minority communities. Caught redhanded were the forces of inaction in their effort to drive yet another wedge, this time involving issues of racial justice and equity, into the climate movement.

WHAT TO DO?

So what can we do to blunt the impact of the deflection campaign currently underway? First of all, realize when you're being played. "Don't feed the trolls" is a popular refrain in social media circles. Learn to recognize trolls and bots and report them when you come across them. They are engaged in a divide-and-conquer strategy against climate advocates, and you become an enabler of that strategy when you get taken in. Be constructive and engage meaningfully with others. Don't let yourself get dragged into divisive spats with those who are on the same side as you. As someone who is actively engaged in social media, I constantly have to remind myself of the very advice I'm giving you right now. When dangerous lies threaten to poison our public discourse, we must do our best to correct the record. But we must avoid traps set by trolls and bots looking to divide us. There's no hard-and-fast rule here. We each must just remain vigilant and use our best judgment.

We should all engage in climate-friendly individual actions. They make us feel better and they set a good example for others. But don't become complacent, thinking that your duty is done when you recycle your bottles or ride your bicycle to work. We cannot solve this problem without deep systemic change, and that necessitates governmental action. In turn, that requires using our voices, demanding change, supporting climate-focused organizations, and voting for and supporting politicians who will back climate-friendly policies—which includes putting a price on pollution—the topic of the next chapter.

CHAPTER 5

Put a Price on It. *Or Not.*

The stock market is roaring and planet Earth is wailing.
—STEVEN MAGEE

AS MY FRIEND BILL MCKIBBEN LIKES TO POINT OUT, THE FOSSIL fuel industry has been granted the greatest market subsidy ever: the privilege to dump its waste products into the atmosphere at no charge.[1] That's an unfair advantage over climate-friendly renewable energy in the playing field that is the global energy marketplace. We need mechanisms that force polluters to pay for the climate damage done by their product—fossil fuels—tilting the advantage to those forms of energy that aren't destroying our planetary home.

Such mechanisms can take the form of tradable emissions permits, also known as *cap and trade*. In this policy, government allocates or sells a limited number of permits to pollute, and the polluters can buy and sell these permits. This strategy limits pollution by providing economic incentives for polluters to reduce emissions. Another policy is a *carbon tax*, wherein a tax is levied at the point of sale on the carbon content of fuels or any other product yielding greenhouse emissions. Additionally, *carbon credits* can be granted for activities that take carbon out of the atmosphere and bury or store it, thus offsetting carbon emissions.

Fossil fuel interests and right-wing anti-regulation plutocrats have fought tooth and nail against any legislation aimed at pricing carbon emissions, for this would diminish their profits. In 2009, they torpedoed a carbon-pricing bill in the United States and similar legislation in Australia and elsewhere. Moreover, a coalition of petrostate actors, including Russia and Saudi Arabia, joined in the United States by the Trump administration, has also conspired to block carbon-pricing initiatives. Ironically, some environmental progressives are now providing them an unintentional assist.

DISOWNING THEIR OWN

As you may recall, former *Republican* president George H.W. Bush signed a cap-and-trade amendment to the Clean Air Act in 1990 that required coal-fired power plants to scrub sulfur emissions before they exited smokestacks. Between 1990 and 2004, sulfur emissions from coal-fired plants fell 36 percent, even as power output increased by 25 percent. The roughly nine-million-ton cap on sulfur emissions was reached in 2007 and fell to about five million tons in 2010. Lakes, streams, and forests in the Northeast—including the western Adirondacks where my family and I often vacation in the summer—recovered. It was a true environmental success story. You might think Republicans would want to own it—and build on this legacy by tackling the climate crisis using the very same market approach.[2] Instead, the GOP disowned its own brainchild.

Presumably expecting some buy-in from moderate Republicans, a cap-and-trade bill sponsored by Democratic congressmen Henry Waxman of California and Edward Markey of Massachusetts (then in the House) was proposed in 2009 to regulate carbon emissions. It passed, but largely on a party-line vote. Opposition from fossil fuel interests and their front groups—which attempted to brand it "cap and tax"—was perhaps predictable.[3] But there was also opposition from some in the environmental community, who argued that the problem was that it *wasn't* a tax. They favored an explicit carbon tax over a system of tradable emission permits.[4]

Nobel Prize–winning economist and progressive *New York Times* columnist Paul Krugman argued that while either a carbon tax or a cap-and-trade policy could achieve the needed reductions in carbon emissions, "in practice, cap and trade has some major advantages, especially for achieving effective international cooperation." He thought the House bill was likely the best compromise possible given the prevailing politics: "After all the years of denial, after all the years of inaction, we finally have a chance to do something major about climate change. Waxman-Markey is imperfect, it's disappointing in some respects, but it's action we can take now. And the planet won't wait."[5]

Further confusing the politics of the matter was the fact that some Republicans actually supported a carbon tax. But with a catch—it had to be "revenue neutral," which is to say, it couldn't increase the overall taxation on the American people, so other taxes, such as income taxes, would have to decrease. South Carolina congressman Bob Inglis and Jeff Flake of Arizona—both fiscally conservative Republicans—made the case for such a vehicle as an alternative to cap and trade.[6]

Fossil fuel interests and their abettors now faced a grave threat. A climate bill was one house of our bicameral legislative branch away from being the law of the land and even some Republicans supported a price on carbon. The inactivists kicked into high gear. First, the Koch brothers used their tremendous wealth and influence to wage a massive disinformation campaign to defeat the climate bill.[7] They had shills such as Myron Ebell of the Koch-funded Competitive Enterprise Institute misrepresent the cap-and-trade bill as a "tax" bill that would hurt our economy and everyday citizens. Even the *New York Times* was hoodwinked into promoting that interpretation. Describing him as "a strong advocate of the acid rain cap-and-trade program," *Times* reporter John M. Broder quoted C. Boyden Gray, who had been White House counsel during the first Bush administration, saying that "opponents were largely correct in labeling the Waxman-Markey plan a tax."[8] The *Times* failed to note that Gray had worked with the Koch brothers as a member of the board of directors for Citizens for a Sound Economy, a conservative think tank the Kochs had founded in 1984. Citizens for a Sound Economy would lead to Freedomworks and Americans for Prosperity.[9]

Americans for Prosperity was in fact a Koch brothers front group. The Kochs employed it as a vehicle for sponsoring a "hot-air" bus tour around the country promoting climate-change denial and fear-messaging about how regulating carbon emissions would supposedly destroy the economy.[10] They even constructed an Astroturf movement that became known as the "Tea Party" to create the illusion of widespread grassroots opposition to the climate bill, marshaling a rabble of disaffected citizens resentful of a changing fiscal, racial, and social landscape that seemed to have left them behind.[11]

Meanwhile, the Kochs served notice to any Republican legislators who might think about supporting climate legislation by making an example of Congressman Bob Inglis (R-SC), who, as noted earlier, had supported a carbon tax bill. Christopher Leonard, author of *Kochland*, described what happened during Inglis's reelection bid in 2010: "Koch Industries stopped funding his campaign, donated heavily to a primary opponent named Trey Gowdy and helped organize teams of Tea Party activists who traveled to town hall meetings to protest against Mr. Inglis. Some of the town hall meetings devolved into angry affairs, where Mr. Inglis couldn't make himself heard above the shouting. Mr. Inglis lost re-election, and his defeat sent a message to other Republicans: Koch's orthodoxy on climate rules could not be violated."[12]

The Kochs' efforts were successful. Democrats were unable to achieve filibuster-proof support (that is, a minimum of sixty votes) in the Senate, and the bill never went forward to President Obama's desk. Even with both houses of Congress under their control and a president in favor of climate action, Democrats were unable to pass a climate bill. While one might blame them for fecklessness, there is little doubt that cap and trade, as Broder at the *Times* put it, "ran into gale-force opposition from the oil industry [and] conservative groups that portrayed it as an economy-killing tax."[13] Tens of millions of dollars from the Koch brothers and dark-money spending aimed at sinking the bill didn't help matters. Nor did a beleaguered president who had already expended considerable political capital fighting a war with the right over health-care reform. And thus ended, with little fanfare, what had once seemed a promising prospect, finally, for a climate bill in the United States.[14] (A noteworthy postscript: Bob Inglis, who in full disclosure is a personal friend, now leads an organization aimed at bringing Republicans on board with climate action. He travels around to speak to conservative audiences about free-market approaches to pricing carbon, and in 2015 he received the JFK Profile in Courage Award.[15])

Similar episodes played out in the other major industrial nations. Australia provides perhaps the most striking example.[16] In some sense, what transpired down under was even more disillusioning than what had occurred in the United States: The Aussies *did* have a national price on carbon, and they lost it. In 2011, after a long, drawn-out battle that dated back several governments, Julia Gillard, prime minister of the ruling labor government, passed an emissions trading scheme, or ETS (another name for a cap-and-trade system). Drawing from the very same textbook that inactivists used to sink the US cap-and-trade bill, Australia's center-right opposition party (the "Liberals"—who are actually *conservatives*) misrepresented the measure as a "carbon tax" that would hurt individuals. This was particularly problematic for Gillard, who had made a campaign promise not to pass a carbon tax but had not ruled out an emissions trading measure.[17]

The usual suspects—Koch-funded front groups combining forces with coal interests and the Murdoch media (which dominate the Australian media landscape)—went to work, savaging Gillard and the Labor Party.[18] The attacks, as described by the *New York Times*, "coalesced around the promise and the tax." The ETS was portrayed "as a burden that would hurt businesses and cost households, instead of one that would cut pollution and ensure a more secure future for our children." There was only the smallest grain of truth to that claim. In principle, some of the cost to polluters of a cap-and-trade policy can be passed on to consumers. But in practice, these costs would have been minimal.[19]

Foreshadowing the attacks on AOC and the Green New Deal that were detailed in the previous chapter, Gillard's critics made not-so-subtle misogynistic appeals to voters in accomplishing their objectives. The *Times* noted that "the heat, anger and vitriol directed at her as a leader—and as Australia's first woman to be prime minister . . . grew strangely nasty."[20]

Liberal Party fossil-fuel advocate and climate-change denier Tony Abbott won the subsequent general election and was eventually able to revoke the ETS. Today the conservative Liberal National Party (LNP), a coalition of the Liberal Party and the National Party, remains in power, with a like-minded prime minister in Scott Morrison who has coddled coal, played a destructive role in international climate negotiations, and downplayed the impacts of climate change even as Australians have suffered through devastating and unprecedented heat, drought, and bushfire outbreaks. It is worth noting that, as in the United States, not all of Australia's conservative politicians were on the wrong side of the climate issue. Former Liberal prime minister Malcolm Turnbull was attacked by the Murdoch press and ousted from office in 2018 in large part because of his support of carbon pricing. He now plays a similar role in Australia to that played by Inglis in the United States, seeking to convince conservatives to come back into the climate tent.[21]

It is instructive, in light of the timeline of these attacks on climate policy, to reconsider the role played by the manufactured "Climategate" controversy. You may recall how that pseudo-scandal played out in late November 2009, just in time to have a detrimental impact on the all-important Copenhagen Summit that December. But we know that it was several months in the making, which means that the plan was likely hatched around the time the Waxman-Markey bill passed the US House of Representatives (late June 2009). The pseudo-scandal dominated conservative media and even some mainstream outlets, including CNN, well into 2010, as the US Senate was taking up the ill-fated cap-and-trade bill. Pretty darned good timing by the inactivists!

In 2009, the Labor Party was in power in Australia, with Kevin Rudd as prime minister. Rudd had attempted to pass a cap-and-trade measure, slated to take effect in July 2010. But an odd coalition of the opposition Liberals (led by Tony Abbott at the time) and the Greens opposed him. According to *The Guardian*, "the Liberal opposition argued that [consideration of the ETS] should all be put off until after the Copenhagen climate conference scheduled for the end of 2009, a tactic that helped to delay the day of reckoning within the Liberal party room."[22] The tactic accomplished more than that. It also postponed consideration of any climate pricing measure until after the "Climategate" pseudo-scandal had broken. With nothing more to go on, I still wonder if Liberal Party insiders were somehow privy to knowledge the rest of us didn't have.

Rudd had anticipated a more favorable political environment for climate action following the Copenhagen Summit. But that was not to be. The proceedings became mired in disputes between developing nations (including China) and the developed world. And the political atmosphere had been poisoned by the climate inactivists' full-on assault, including the ammunition that trumped-up Climategate rhetoric provided.

As we have seen, two petrostates—Russia and Saudi Arabia—are known to have played an important role in the spread of Climategate propaganda. Indeed, Saudi Arabia attempted to sabotage the entire Copenhagen Summit based on the false Climategate claims. This bloc of climate-denying petrostates has since welcomed two additional members: the United States under Trump, and oil-soaked Kuwait. This "coalition of the unwilling" attempted to thwart the findings of the UN Intergovernmental Panel on Climate Change during the December 2018 UN Climate Change Conference in Poland. The IPCC report concluded that rapid and immediate reductions in global carbon emissions were necessary to avert catastrophic planetary warming. The four countries were the only member nations that refused to support a motion to "embrace" the findings of the new report (instead they agreed only to "note" the report's findings—a far weaker measure that is much easier for policymakers to ignore). The delegate for St. Kitts and Nevis—a West Indian island nation threatened by sea-level rise and increasingly dangerous hurricanes—told the UN plenary that it was "ludicrous" for this minority of countries to hold up the critical proceedings over two words.[23]

Based on recent behavior, the coalition of the unwilling now includes Brazil under Jair Bolsinaro and Australia under Scott Morrison. Russia, by far, though, remains the most active member of the coalition of inactivist states. As we have already seen, it was implicated in efforts to influence recent US elections in a manner that was disadvantageous for climate policy. It also appears to have interfered in recent elections in the United Kingdom, working with the climate-change-denying UK Independence Party (UKIP), for example, to pass "Brexit" (the withdrawal of the United Kingdom from the European Union). Brexit is expected to erode the power of the European Union—including its influence on climate policy.[24]

Russia is also believed to have played a role in instigating the 2018 "Yellow Vest" revolts in France that sabotaged governmental efforts to introduce a carbon tax there.[25] In that movement, Russian trolls helped incite protests and rioting in the streets using messaging that played upon class conflict and perceived economic injustice. Ironically, although most of the protesters actually supported action on climate, they opposed a proposed fuel tax, which they were led to believe would be financed by the working class and poor to the benefit of multinational corporations.[26]

Russia has also tampered in Canadian politics. Russian bot farms have been used, for example, in an effort to convince environmental progressives in Canada that Prime Minister Justin Trudeau, who supported carbon pricing, was in fact against taking meaningful action on climate.[27] Trudeau's environment minister, Catherine McKenna, who was responsible for implementing Canada's new carbon tax program, has been subject to an onslaught of Russia-style troll- and bot-based social media attacks since taking office in 2015. Many are tinged with misogyny, dismissing her as a "climate Barbie," and ridden with slurs like "bitch," "c—t," "slut," and "twat."[28] Going into the 2019 Canadian federal election, Russian Twitter trolls attempted to stoke anger against the Trudeau government by focusing on issues such as immigration, employment, the economy, and, of course, climate policy.[29]

What might Russia and other petrostate bad actors be trying to accomplish through these sorts of activities? For one thing, a few early carbon-pricing political disasters in countries like France and Canada might cause other governments considering climate policy to get cold feet, much as the failure in the 1970s and 1980s of many of the efforts to pass bottle bills in individual states in America sank any chance of a national bottle bill. So the theory might be to nip any promising new efforts at carbon pricing in the bud before they have a chance to succeed. And to make a price on carbon toxic, all they have to do is associate it with social unrest, disruption, and economic pain.

We can see how these efforts have paid off for the inactivists when it comes to recent climate policy efforts in the United States. Consider the defeat of a climate tax initiative by voters in Washington state in November 2016. Sure, there was massive opposition and a flood of advertising from fossil fuel interests. But ironically, those opposing the initiative got an assist from environmental organizations such as the Sierra Club, which argued that the carbon tax would violate principles of social justice. This leads to our next discussion: the ironic alienation of environmental progressives from pricing carbon.[30]

PIPELINES, NOT PRICING

From a market vantage point, the fossil fuels we burn are a consequence of both supply and demand. And so there are two basic, complementary approaches to regulating fossil fuels: control *supply* and/or control *demand.* Pricing carbon (or, alternatively, incentives for renewables) reflects an effort to diminish demand, while fossil fuel divestment campaigns and opposition to pipelines, offshore oil drilling, or mountain-top-removal coal mining constitute efforts to diminish supply. Leading climate advocates like Bill McKibben and Senator Bernie Sanders of Vermont at least originally endorsed both approaches.[31]

Despite the natural duality between demand-side and supply-side measures, there is also an asymmetry—at least when it comes to political organizing. It's easy to motivate activists to protest a pipeline or mountain-top removal. Or to attend demonstrations at college campuses demanding that administrators divest of fossil fuel holdings. These events are visual, involve conflict, bring out A-list celebrities, and generate front-page headlines and graphic photos. Think the Dakota Access Pipeline demonstrations at the Standing Rock Indian Reservation, or Darryl Hannah and James Hansen being arrested protesting Massey Energy's coal processing plant in West Virginia.[32] Or Harvard and Yale student protesters joining forces to disrupt the 2019 Harvard/Yale football game, demanding that both institutions divest of fossil fuel holdings.[33]

Carbon pricing, by comparison, seems wonkish and abstract, and it's hard to capture it in a front-page image or on a television screen. Moreover, while both carbon pricing and pipeline protests reflect efforts to influence the underlying market economics of fossil fuel use, carbon pricing is more readily seen as buying into market economics. As a result, carbon pricing has been vulnerable not just to attacks from the right but also to attacks from the left. We've seen how conservatives have been led to oppose carbon pricing—by fear messaging that warns of infringements to personal liberty and heavy-handed governmental mandates. But progressives have also been led to oppose carbon pricing—for them, it has been portrayed as an ostensible mechanism of neoliberal economics that discounts social justice.

One argument that seems to have resonated with the environmental left is that a price on carbon amounts to a regressive tax that selectively hurts low-income workers. This was the claim that was used to foment the Yellow Vest uprising.[34] It is telling that Donald Trump, in his role as patsy for the fossil fuel interests that write his energy and environmental policies, insisted that the Yellow Vest violence was proof that people oppose environmental protection (as noted earlier, it showed nothing of the sort).[35]

In reality, whether a carbon tax is progressive or regressive depends on how it is designed. A fee-and-dividend method, for example, returns any revenue raised back to the people. Such a plan could be designed to be progressive, returning revenue to the poor and those most impacted through an appropriately constructed dividend.

In fact, the carbon-pricing schemes that have been successfully instituted have been *progressive* in nature. With the ETS scheme implemented by Australian prime minister Julia Gillard, the government compensated low-income earners, who ended up *benefiting* financially. Under Canada's carbon tax-and-rebate system, most households actually save money.[36] No less than Pope Francis, a champion of social justice and a true advocate for the poor and downtrodden, has called carbon pricing "essential" for tackling the climate-change "emergency."[37]

Another argument is that carbon pricing would represent a sort of political zero-sum game for climate action, with any carbon tax coming at the expense of losing legal avenues for holding polluters accountable. More specifically, some in the climate movement believe that passage of a carbon tax would shield fossil fuel companies from legal liability for their actions. This simply isn't true.

Much as the tobacco industry was finally held liable for its efforts to hide the dangers of its products from the public, so, too, are there efforts today to use the legal system to bring polluters to justice for hiding the dangers of their product—fossil fuels—to the entire planet.[38] A number of lawsuits against fossil fuel companies are currently working their way through the legal system.[39] Two states have launched fraud investigations targeting ExxonMobil (one went to trial in 2019 and failed). Nine cities and counties, including New York and San Francisco, have used the courts to seek compensation from fossil fuel companies for the climate damages they have caused. Perhaps best known, however, is *Juliana v. U.S.*, brought by twenty-one children who sued the federal government for violating their right to a safe climate. The suit was thrown out but is currently under appeal.[40]

The belief that a carbon tax would somehow end legal liability on the part of fossil fuel interests is premised on mistaking what fossil fuel interests might *want* for what they're actually going to *get*. Some climate activists have breathlessly warned that climate pricing legislation is a "fossil-fuel-funded Trojan Horse" that would amount to "letting oil, gas, and coal companies off the hook" by "exempting fossil fuels companies from . . . lawsuits."[41] While fossil fuel companies have lobbied for a bill that would do just that, none of the climate bills that have been introduced in Congress have proposed to absolve fossil fuel companies of liability.[42] It is simply a fallacy to equate carbon pricing with releasing fossil fuel interests from legal liability.

Another argument frequently made by progressive critics is that a carbon tax cannot achieve the needed emissions reductions. But that depends on the magnitude of the tax.[43] Consider, for example, what transpired in Australia between 2012 and 2014, when Gillard imposed a modest price on carbon through the ETS that ended up costing polluters about $23 per metric ton of emitted CO_2. Emissions in the electricity sector dropped more than 9 percent during the first six months of implementation. And what happened when the Abbott government repealed the ETS in 2014? Emissions recorded their single greatest annual gain (more than 10 percent).[44] Of course, a carbon tax is just one tool in the climate action toolbox and must be combined with other demand-side and supply-side measures in any comprehensive climate plan.

Nonetheless, because of objections from some on the environmental left, the version of AOC's Green New Deal endorsed by leading environmental organizations advocates *against* a price on carbon. A letter signed by 626 groups, including Greenpeace and 350.org, was delivered to every member of Congress in early 2019 laying out support for a Green New Deal, while stating that the groups "will vigorously oppose any legislation that . . . promotes corporate schemes that place profits over community burdens and benefits, *including market-based mechanisms . . . such as carbon and emissions trading and offsets*" (emphasis added).[45] There are other recent cases in which environmental progressives and green groups have opposed carbon-pricing efforts. As we learned earlier, for example, the Sierra Club helped defeat a 2016 climate tax initiative in Washington because its leaders felt it didn't satisfy principles of social justice.[46]

Then there's the Carbon Pollution Reduction Scheme (CPRS) that former Australian Labor prime minister Kevin Rudd proposed back in 2009. Rudd's government had negotiated a package with climate-policy-friendly Liberal leader Malcolm Turnbull that could pass Parliament. Turnbull, however, was replaced as Liberal leader by fossil fuel flack Tony Abbott. On Abbott's first full day as Liberal leader, the members of Parliament (MPs) for the *Green Party*—yes, the party whose very name bespeaks ostensible prioritization of environmental preservation—voted with Abbot *against* the CPRS, purportedly because its members wanted *more ambitious* reduction targets. This fateful decision by the greens, as Mark Butler explained in *The Guardian*, "allowed Abbott to begin to build the momentum that has hamstrung long-term climate action for almost a decade." According to Butler, "had the CPRS passed the parliament in 2009, an emissions trading scheme would likely have been operating for some years before Abbott was able to become prime minister. And it's likely that Abbott would not have been able to build a platform to tear down such a large reform after that time."[47]

Prominent spokespeople within the scientific community, too, sometimes fan the flames of progressive opposition to carbon pricing. Consider the words of Australian environmental scientist Will Steffen, executive director of the Australian National University Climate Change Institute and lead author of a controversial "Hothouse Earth" commentary in the *Proceedings of the National Academy of Sciences*.[48] Asked what could be done to prevent a Hothouse Earth scenario, Steffen said the "obvious thing we have to do is to get greenhouse gas emissions down as fast as we can. . . . *You have got to get away from the so-called neoliberal economics* . . . [and shift to something] more like wartime footing [to decarbonize society] at very fast rates" (emphasis added).[49] While Steffen is no doubt an expert in environmental science, his statements about economics and policy here are ill-informed. If we are to achieve rapid decarbonization of our economy, carbon pricing (which one suspects he is lumping in with "neoliberal economics") is essential—it's the main lever arm we have available to us in a market economy.[50]

Among the most market-economics-averse of proponents of a Green New Deal is social activist Naomi Klein, who has long argued that modern-day capitalism—which is to say, neoliberal market economics—is fundamentally at odds with basic human rights and environmental sustainability. According to Adam Tooze, in his article "How Climate Change Has Supercharged the Left" in *Foreign Policy* magazine, "the denunciation of neoliberalism in Naomi Klein's *This Changes Everything* gave a manifesto to the new green left."[51]

I published a commentary in *Nature* that recommended Klein's latest book on the GND but questioned her critique of market mechanisms, pointing out that—as we've already seen—there is no reason that carbon pricing has to be either regressive or inadequate.[52] Her followers immediately took to social media to expressly denounce me. I can understand that some of her supporters might have been disappointed that I had some points of disagreement with her and didn't endorse her precise vision of the Green New Deal. But we are on the same side. And I didn't expect the vitriolic personal attacks of the sort I'm used to getting from the climate-denying right coming instead from the left.

One reader dismissed my commentary as "mansplaining trash from myopic white bros who do not speak for those on the front lines." Now, I'll humbly submit that I *do* know a thing or two about being on the front lines. For two decades I've been in the cross hairs of the attack machine funded by the fossil fuel industry, and I have devoted my professional life to study and activism relating to climate change.[53] Eric Holthaus jumped in to express his disapproval as well, tweeting "Ladies, does he . . . leverage his platform to write op-eds in prominent magazines disparaging the Green New Deal? He's not your climate hero, he's a gatekeeper."[54] These responses—from both strangers and people who are ostensibly on the same side of the issue as I am, seemed to exemplify once again the divisive way that race, gender, and callout culture are being used to divide the climate movement.[55]

The takeaway message from this particular episode, however, is that there is a fairly aggressive effort underway by some on the environmental left to turn support for the GND *in its current form* (including *opposition to carbon pricing*) into a purity test. Even questioning it can lead to massive, mob-like online assaults and ugly accusations that somehow become framed in identity politics and tinged with issues of race, gender, and ageism. We have already seen that the inactivists seize upon such internal conflict and amplify it to sow dissent and divide the climate community. They are surely doing that here. Fortunately, as we've seen, there are also many committed climate advocates who recognize this threat and are willing to push back against needlessly divisive rhetoric. That will remain critical if we are to find some degree of common ground, as a society, when it comes to climate action—including carbon pricing. That leads us to our next topic.

PRICING AIN'T PARTISAN

Despite the divisiveness that has arisen around the role of carbon pricing, there is nothing intrinsically divisive or partisan about it. As we have seen, market mechanisms for dealing with pollution actually have their origins in the Republican Party. Carbon pricing is supported by all former Republican chairs of the president's Council of Economic Advisers. But carbon pricing is also widely supported by Democrats. Nine of the ten leading candidates for the Democratic presidential nomination supported it as of July 2019. The one exception as well as one subsequent major "flip" are rather interesting, and we will discuss them later.[56]

It is only relatively recently, as efforts to implement carbon pricing have actually started to move forward—that we've seen support for carbon pricing start to erode on *both* sides of the political spectrum. That's convenient for fossil fuel interests, whose spokespeople might publicly claim, for public relations purposes, that their companies and organizations support carbon pricing, but behind the scenes still fund groups working to undermine it.[57]

It's hardly surprising that Donald Trump, who has outsourced his policymaking to polluting interests, is dismissive of carbon pricing, which he has derided as "protectionism."[58] But the fact that some environmental *progressives* have grown apprehensive of carbon pricing has almost certainly influenced recent decisions by other climate-friendly politicians to steer clear of it. Consider New York governor Andrew Cuomo. Cuomo has been a leader in many respects when it comes to climate action. He has supported supply-side measures to restrict fossil fuel extraction, becoming only the second governor to ban natural gas drilling via hydraulic fracturing (fracking).[59] And he has promoted at least *one* type of demand-side measure, namely, governmental incentives for renewable energy (the topic of the next chapter). What he has proposed for New York is that it require 70 percent of the state's electric power supply to come from renewable energy sources by 2030 and mandating that it be free of carbon emissions by 2040. But Cuomo has not endorsed a price on carbon—as yet.[60]

Others have nonetheless called upon him to do so. Richard Dewey is the president and CEO of the New York Independent System Operator (NYISO), a not-for-profit corporation responsible for operating New York State's bulk electricity grid, administering its competitive wholesale electricity markets, conducting comprehensive long-term planning for its electric power system, and advancing the technological infrastructure of its electric system.[61] Dewey has insisted that Cuomo cannot achieve these goals without imposing a price on carbon: "These goals are really going to come fast," he has stated, adding that carbon pricing "is a necessary element in meeting them."[62]

The conclusion that we need carbon pricing is also supported by the International Monetary Fund (IMF), hardly a left-leaning organization. The IMF exists to "secure financial stability, facilitate international trade," and "promote high employment and sustainable economic growth."[63] It has estimated that there is an effective global average price of roughly $2 per metric ton, given the various carbon-pricing systems that are in place around the world. It has warned, however, that the world needs an average price of $75 per metric ton if we are to meet the Paris Agreement goal of keeping warming below 2°C (3.6°F). (An even higher price would be needed to keep warming below 1.5°C [2.7°F]—a level of warming increasingly considered to constitute dangerous climate change.[64])

These are examples of objective, moderate, nonpartisan institutions, with no particular axe to grind, that have called for carbon pricing. There are both Democrats and Republicans who support carbon pricing. Why is it proving so difficult to find political common ground here? Part of the answer, of course, is that fossil fuel interests, and the forces of inaction doing their bidding, have worked hard to poison the well (look no further than Donald Trump's threats to retaliate against the European Union over its proposed carbon tax[65]). But frankly, progressive scientists and thought leaders have at times made it easy for them, helping to create a political economy that is toxic for bipartisan compromise.

Let me relate an episode involving David Mastio, the deputy editorial page editor of *USA Today* and a self-avowed "libertarian conservative." In June 2019, I coauthored an op-ed about the dangers of the new climate "deflection campaign" discussed in this book.[66] I was sure the *New York Times* would publish it, but it did not. I was sure the *Washington Post* would then publish it. It didn't. I then went to *USA Today*. David not only embraced the piece and offered to publish it, but encouraged me to keep *USA Today* in mind for any future op-eds. He's precisely the sort of conservative we need on board.

Well, I was crestfallen to read a controversial statement David made some months later when he tweeted this: "Why I remain skeptical of the climate change consensus. If this was a real emergency, the scientists would be in favor of mobilizing the power of capitalism, not government control."[67] I wondered what could have set him off? Clicking through, I saw that it was a tweet paraphrasing a letter signed by eleven thousand scientists: "11,000 scientists have declared we are in a climate emergency. Among other things, *we need to move away from capitalism* . . ."[68] I've intentionally eliminated the rest of the tweet (you can find it in the endnotes) because I want you to read only as far as David would have had to read before becoming suspicious that the declaration of a climate emergency is just a tool—at least to some—for overthrowing capitalism. The "watermelon" fears, revisited.

A parochiality has emerged among environmental progressives that is unhelpful to the process of building consensus for climate action. Here's an example. In January 2020, George P. Shultz, secretary of state under President Ronald Reagan, and Ted Halstead, chairman and chief executive of the nonpartisan Climate Leadership Council, coauthored an op-ed in the *Washington Post* titled "The Winning Conservative Climate Solution."[69] In it, they advocated for a revenue-neutral carbon tax, or, more specifically, a fee-and-dividend system, similar to what is advocated by the nonpartisan Citizens Climate Lobby. In such a system, a fee is charged to carbon polluters, and the revenue is distributed, through a dividend, to the people (for example, in the form of quarterly checks sent by the government to individuals).

Now consider the response to the op-ed by David Roberts, a writer for *Vox*. Roberts tweeted, "I'll never get used to the bizarre convention of calling a policy that the GOP has repeatedly rejected & the vast bulk of conservatives oppose . . . a 'conservative solution.'" He went on to add, "The conservatives who are actually attracted to this policy are conservative centrists & conservative Democrats. This is an intra-left dispute in which one side is fraudulently claiming to be able to count on the right's support."[70]

Roberts often has keen insights into climate politics. But here, he is misguided. He fails to distinguish between traditional conservatives—that includes Reagan conservatives, like George Shultz, who, as we have seen, not only supported but actually *gave us* market-based approaches to reducing pollutants—and the current-day Republican Party, which has indeed been cowed into complicity with the Koch brothers, the Murdoch media, and the fossil fuel industry.

These old-school conservatives—George Shultz, Hank Paulson, Bob Inglis, Arnold Schwarzenegger, or, in the United Kingdom, former prime minister David Cameron—not only support climate action, but are passionate about it. Nevertheless, they are apprehensive about what they perceive to be heavy-handed governmental regulatory approaches, including the GND in its current form. As Shultz and Halstead put it, "the climate problem is real, the Green New Deal is bad."[71] According to Schwarzenegger, who as governor of California led efforts to cut back carbon emissions, and has roundly criticized Donald Trump's efforts to roll back environmental protections, the Green New Deal is "a slogan" and "marketing tool" that is "well intentioned" but "bogus."[72] Cameron has implored his fellow conservatives not to abandon the matter: "Don't leave the issues of climate and the future of the planet . . . These are natural conservative issues, don't leave this to the left or you'll get an anti-business, anti-enterprise, anti-technology response."[73]

We are unlikely to see a climate bill resembling the current version of the GND pass both houses of Congress in the United States. There will need to be some degree of bipartisan compromise, which means bringing along moderate conservatives. Rather than alienating them through partisan rhetoric, we need to create space for them and welcome them into the fold. There *is* a legitimate wedge to be formed, and it's between moderate conservatives, who are on board with climate action, and the recalcitrant deniers, delayers, and deflectors.

Nobody said it would be easy to pass climate legislation with the fossil fuel interests and the Koch brothers doing their best to enforce Republican Party purity. But fissures are starting to form, particularly as a result of generational shifts that favor action. Republican pollster Frank Luntz found that Republican voters under the age of forty favor a fee-and-dividend carbon-pricing policy by a whopping six-to-one margin.[74] The same generational trends that led to a tipping-point-like response on marriage equality during the Obama years will soon reach a tipping point on climate, too. But we don't have a decade to wait, and the most viable path forward toward comprehensive climate legislation in the United States involves market mechanisms, including carbon pricing. It would be sadly ironic—and indeed tragic—if progressives, rather than conservatives, became the greatest obstacle to climate progress by refusing to engage in compromise, cooperation, and consensus building.

Ironically, not only is there in an increasing tendency among progressives to oppose seeking a middle ground when it comes to climate policy, but we've arrived in a "bizarro" world where the climate-change talking points employed on the political left are sometimes virtually indistinguishable from those on the political right. Adam Tooze reported in *Foreign Policy* what transpired at a conference of the UK Labour Party in September 2019: "The general secretary of the GMB trade union, Tim Roache, warned that a crash program of decarbonization would require the 'confiscation of petrol cars,' 'state rationing of meat,' and 'limiting families to one flight for every five years.' He concluded: 'It will put entire industries and the jobs they produced in peril.'"[75] Other labor leaders have an arguably more enlightened view of carbon pricing. In March 2020, James Slevin, president of the Utility Workers Union of America, coauthored an op-ed with Senator Sheldon Whitehouse (D-RI) articulating the case for carbon pricing. They advocated measures to ensure that the revenue raised is rebated to consumers and used to help individuals and communities—particularly coal workers and their families—with support for health plans, pensions, and educational opportunities.[76]

Or consider Kevin Anderson, a climate scientist in the United Kingdom who has criticized the mainstream climate research community for understating the degree of the threat posed by climate change and overstating the progress that has been made. In critiquing a report by the Committee on Climate Change (an independent committee created to advise the UK government on matters of climate mitigation) on what measures are required to meet commitments under the Paris Agreement, Anderson stated that "it is designed to fit with the current political and economic status quo." Then he went further, accusing the entire climate research community of complicity: "The overall framing is firmly set in a politically-dogmatic stone with academia and *much of the climate community running scared of questioning this for fear of loss of funding,* prestige, etc." (emphasis added).[77] That charge is virtually indistinguishable from the shopworn accusation by climate-change deniers that climate scientists invented the climate crisis to bring in loads of grant money.[78]

Indeed, the prevailing politics of climate change today sometimes resemble the metaphorical snake biting its tail, with some on the left end of the spectrum promoting the positions on climate typically found on the right. Consider this characterization of Democratic presidential candidate Tulsi Gabbard by Brian Boyle in the *Los Angeles Times*: "Gabbard is a tricky candidate to pin down. Her domestic policy positions graft rather cleanly with Bernie Sanders' and Elizabeth Warren's progressive platforms—in fact, she was one of Sanders' fiercest supporters in 2016."[79] Sounds "left" doesn't it? But Boyle goes on to point out that Gabbard has taken curiously pro-Russian positions on any host of issues, and indeed, her candidacy was promoted by Russian bot armies. Is it a coincidence that she also happens to be the one Democratic candidate who went on record during the primaries to oppose a price on carbon—a position that aligned suspiciously with Putin's Russia and the Trump administration?[80] This contradiction speaks to the breakdown in our conventional descriptions of "right" and "left" in the current geopolitical environment.

An even more extreme example of the blurring of the political boundaries is the British Internet magazine *Spiked*, which purports to reflect the views of the Marxist far left. *Spiked* frequently engages in what it sees as "pushback against the protected hysteria of modern environmentalism," including rejection of climate science (for example, dismissing IPCC reports as "often over-the-top" and "scare mongering").[81] The magazine also promotes caricatures of the climate movement. It insists, for example, that climate advocates claim "that we have 12 years to save the planet."[82] This is a bastardization of the scientifically backed estimate that we only have around twelve years to bring carbon emissions down (by a factor of two) if we are to avert a dangerous 1.5°C (2.7°F) warming.[83] *Spiked* also promoted Brexit, which, as we know, will help derail EU climate pricing efforts. It was a confusing mix of positions for a far-left magazine, but it all became clear thanks to the work of British columnist George Monbiot. In an exposé for *The Guardian*, Monbiot revealed that among the funders of *Spiked* is in fact the foundation of fossil fuel billionaire (and apparently secret Marxist) Charles Koch.[84] Far right posing as the far left? Can you say *Manchurian Candidate*—backward? If there's a lesson in all of this, it's that inactivists are working hard to generate conflict within the climate movement, literally infiltrating the environmental "left" in an effort to turn climate identity politics on their head. They'll seemingly stop at nothing in their efforts to block climate progress and carbon pricing. Forewarned is forearmed.

ACCELERATING THE TRANSITION

Climate action requires a fundamental transition in our global economy and massive new infrastructure, but there is no reason to think we can't accomplish it—and accomplish it rapidly—with the right market incentives. Those incentives, as we've seen, must involve both supply-side and demand-side measures.

Supply-side measures take the form of blocking pipeline construction, banning fracking, stopping mountain-top-removal coal mining, divesting in fossil fuel companies, and putting a halt to most new fossil fuel infrastructure. These actions obviously lend themselves to activism, protests, and media-ready conflict and publicity. But they *can* also have a material impact. Consider, for example, the Keystone XL Pipeline, which promised to deliver huge amounts of the dirtiest, most carbon-intensive petroleum from the Canadian tar sands to the open market. It's a scenario that climate scientist James Hansen exclaimed would be "game over for the climate."[85] In response to massive protests and pressure from environmental organizations, former president Obama ultimately blocked the construction of the pipeline in 2015, arguing that it would "undercut" his administration's "global leadership" in "taking serious action to fight climate change."[86] Combined with the clean power plan and tighter fuel-efficiency standards imposed by his administration, blocking Keystone XL gave Obama a strong hand in negotiating a bilateral climate agreement with China in 2015 that would, in turn, lay the groundwork for the monumental Paris Agreement later that year.[87]

But, just as personal action is no substitute for systemic change, supply-side efforts are no substitute for demand-side approaches. Both are necessary. Demand-side measures attempt to level the playing field, so that climate-friendly energy, transportation, and agricultural practices outcompete fossil fuels in the marketplace. Carbon pricing is one of the most powerful tools we have to do that. Taking it off the table would constitute unilateral disarmament in the climate wars.

That is literally what happened in Australia. A successful carbon-pricing program that both progressives and conservatives initially supported was nixed by a climate-change-denying, fossil-fuel-flacking prime minister in Tony Abbott. Fatefully, Australia, in the record hot, dry, bushfire-plagued summer of 2019/2020, morphed into a dystopian hellscape resembling a scene from the 1979 Australian film *Mad Max*. Once a shining example of climate leadership in the industrial world, Australia has now become a poster child for the cost of climate inaction. Yet it is not too late for Australians to reclaim leadership by voting in a government that promises to act on climate in the next election.

Nor is it too late in the United States. As I write, the fate of carbon-pricing remains uncertain. The election of Donald Trump in 2016 was a major setback. A Biden presidency would put carbon pricing back on the table. Still there are signs, as we've seen, that some on the political left are also hostile to this policy. During the 2020 Democratic primaries, for instance, Bernie Sanders flipped on the issue of carbon pricing sometime between July 2019, when he supported it (albeit with qualifications), and November 2019, when, in response to direct questioning by the *Washington Post*, he indicated he no longer favored such policies. A cynic might imagine that this concession reflected an effort to wrest carbon-pricing-averse Green New Deal supporters from his chief primary campaign challenger, Elizabeth Warren. The great irony is that, as a result of this flip-flop, *both* major party candidates for the 2020 presidency could have ended up opposed to this important mechanism for climate action.[88]

Of course, a truly comprehensive strategy for leveling the playing field involves more than simply forcing corporate polluters to pay for the damage they're causing. That's the stick. But we need the carrot, too. That means incentives for energy providers to replace fossil fuels with cleaner, safer, carbon-free energy (and, conversely, eliminating the perverse existing subsidies that are provided to fossil fuel energy producers). The inactivists, naturally, as detailed in the next chapter, have opposed these measures, too.

CHAPTER 6

Sinking the Competition

We are like tenant farmers, chopping down the fence around our house for fuel, when we should be using nature's inexhaustible sources of energy—sun, wind, and tide.

—THOMAS EDISON

WE SAW IN THE PREVIOUS CHAPTER THAT CARBON PRICING IS A means of leveling the playing field in the energy market, so that those sources of energy that are not warming the planet (i.e., renewable energy) can compete fairly against those that are (i.e., fossil fuels). A complementary approach is to introduce explicit incentives for renewable energy (and eliminate those for fossil fuels). Here again, the inactivists have put their thumbs on the scale by promoting programs that favor fossil fuel energy while sabotaging those that incentivize renewables, and engaging in propaganda campaigns to discredit renewable energy as a viable alternative to fossil fuels.

SELECTIVE SUBSIDIES

The fossil fuel industry loves subsidies and incentives. When *they* receive them. According to the International Monetary Fund, the industry receives about half a trillion dollars globally in explicit subsidies, such as in the form of assistance to the poor for the purchase of fossil-fuel-generated electricity, tax breaks for capital investment, and public financing of fossil fuel infrastructure. It's a lot of money.

But when *implicit* subsidies are included—that is to say, the health costs and damage born by citizens for the associated environmental pollution, including the damage done by climate change—the estimate rises to a whopping $5 trillion.[1] These perks didn't arise by accident—the industry used its immense wealth and influence to obtain them. In the 2015–2016 election cycle alone, fossil fuel companies spent $354 million in campaign contributions and lobbying.[2]

Fossil fuel interests have also done everything possible to *block* subsidies and incentives for their competition—renewable energy—and they've had a lot of success doing so. That has led to a perverse incentive structure in the energy marketplace through which we are artificially boosting the very energy sources that are hurting the planet, while devaluing those that can save it. Industry front groups like the American Legislative Exchange Council (ALEC) and the Heartland Institute have been particularly active in sabotaging efforts at the national and state levels to promote renewable energy.

The watchdog group SourceWatch describes ALEC as a "corporate bill mill" through which "corporations hand state legislators their wish lists to benefit their bottom line."[3] In recent years, fossil fuel corporations such as ExxonMobil, Shell, and BP have pulled out of ALEC, concerned about increased public scrutiny of their funding activities. But the privately held fossil fuel giant Koch Industries has remained steadfast in its funding of the group.[4] In one year alone, ALEC helped push through seventy bills in thirty-seven states designed to disadvantage clean energy. ALEC has proposed legislation that would undermine state policies mandating that a fraction of the energy produced come from renewable sources (so-called Renewable Portfolio Standards).[5] One bill sponsored by Wyoming Republicans in 2020 was a caricature of these efforts. It would have required utilities to provide 100 percent of electricity from coal, oil, and natural gas by 2022. It failed.[6]

ALEC has also promoted legislation that penalizes those who choose to install solar panels on their homes. This would be accomplished by placing a surtax on homeowners with solar panels who attempt to sell power they don't need back to electric utilities.[7] Such

efforts, ironically, managed to earn the Koch brothers—apparently against intrusive state interference only until their bottom line is threatened—the ire of members of the Tea Party they helped create.

The Koch-funded Heartland Institute has been engaged in similar attacks on renewable energy.[8] Beginning in 2012, it sponsored ALEC's Electricity Freedom Act, model legislation aimed at repealing state renewable energy standard programs. Fortunately, these efforts have largely failed at the state level—with only Ohio halting its program, and only for one year (2014). These efforts have failed at the national level as well. Heartland has also tried to block state-level programs incentivizing solar energy.[9]

The goal of these efforts is to undermine the decarbonization of the power sector. But no assault on renewable energy would be complete without an attack on electric vehicles (EVs), for they are the path to decarbonizing the transportation sector as well. If you get your electricity from renewables and charge your car off an outlet in the garage, you're no longer driving off fossil fuels. That's a threat to the oil industry, which profits off the sale of gasoline, and to Koch Industries, which profits off the refining and distribution of oil and gasoline. Recognizing the threat to their bottom line, agents of the Koch brothers met with oil-refining and marketing companies in 2015 to pitch a "multi-million-dollar assault on EVs."[10]

Central to the plan was one of their bought-and-sold politicians, Republican senator John Barrasso of Wyoming, who was the third-highest recipient of Koch brothers dollars during the 2018 election cycle.[11] Barrasso, as chair of the Senate Environment and Public Works Committee, introduced the Fairness for Every Driver Act in 2019. It would not only end federal tax credits for EVs, but in addition would create an annual "highway user fee" for all "alternative fuel vehicles." It might not surprise you to learn that Barrasso, in his efforts to sell this bill to voters, used talking points that were taken directly from Koch brothers propaganda (for example, that the tax credit "disproportionately subsidizes wealthy buyers," and that "hard-working Wyoming taxpayers shouldn't have to subsidize wealthy California luxury-car buyers"). He and his fellow Republican proponents also used talking points manufactured by the Koch-funded Manhattan Institute (for example, the bogus claim that ending the electric vehicle tax credit would save roughly $20 billion in taxpayer funds over the next decade). These arguments have been characterized as resting on "every conceivable kind of error: data dredging, wishful thinking, truculent dogmatism, and, now and again, outright fraud."[12]

Tesla may be the greatest threat of all to the fossil fuel industry. Not only do Teslas compete with the sleekest of conventional automobiles performance-wise, but Elon Musk and his company have also literally redefined what an electric automobile can be. In North Carolina, American-made Teslas were outselling high-performance conventional vehicles, including foreign brands like BMWs, Mercedes, and Audis. The company's success was a triumph of American innovation, industry, and free markets! So the Republican state senate stepped in and tried to pass a bill that would prohibit the sale of Teslas.[13] (While the bill failed, Tesla sales were nonetheless banned in one major city, Charlotte.[14]) Soon thereafter, Republican governor Chris Christie tried to do the same thing in New Jersey.[15] Other red states—Texas, Utah, West Virginia, and Arizona—followed suit.[16] So much for "free-market" Republicans!

Meanwhile, the conservative media, doing the bidding of fossil fuel interests, have promoted mythologies designed to undermine public support for renewable energy. Solyndra was a California manufacturer of thin-film solar cells that used unusual, innovative technology. Plummeting silicon prices, however, led to the company being unable to compete with conventional solar panels, and it went bankrupt in September 2011.[17] The company defaulted on a $535 million loan it had received from the US Department of Energy under President Barack Obama's 2009 economic stimulus package. The vast majority (98 percent) of the funds provided under the federal program went to companies that have *not* defaulted on their loans; in fact, the Department of Energy projects a profit of more than $5 billion over the next two decades, with twenty of the program's thirty enterprises operating and generating revenue.[18]

The overall success of the program notwithstanding, inactivists have sought to make Solyndra the poster child for the supposed failings of renewable energy. They also used Solyndra scandal-mongering to attack Obama's proposed budget in 2015. Presumably what they *really* didn't like about the budget was that it would repeal nearly $50 billion in tax breaks for the oil, natural gas, and coal industries.[19] So in a masterful display of propagandistic jujitsu, Fox News and the *Daily Caller* (a Koch brothers front group masquerading as a media outlet), among others, sought to use Solyndra to tie the Obama budget to an ostensibly failed renewable energy agenda.[20] Despite what they claimed, Solyndra had not received the clean energy tax credits included in the president's 2015 budget. The budget didn't even increase funding for the largely successful loan guarantee program that had supported Solyndra in 2009.[21] But facts be damned when there's an opportunity to simultaneously both smear renewables and protect fossil fuel subsidies.

CROCODILE TEARS

Another line of attack by the inactivists is to cry crocodile tears over the purported threat posed by renewable energy. It's once again the classic tactic of dividing the environmental community, in this case by convincing them that renewables—which actually promise environmentally safe and reliable energy—are instead somehow a threat to our health and the environment.

So we get myths and distortions that seek to create a false dilemma for the environmentally minded, namely, that decarbonizing our economy will somehow come at the expense of environmental peril. None is more prominent than the supposed threat wind turbines pose to birds. Robert Bryce of the aforementioned Koch-funded Manhattan Institute has been out in front promoting this myth, both on the editorial pages of the *Wall Street Journal* and in ultra-right-wing venues like the *National Review*.[22] Do we really think that Bryce cares a feather about the birds whose supposed turbine-driven demise he laments? More birds are killed every year by housecats. Why aren't Bryce and the Murdoch media crusading to rein in our felines? Might they—and other fossil fuel water carriers advancing the "wind is a threat to birds" myth—be crying crocodile tears?

When it comes to the welfare of our feathered friends, I put more trust in the Audubon Society, whose *actual* mission is to "protect birds and the places they need, today and tomorrow." The Audubon Society has stated that climate change is a far greater threat than wind turbines. According to an Audubon Society report, hundreds of bird species in the United States—including our national symbol, the bald eagle—are at "serious risk" due to climate change, with the ranges for some species predicted to be diminished by 95 percent by 2080. Bird catch by wind turbines can be minimized by siting wind farms away from bird migration routes. Accordingly, Audubon supports "properly sited wind power as a renewable energy source that helps reduce the threat posed to birds and people by climate change."[23]

The inactivists have even managed to invent an imaginary health affliction in their efforts to scare people away from wind power—"wind turbine syndrome." Anti-wind advocates have claimed that a whole array of afflictions, including lung cancer, skin cancer, hemorrhoids, and both the gain and loss of weight, are somehow caused by proximity to wind farms. It is just one example of how Sagan's worst fears about "pseudo-science" have come to pass.[24] With absolutely no scientific evidence behind the phenomenon, the fact that some honest actual individuals have claimed to suffer from the imaginary syndrome is a classic example of a "communicated disease"—that is to say, people who might be experiencing any number of maladies and happen to live near a wind farm hear others talk about the putative syndrome and, looking for someone or something to blame, embrace this pseudoscientific but seemingly plausible explanation.[25]

It should come as no surprise that Koch-affiliated groups, fossil fuel interests, and the Murdoch media empire have sought to spread the myth of "wind turbine syndrome" far and wide.[26] Consider the utterings of Fox Business network's Eric Bolling: "Turbines are popping up all across America, as the demand for the usage of wind energy is increasing. But at what cost? Residents close to them have reported everything from headaches to vertigo to UFO crashes."[27] Yes, you read that right: "UFO crashes," too! The anti-wind brigade even managed to recruit President Donald Trump to the cause. Among his long list of ridiculous claims about wind turbines, he suggested they "cause cancer."[28]

Trump, in fact, used a fundraising address on April 2, 2019, to promote fears that allowing wind farms in communities causes financial damage, warning Americans, "If you have a windmill anywhere near your house, congratulations, your house just went down 75 percent in value."[29] Actual studies have found no evidence for the claim that wind turbines affect property values.[30]

Crocodile tears have also been shed over the supposed environmental impact of solar energy. That isn't to say that solar farms and solar panels have no environmental footprint—there are valid issues regarding land use and habitat loss, water use, and the potential release of hazardous materials in manufacturing.[31] But that footprint is tiny compared to the environmental impact of coal, natural gas, and petroleum. And that's not even considering the damages from climate change!

Enter the so-called Breakthrough Institute (BTI), a group originally linked to fossil fuel interests that has more recently been called a "nuclear [industry] front group."[32] Public ethics expert Clive Hamilton has accused BTI of "misrepresenting data on the energy savings of investment in energy efficiency, [criticizing] almost every proposed measure to reduce America's greenhouse gas emissions [and allying] with anti-climate science organizations."[33] Thomas Gerke, writing for *Clean Technica*, noted BTI's propensity for articles "discrediting renewable energy on the one hand and on the other preaching about nuclear energy as the solution for the global energy crisis of the 21st century."[34]

BTI cofounder Michael Shellenberger promotes the myth that solar energy poses a major threat to the environment. In May 2018 he penned a column for *Forbes* soaked with plaintive tears over the supposed toxicity of chemicals in solar photovoltaic cells.[35] Curiously unmentioned in his piece is the fact that (1) solar panel manufacturers in the United States must follow laws to ensure that workers are not harmed by exposure to toxic chemicals, and that chemical waste products are disposed of properly, and (2) manufacturers have a strong financial incentive to ensure that valuable and rare materials are recycled rather than disposed of.[36]

Just months later, Shellenberger followed up with another *Forbes* piece in which he asserted, presumably in all seriousness, that "nuclear is the safest source of electricity," that "low levels of radiation are harmless," and that "nuclear waste is the best kind of waste."[37] You see, nuclear = safe, solar = dangerous. Black = white. Up = down. Welcome to the bizarro world of soft denial.

Fox News has regularly subjected its viewers and readers to anti-solar propaganda warning of the dire environmental threats posed by solar energy. It has given us headlines like "Solar Energy Plants in Tortoises' Desert Habitat Pit Green Against Green."[38] It's an inactivist two-fer, combining feigned environmental concern with environmentalism wedge creation, all in one headline! Other examples include "Environmental Concerns Threaten Solar Power Expansion in California Desert," "Massive East Coast Solar Project Generates Fury from Neighbors," and my favorite: "World's Largest Solar Plant Scorching Birds in Nevada Desert."[39] It's touching to behold once again Rupert Murdoch's deep and abiding empathy for our avian cousins. Which makes total sense when you realize that birds are the modern descendants of dinosaurs.

Oddly, though, I don't recall seeing any Fox News headlines like "Mountain-Top-Removal Coal Mining Kills Off Fish and Amphibians," or "Deep Oil Drilling Destroys the Gulf of Mexico," or "Our Dependence on Fossil Fuels Is Scorching the Planet." Fox News and conservative media display curiously selective outrage over impacts on people and the environment where renewable energy, rather than fossil fuels, is concerned.

Some of the solar scare tactics used by the right-wing media border on the comical. Just as wind turbines supposedly cause cancer, solar panels will apparently cause you to freeze to death in cold climates. Or so claimed Fox News host Jesse Watters as he attempted to discredit the Green New Deal and its architect Alexandria Ocasio-Cortez: "They have this new green deal or whatever. Ok, where they want to eliminate all oil and gas in 10 years. If you're in the polar vortex, how are you going to stay warm with solar panels?"[40]

Of course, the fine art of scaring the public about renewables isn't confined to the United States. Australian prime minister Scott Morrison, known, among other things, for having brandished a lump of coal on the floor of Parliament as a testament to his idea of "clean energy," has also demonstrated some facility in this department. In April 2019, Morrison launched an attack on the Labor Party's proposed target that EVs constitute 50 percent of all new car sales by 2030. Admonishing Labor leader Bill Shorten, he said that pro-EV policies would "end the weekend" for Australians. Morrison warned, "You've got Australians who love being out there in their four-wheel drives. [Shorten] wants to say see you later to the SUV when it comes to the choices of Australians." Ironically, Morrison's own government (the Liberal-National coalition) had proposed policies that were only slightly less bullish on electric vehicles, setting a goal that 25 percent of all new car sales by 2030 be EVs. Noting the irony, Shorten responded that Morrison and the coalition government were "so addicted to scare campaigns, they're even scaring you with their own policies."[41]

"LET THEM BURN COAL"

If the inactivists have shed a few crocodile tears when it comes to the supposed threat posed by renewable energy to our health and the environment, they've cried a whole river when it comes to their supposed concern for the plight of the poor. They've appealed to the logical fallacy known as "you can't chew gum and walk at the same time," or, to be more specific, the idea that promoting renewable energy over ostensibly cheaper fossil fuel energy will somehow divert essential resources from efforts to fight third-world poverty. Welcome to the contrived concept of "energy poverty."

The energy-poverty conceit rests on the flawed premise that lack of access to energy (rather than to, say, food, water, health care, and so on) poses the primary threat to people in the developing world, and, moreover, that fossil fuels are the only viable way to provide that energy. In other words, if you are concerned about the disadvantaged of the world, you should be promoting fossil fuels. It's a truly brilliant, if cynical and manipulative, strategy by fossil-fuel-promoting inactivists to recruit political progressives and moderates to their cause.

Among the promoters of the concept is the aforementioned BTI, whose mission, as stated on its website, is "[to make] clean energy cheap through technology innovation to deal with both global warming and energy poverty."[42] Also among the ranks of energy-poverty adherents are Microsoft CEO Bill Gates and former ExxonMobil CEO Rex Tillerson. Tillerson once posed, without any apparent sense of irony, the question, "What good is it to save the planet if humanity suffers?"[43]

Indisputably the most enthusiastic of energy-poverty crusaders, however, is Bjorn Lomborg. A self-styled "skeptical environmentalist," Lomborg is neither—skepticism, remember, involves good-faith scrutiny of tenuous-seeming claims, not indiscriminate rejection of well-established science. The charismatic Lomborg brandishes a Greenpeace T-shirt to prove his environmental bona fides.

Dig a bit deeper, however, and a rather different story emerges. Lomborg's Copenhagen Consensus Center has been funded by the Randolph Foundation, whose main trustee, Heather Higgins, is also the president of the Koch-funded International Women's Forum.[44] The center is in fact a virtual entity, with an official address at a Lowell, Massachusetts, parcel service. The conservative Abbott government in Australia attempted to provide it with a permanent home, offering $4 million in taxpayer funds to the University of Western Australia if it would provide a home for the center. The university ultimately walked away from the offer.[45]

Lomborg frequently pens commentaries in leading newspapers, including the *Wall Street Journal*, the *New York Times*, and *USA Today*, downplaying the impacts of climate change, criticizing renewable energy, and promoting fossil fuels. With a smile and a professed concern for the environment and the poor, he scolds those who would misguidedly wean us off fossil fuels and promote clean energy.[46]

For someone with such professed sympathy for the plight of the developing world, Lomborg displays a remarkable dismissiveness toward those most vulnerable to the devastating impacts of climate change. In one op-ed he warned that "a 20-foot rise in sea levels . . . would inundate about 16,000 square miles of coastline, where more than 400 million people currently live." An alarming fact. But Lomborg couldn't quit while he was ahead. He continued: "That's a lot of people, to be sure, but hardly all of mankind. In fact, it amounts to less than 6% of the world's population—which is to say that 94% of the population would not be inundated."[47]

Conservatives apparently now study Lomborg's talking points. This type of "big picture" thinking cropped up again in the middle of the coronavirus crisis of early 2020. Take, for example, right-wing Wisconsin senator Ron Johnson's message to his constituents over the Trump administration's failure to take meaningful actions in the early stages of the pandemic. "Right now, all people are hearing about are the deaths," Johnson complained. "Sure the deaths are horrific," he conceded, but "the flip side of this is the vast majority of people who get coronavirus do survive." He cheerily added that, in the end, the coronavirus would kill "no more than 3.4 percent of our population."[48] What's a few hundred million people among friends, after all, Bjorn/Ron?

When it comes to the plight of the poor, I must confess that my own bias is to take Pope Francis more seriously than Bjorn Lomborg. And the pope has rejected the energy-poverty myth, pointing out that distributed, renewable energy in the form of solar power and hydropower is far more practical than fossil fuel use in most of the developing world.[49] Even the fossil-fuel-friendly *Wall Street Journal* has acknowledged as much, noting that "renewable energy could offer a . . . solution for remote areas, because it is created and consumed in the same region and doesn't require massive power plants and hundreds of kilometers of power lines."[50] If you've lost the *Wall Street Journal*, Bjorn, well . . .

There is an even deeper problem, of course, with the premise that climate action detracts from the concerns of the poor. As Pope Francis emphasized in his papal encyclical on the environment, climate change *aggravates* other societal challenges—food, water and land scarcity, health, and national and international security. The US Department of Defense agrees.[51] The irony of the energy-poverty myth is that climate-change impacts will actually place far more people in poverty than are in poverty today. In a scenario of climate collapse, there *is* no economy. Don't take my word for it, though. A World Bank study from 2015 concluded that climate change could "thrust 100 million into deep poverty by 2030." Even Fox News reported it.[52]

IT'S THE JOBS, STUPID!

Another tactic the inactivists use is to scare people into thinking that climate action and renewable energy will take away their jobs. A group connected to the Koch Foundation that calls itself Power the Future has sought to blame Tom Steyer—a climate activist and philanthropist, and perhaps not coincidentally, from the standpoint of being an eligible boogeyman, a Jewish billionaire—for the steady, decades-long decline of the coal industry and the demise of coal communities across America. The organization has even attempted to brand collapsing coal towns as "Steyervilles." Their "proof" is the fact that Steyer's philanthropic spending has increased as coal jobs have decreased—not exactly the sort of iron-clad argument that would pass muster in the peer-reviewed literature, or the pages of a reputable newspaper, or even a fortune-cookie fortune.[53]

Yes, coal jobs are disappearing. And there are now far more jobs in the burgeoning renewable energy industry (hundreds of thousands in solar alone) than there are in the dying coal industry (which currently has less than fifty thousand coal-mining jobs).[54] But these job losses have more to do with increased mechanization and automation of coal mining and competition from cheaper fossil fuels (namely, natural gas) than they do with competition from renewable energy, let alone climate activism itself.

Despite job retraining programs and other efforts to help those displaced by the demise of coal, there are inevitably those—especially older workers—who will encounter difficulty finding subsequent employment. Labor leaders representing the energy sector, such as James Slevin, president of the Utility Workers Union of America, have thus argued that climate policies must include measures to help coal workers and their families by providing financial support for their health plans, pensions, and educational opportunities.[55]

Technological transitions are never easy, and there are always winners and losers. But it is no more appropriate to blame the renewable energy industry for lost coal jobs than it is to blame the fossil fuel industry for destroying the whaling industry, which provided much of the lamp oil that was replaced by kerosene and then coal-powered electrical lighting.

ET TU, MICHAEL MOORE?

File this one under the category of "with friends like this . . ." None other than liberal icon Michael Moore has now joined the ranks of the renewable energy bashers. Working with director Jeff Gibbs, his longtime collaborator on left-of-center polemics like the anti-NRA *Bowling for Columbine* and the anti-Bush, anti–Iraq War film *Fahrenheit 9/11*, Moore, in his 2020 film *Planet of the Humans* (*POTH*), has promoted a full-on assault on renewable energy. Though Gibbs directed the documentary, Moore put the full weight of his celebrity into the project, doing the talk-show circuit and flacking the film like next month's rent depended on it.[56]

POTH had no sooner been screened at film festivals when the negative reviews started to come in.[57] The film, in fact, proved to be so toxic that Moore couldn't get a major distributor to adopt the film. Nor would Netflix or any other major streaming platform show it. So he ended up posting it for free on YouTube on Earth Day 2020, as if his intention were to launch a hand grenade that would produce maximum collateral damage to action on climate.[58]

The fatal flaws in the film, enumerated in excruciating detail by a number of energy and climate experts, comprise a laundry list of deceptive facts and bad-faith arguments.[59] They include: (1) the misleading use of data, photographs, and interviews that are a decade old to dramatically overstate the limitations of renewable energy and understate the efficiency and capacity of current-day renewable energy sources and storage technology; (2) complaints that a still largely fossil-fuel-driven electricity grid is used in the construction of solar panels and wind turbines, without noting that the life-cycle carbon emissions are tiny compared to either coal or gas, and that decarbonization of the grid is precisely what the renewable energy transition is about; and (3) grossly inflated estimates of the carbon footprint of biofuels and biomass (which is tiny compared to that of fossil fuels), while failing to note that biomass accounts for only 2 percent of domestic electricity generation (though Moore and Gibbs spend about 50 percent of the film complaining about it).[60]

The film, disappointingly, promotes the sorts of myths about renewable energy that one expects to hear on Fox News rather than in a Michael Moore–produced film. For example, it decries electric vehicles as not being green because they're fueled off the grid, which is still driven substantially by fossil fuel energy. But this argument neglects the fact that a fundamental component of any meaningful green energy transition is the electrification of transport in concert with the _decarbonization of the electric grid_.[61] To focus on the former without acknowledging the latter is to entirely miss the point, unintentionally or otherwise.

We are treated once again to the now familiar crocodile tears over the ostensible horrible environmental impacts of renewable energy—the large tracts of land required for solar and wind farms, the reliance on mining for metals used in solar panels, and so on. It's odd that Michael Moore seems far more concerned by fields dotted with wind turbines and solar panels than by his newfound concern about climate change. Shortly after the release of the film, he tweeted that "the public knows we're losing the climate battle, thanks to profit & greed & leaders who led us wrong."[62] First of all, we're not "losing the climate battle." As we will see later, substantial progress is now being made. And while profit and greed are certainly part of the problem, so, too, are misguided attacks on renewable energy and the false prophets who bear them. Which brings us back to Michael Moore and Jeff Gibbs.

They are shocked, for example, to learn that the United States gets some of its renewable energy from the burning of biomass (mostly, organic refuse). But in what stands out as a blatant untruth in an already a gratuitously error-ridden film, they claim that power generation from biomass exceeds that of solar and wind. The actual numbers indicate just the opposite, with biomass providing only 1.4 percent and solar and wind providing 9.1 percent of total power generation.[63] Adding insult to injury, they repeat the outrageously misleading claim that "biomass releases 50 percent more carbon dioxide than coal and more than three times as much as natural gas." The erroneous claim is the by-product of the very same bad math we encountered in an earlier chapter with the 2014 film *Cowspiracy*.

Cowspiracy, as readers may recall, falsely asserted that livestock are responsible for 51 percent of carbon emissions. This figure is based on bad accounting coupled with poor scientific understanding. The scriptwriters appear to have been unaware of the simple fact that the carbon produced by cows when they exhale (in the form of carbon dioxide, through what we call "respiration") comes from consumed plant matter that had extracted the carbon from the atmosphere in the first place (through the process of "photosynthesis"). When cows, or any animals—including us—exhale, we're not adding net carbon dioxide to the atmosphere, we're simply helping circulate the carbon through the atmosphere/biosphere system.[64] The actual contribution of livestock to carbon emissions comes from entirely different processes: fermentation, manure management, feed production, and energy consumption. Cows do also belch *methane*, which is itself a potent greenhouse gas, but its lifetime in the atmosphere is much shorter than that of CO_2. The true net contribution to carbon emissions from livestock (15 percent), curiously enough, corresponds to a simple reversal of the two digits in the number (51 percent) cited in *Cowspiracy*.

Moore and Gibbs make essentially the same error in *POTH*, failing to inform their audience that the carbon dioxide produced by burning biomass (with the exception of old-growth forests) is carbon dioxide that recently came from the atmosphere anyway. Biomass is therefore largely "carbon neutral"—far from perfect when we are trying to reduce the amount of carbon in the atmosphere, but still better than releasing CO_2 from the Carboniferous era, as we do when we burn coal or gas. Burning biomass itself doesn't increase carbon dioxide levels in the atmosphere. There are, of course, some carbon emissions associated with processing and transportation, and that's simply a result of the fact that much of our basic infrastructure still relies upon a fossil-fuel-energy economy—a fact that is less true every day as a *result* of the renewable energy revolution! But the carbon emissions are tiny—about ten grams of carbon pollution per kilowatt-hour. For comparison, natural gas yields about five hundred grams and coal nine hundred grams per kilowatt-hour! Much as animal rights activists have overstated the role of meat-eating in climate change to advance their (admittedly worthy) agenda of decreasing meat consumption, so, too, have some forest preservation activists overstated their (admittedly worthy) goal of stopping deforestation.[65]

It's important to get the facts right. The wood chips used in biomass are generally a by-product of already-existing forestry practices, not the result of cutting down trees for fuel as some imply. And biomass is a broad category. While we certainly shouldn't be turning forests into wood chips for burning, it does make sense to burn some forms of organic waste, which can provide a near carbon-neutral source of energy, while we transition to cleaner renewable energy.

POTH reinforces so many of the tropes we've encountered that it almost serves as a poster child for the new climate war. One challenge we face in this new war on climate action is, as we saw in the previous chapter, the wedge that has emerged within the climate movement itself when it comes to market-driven climate solutions. Moore and Gibbs attempt to pry that wedge wide open. The fact that wind and solar energy are increasingly profitable is somehow an indication, to them, that they're "bad." In the words of the editorial board of the *Las Vegas Review-Journal*, Moore seems "particularly aghast to discover that . . . any transition to green energy will require massive investment from evil industrialists and capitalists who might turn a profit. Who knew?"[66]

So heroes become villains—and villains, ironically, become heroes. Climate champion Bill McKibben is vilified for having once, long ago, supported the limited use of biomass energy.[67] Al Gore is attacked for supposedly being "more focused on cashing in than saving the planet."[68] (Couldn't a similar argument be made about Michael Moore and his $50 million net worth?[69]) Moore and Gibbs were apparently "shocked to find a company owned by Charles and David Koch receiving solar tax credits." Now, there are *many* reasons to dislike the Koch brothers—but the fact that they invested in solar energy is not one of them. Only in the Trumpian era of gaslighting could a progressive filmmaker produce a polemic premised on the absurd notion that ultra-right-wing plutocrats are secretly behind the effort to end our dependence on fossil fuels. And get progressives to actually fall for it.

Then there is defeatism and despair-mongering (a topic we'll explore in detail in Chapter 8). As *The Guardian* put it, "most chillingly of all, Gibbs at one stage of the film appears to suggest that there is no cure for any of this, that, just as humans are mortal, so the species itself is staring its own mortality in the face."[70] Writing for *Films for Action*, an award-winning longtime environmental filmmaker, Neal Livingston, had an even harsher critique: "SHAME on these filmmakers for making a film like this, full of misinformation and disinformation, to intentionally depress audiences, and make them think there are no alternatives. . . . Let me make it absolutely clear that the new documentary, *Planet of the Humans*, by Jeff Gibbs—with executive producer Michael Moore, is inaccurate, misleading and designed to depress you into doing nothing."[71] Doomism and the loss of hope can lead people down the very same path of inaction as outright denial. And Michael Moore plays right into it.

Then there is the classic deflection of the sort we've encountered before. Technically, Moore and Gibbs do advance one "solution." Rather than focusing on the systemic source of the problem—our reliance on fossil fuels, they deflect attention toward individual behavior, which, as we have seen, is a classic new-climate-war tactic. The twist here is that it's all about the behavior of *others*. Environmental author Ketan Joshi remarks that Moore "ends up at population control—a cruel, evil and racist ideology that you can see coming right from the start of the film."[72] Brian Kahn, writing in *Earther*, noted, "Over the course of the movie, [Gibbs] interviews a cast of mostly white experts who are mostly men to make that case. . . . There's a reason that Breitbart and other conservative voices aligned with climate denial and fossil fuel companies have taken a shine to the film. It's because it ignores the solution of holding power to account and sounds like a racist dog whistle."[73] It is worth noting, by the way, that people in the developing world, where the main population growth is taking place, have a tiny carbon footprint in comparison with those in the industrial world. The world's richest 10 percent produce half of global carbon emissions.[74] The problem isn't so much "too many people" as it is "too many people who burn a lot of carbon." As environmental sociologist Grant Samms put it, Moore and Gibbs spend the entire film oscillating between "ecological nihilism and ecological fascism."[75]

Conservative foundations and media outlets, on the other hand, loved Moore's film. And it wasn't just Breitbart News that was "full of gratitude and admiration that they should have made this bold, brave documentary."[76] Fossil-fuel-funded groups like the Competitive Enterprise Institute and the Heartland Institute (and their paid attack-dog Anthony Watts) lapped it up.[77] CEI encouraged people to "Hurry, see *Planet of the Humans* before it's banned," while the Heartland Institute promoted the film in a podcast series.[78] Watts advertised it as an "Earth Day Epic," linking to it directly on his blog.[79] Industry-funded denier-for-hire Steve Milloy insisted that "EU politicians should be forced to watch Michael Moore's *Planet of the Humans* . . . with their eyes clamped open if necessary."[80] Other fossil-fuel-industry shills, including Marc Morano of the Committee for a Constructive Tomorrow (CFACT), promoted the film and attacked its critics on Twitter, which also became a predictable venue for manufactured outrage by right-wing trollbots.[81] And yes, even the Koch brothers got in on some of the action. An anti-renewables Koch brothers front group known as the American Energy Alliance spent thousands of dollars promoting the film.[82]

We are left, in the end, to wonder why Michael Moore ever produced this film. Politics can make for strange bedfellows. Moore was a huge supporter of Bernie Sanders during his campaign for president. Sanders made his support for the Green New Deal a centerpiece of his platform, and the GND, at its core, supports renewable energy. But Moore has also been a supporter of Julian Assange for years.[83] The WikiLeaks leader has collaborated closely with Russia in its efforts to attack climate science and undermine action on climate.

Moreover, Moore has been a longtime advocate for blue-collar workers and the unionization movement, beginning with his breakout 1989 film *Roger and Me*, which denounced General Motors' crackdown on union workers. It is hardly unprecedented for the labor left to find itself in conflict with the environmental left. Recall from Chapter 5 that the general secretary of the GMB trade union, Tim Roache, warned that climate action would lead to the "confiscation of petrol cars," "state rationing of meat," and "limiting families to one flight for every five years," placing "entire industries and the jobs they produced in peril."[84]

Does Moore see decarbonization of our economy as a threat to workers? Had Moore struck a secret deal with the fossil fuel industry? Or had he simply lost his mind? Had the Trump presidency somehow caused him to "flip"? Or did Moore simply care more about being provocative than about being right? With his most successful films now more than a decade behind him and his relevance increasingly in question, was he simply looking for a dramatic way to attach himself to the defining issue of the day? Once a polemicist, after all, always a polemicist.

Maybe this is simply a manifestation of what environmental journalist Emily Atkin has referred to as the phenomenon of "first-time climate dudes."[85] It's the tendency for members of a particular, privileged demographic group (primarily middle-aged, almost exclusively white men) to think they can just swoop in, surf the Internet, interview a few hand-selected "experts," and solve the great problems that others have spent decades unable to crack. It is almost inevitable that the product, in the end, is a hot mess, consisting of fatally bad takes and misguided framing couched in deeply condescending mansplaining. On climate change, we've seen it with Bill Gates, *FiveThirtyEight*'s Nate Silver, and now with Michael Moore.[86]

The fact is that we may never know the motives behind this ill-premised, intellectually dishonest stunt by Michael Moore and Jeff Gibbs. What we *do* know is that their misguided polemic furthers the agenda of fossil fuel interests and their tactic of denial, delay, distraction, and deflection by buying into misleading and false narratives about renewable energy. It appears they will go down in history as having ironically sided with wealthy, powerful polluters, rather than "the people" they purport to care about, in the defining battle of our time.

"YOU'RE NOT GONNA HAVE IT!"

Finally, when all other arguments fail, we're left with "Well—it just won't work. You can't do it!" Inactivists in fact twist themselves into veritable pretzels to explain why there's no way we can possibly power our economy with renewable energy. There are fundamental obstacles, they say. Intermittency! Insufficient batteries!

Yes, the wind isn't always blowing, and the sun isn't always shining. And batteries don't have infinite storage capacity. But these challenges are, if you will forgive the pun, overblown. Smart grid technology that adaptively combines various renewable energy sources can overcome these limitations—not in the future, but right now. Utility-scale "big battery" systems like those produced now by Tesla are outperforming and outcompeting fossil fuel generators in providing grid stability to blackout-prone regions like South Australia.[87]

Peer-reviewed research demonstrates authoritatively that even without any technological innovation—that is, using current renewable energy and energy-storage technology—we could meet up to 80 percent of global energy demand by 2030 and 100 percent by 2050. This would be accomplished through increased energy efficiency, electrification of all energy sectors, and decarbonization of the grid through a mix of generation sources, including residential rooftop solar and solar plants, onshore and offshore wind farms, wave energy, geothermal energy, and hydroelectric and tidal energy. The precise mix of technologies would depend on the location, season, and time of day.[88] Sorry, Bill Gates, but we don't "need a miracle."[89] The solution is already here. We just need to deploy it rapidly and at a massive scale. It all comes down to political will and economic incentives.

A renewable energy transition would create millions of new jobs, stabilize energy prices in the absence of fuel costs, reduce power disruption, and increase access to energy by decentralizing power generation.[90] But that's not what we hear from Koch-funded groups like the Heartland Institute. Instead we get supposed experts like coal-industry shill and climate-change denier David Wojick penning pieces with titles like "Providing 100 Percent Energy from Renewable Sources Is Impossible."[91] In dismissing the viability of a renewable energy transition, Wojick engages in a classic game of denial bingo, harping on the ostensible fatal problems of "intermittency" (largely already solved, as discussed earlier), "scalability" (that's simply a matter of government incentives—the very incentives that Wojick's bosses, the Kochs, have worked so hard to game in favor of the fossil fuel industry and against renewable energy), and "expense" (he grossly overestimates battery storage costs; ignores that there are multiple storage options aside from batteries, like pumped-storage hydroelectric power; and pretends that places like Colorado have no sun).

Wojick ends by offering us some revisionist history, dismissing as "false claims" the dramatic success stories that have been told of towns and municipalities that have already transitioned to 100 percent renewable energy. Pay no attention to Greensburg, Kansas—the town that was leveled by an EF5 tornado and rebuilt 100 percent renewable by its conservative Republican mayor.[92] Really, it doesn't exist! Fake news! The critics have gone beyond denial of climate change to denial of reality itself.

Speaking of denial of reality, let's again talk about Fox News and its take on solar energy in the United States. In a 2013 segment attacking the Obama administration's support for renewable energy, Fox News host Gretchen Carlson questioned Fox business reporter Shibani Joshi on why solar power was so much more successful in Germany than in the United States. "What was Germany doing correct?" Carlson asked. "Are they just a smaller country, and that made it more feasible?" Carl Sagan surely rolled over in his grave after hearing the response: "They're a smaller country," Joshi said, "*and they've got lots of sun. Right? They've got a lot more sun than we do*" (emphasis added). Perhaps sensing she had just said something absurd, Joshi doubled down in an effort to explain herself. "The problem is it's a cloudy day and it's raining, *you're not gonna have it*" (emphasis added). Conceding that California actually gets just a bit of sunlight now and then, she elaborated, "Here on the East Coast, it's just not going to work."[93]

Of course, it's only in the mythological universe of Fox News where the East Coast of the United States gets less sun than Germany. As Media Matters pointed out in its response to the segment, estimates from the US Department of Energy National Renewable Energy Laboratory (NREL) show that nearly the entire continental United States gets more sun on average than even the most sunladen regions of Germany.[94] In fact, as one NREL scientist pointed out, "Germany's solar resource is akin to Alaska's." (Alaska receives by far the least average sunlight of any US state.[95]) But, returning to Carlson's original question: What's the real reason that German's solar industry is doing so much better than the solar industry in the United States? Simple: It doesn't have Fox News, the rest of the Murdoch media, the Koch brothers, and fossil fuel interests all joining forces to destroy it.

FALSE SOLUTIONS

We have seen that there is a dual attack underway by inactivists in the form of efforts to both block carbon pricing and blunt or at least slow the renewable energy transition now underway. Fight back. When you encounter myths about the supposed environmental threat of wind turbines and solar panels, push back against them. Correct the misinformation. If you have friends or family or colleagues who have been taken in by the crocodile tears, hand them a handkerchief and explain to them they've been had. When someone cites "energy poverty" or "lost jobs" as arguments against renewable energy, point out that the opposite is true: the safest and healthiest path to economic development in the third world is access to clean, decentralized, renewable energy, and the greatest opportunity for job growth in the energy industry comes with renewables, not fossil fuels.

But also be prepared for the next line of attack: There is yearning now among the public for a meaningful climate solution. If it's not renewable energy, it must be something else. So inactivists seek to fill that void with reassuring, plausible-sounding alternative "solutions" that do not pose a threat to the fossil fuel juggernaut. And they have done so by introducing a new, seemingly empowering lexicon: "geo-engineering," "clean coal," "bridge fuels," "adaptation," "resilience." Welcome to our next chapter—the *non-solution solution*.

CHAPTER 7

The Non-Solution Solution

It is a wholesome and necessary thing for us to turn again to the earth and in
the contemplation of her beauties to know the sense of wonder and humility.
—RACHEL CARSON

When I am working on a problem, I never think about beauty but when I have
finished, if the solution is not beautiful, I know it is wrong.
—R. BUCKMINSTER FULLER

THE INACTIVISTS HAVE SOUGHT TO HIJACK ACTUAL CLIMATE
progress by promoting "solutions" (natural gas, carbon capture, geoengineering) that aren't real solutions at all. Part of their strategy is
using soothing words and terms—"bridge fuels," "clean coal," "adaptation," "resilience"—that convey the illusion of action but, in context, are empty promises. This gambit provides plausible deniability:
inactivists can claim to have offered *solutions*. Just not good ones.
They are delay tactics intended to forestall meaningful action while
the fossil fuel industry continues to make windfall profits—what
noted climate advocate Alex Steffen has referred to as "predatory
delay."[1] It is essential that we recognize and expose these efforts for
the sham they are, for the clock is ticking. We cannot afford any
further delay when it comes to the climate crisis.

A BRIDGE TO NOWHERE

Let me sell you a bridge to a fossil-fuel-free future. Beware of a bait-and-switch, however, for it is actually a bridge to nowhere. It's called natural gas, a naturally occurring gas composed primarily of methane—the same methane that, as we learned earlier, is belched by cows, contributing to greenhouse gas emissions. This particular source of methane isn't biogenic, however. It is a fossil fuel formed from ancient organic matter—plants and animals that died and were buried beneath Earth's surface millions of years ago. They eventually made it down deep into Earth's crust, where, subjected to great pressure and heating, they eventually turned into an admixture of hydrocarbon molecules residing in either the solid, liquid, or gaseous state (coal, oil, or natural gas, respectively). Like other hydrocarbons, natural gas is energy rich, and it is readily burned for heating, cooking, or electricity generation. Or it can be cooled into a liquid (liquefied natural gas, or LNG) that can be used as a fuel for transportation.

Natural gas reservoirs can be found in sedimentary basins around the world, from Saudi Arabia to Venezuela to the Gulf of Mexico, from Montana and the Dakotas to the Marcellus Shale spanning the Appalachian Basin. That includes my home state of Pennsylvania, where the discovery of extensive natural gas deposits has led to an explosion in natural gas drilling over the past decade and a half. Pennsylvania is now responsible for more than 20 percent of all the natural gas produced in the United States.

The fracking boom has generated billions of dollars in revenue for the state. It has also generated a heated debate, forgive the pun, about the role Pennsylvania should be playing in expanding fossil fuel extraction at a time when we are increasingly dealing with the negative impacts of climate change (and that's not even accounting for the other serious potential environmental threats from natural gas extraction, including the impact of fracking chemicals on the safety of water supplies).[2]

The debate is playing out over an increasingly large stage. Australia's natural gas boom is threatening its agreed-upon carbon emissions targets.[3] Indeed, before the devastating bushfires of the summer of 2019/2020 had even ended, Australia's conservative, pro-fossil-fuel prime minister, Scott Morrison, had eagerly announced a $2 billion plan to boost the domestic natural gas industry.[4] The tragic irony was apparently lost on him.

The Trump administration, meanwhile, heavily promoted natural gas in the United States, attempting to improve its image by re-branding it as "freedom gas."[5] The implication that it will somehow help spread freedom evokes propaganda campaigns from days of yore. The tobacco industry used the phrase "Torches of Freedom" in the early twentieth century in an effort to encourage women to smoke, convincing them it was a source of empowerment during the first wave of feminism in the United States.[6]

Natural gas has often been characterized as a bridge fuel, a way to slowly wean us off more carbon-intensive fuels like coal and gently nudge us toward a renewable energy future. The rationale is that, nominally, natural gas produces about as half as much carbon diox-ide as coal for each watt of power generated. Indeed, the "coal to gas switch," as it's called, is partly responsible for the flattening of global carbon emissions as natural gas displaces more carbon-intensive coal. In the United States, for example, it has been tied to a 16 per-cent decrease in carbon emissions from the power sector during the 2007–2014 period.[7]

What is unique about natural gas among fossil fuels, however, is that it is not only a fossil fuel. It's also a greenhouse gas. In fact, methane is nearly one hundred times more potent as a greenhouse gas than carbon dioxide on a twenty-year time frame.[8] That means it can cause warming not only when we burn it for energy, and it re-leases carbon dioxide, but when the methane itself escapes into the atmosphere. The process of hydraulic fracturing, or fracking, that is used to break up the bedrock to get at natural gas deposits inevitably allows some of the methane to escape directly into the atmosphere (what's known as "fugitive methane").

The Obama administration sought to limit fugitive methane emissions by requiring natural gas interests to curb methane releases from drilling operations, pipelines, and storage facilities. The Trump administration disbanded these regulations, claiming it would save industry millions of dollars.[9]

The rest of us pay the price. Research from 2020 has demonstrated that the spike in atmospheric methane levels in recent decades is coming from natural gas extraction (as opposed to farming and livestock, or natural sources such as peat bogs and melting permafrost).[10] Moreover, the rise in methane is responsible for as much as 25 percent of the warming during this period.[11] Connecting the dots, it is reasonable to say that fugitive methane emissions from fracking are contributing substantially to warming—enough that they may well offset, at least in the near term, the nominal decrease in carbon dioxide emissions from the coal-to-gas switch.

There are other problems with the bridge-fuel framing. Perhaps the most obvious is that we don't have decades to get this right. If we are to avert warming beyond the 1.5°C (2.7°F) danger limit, we've got *one decade* to decrease global carbon emissions by a factor of two.[12] That's a very short bridge. And increased use of natural gas for power generation is likely to crowd out investment in a true, zero-carbon solution in the power sector: renewable energy. Ultimately, the predicament with natural gas is that the solution to a problem created by fossil fuels cannot be a fossil fuel.

UNCLEAN COAL

Why not just gather the carbon dioxide released from coal burning at a coal-fired power plant before it makes it to the atmosphere? Then contain it, burying it somewhere beneath Earth's surface (or below the ocean floor)? There's a name for that—it's called carbon capture and sequestration, or CCS and it's already being implemented.

As I first drafted the paragraph above, the TV was on in the background. Playing was an ExxonMobil commercial promoting CCS. The advertisement conjured an enticing vision of technology overcoming our problems: coal power without carbon pollution—at last, the promise of "clean coal"! Problem solved, right? Not quite. There are in reality a number of fundamental problems with the feasibility, cost, and reliability of CCS.

With CCS, typically, the carbon dioxide released during the burning of coal is scrubbed from emissions and captured, compressed, and liquefied. It is then pumped deep into the Earth, several kilometers beneath the surface, where it is reacted with porous igneous rocks to form limestone. This approach mimics the geological processes that bury carbon dioxide on geological time scales and provides a potential means of long-term geological sequestration of carbon dioxide.

The first full-scale proof of concept for CCS was built in Illinois. Called FutureGen, it was designed to provide data about efficiency, residual emissions, and other matters that would enable scientists to evaluate CCS performance. If CCS were to be deployed commercially at a larger scale in the future, that data would be vital. The project was funded by an alliance of the US Department of Energy and coal producers, users, and distributors. It was ultimately canceled in 2015 as a result of difficulties acquiring public funds.[13] Other CCS projects followed, however, including the large-footprint Petra Nova project in Texas.

Despite its failure, FutureGen did provide some useful insights into the viability of CCS. The scientists involved in the project estimated that they could bury roughly 1.3 million tons of carbon dioxide annually, equivalent to roughly 90 percent of the carbon emitted by the plant's coal burning.[14] But the FutureGen site was chosen in part for its favorability, as it is located above geological formations that are suited to carbon sequestration. This might not be true for many existing coal-burning sites.

The Global CCS Institute reports that there are today fifty-one CCS facilities globally in some stage of development that plan to capture nearly 100 million tons of carbon dioxide per year. (Nineteen facilities are currently in operation, and another thirty-two are either under construction or in development.) Of these, eight are in the United States.[15]

CCS might sound like a foolproof way to mitigate coal-based greenhouse emissions, but there are real questions about its scalability. It simply isn't feasible to bury the *billions* of tons per year of carbon pollution currently produced by coal burning. Many coal-fired power plants are not located at CCS-favorable sites. Moreover, given unforeseen factors, such as earthquakes and seismic activity, or groundwater flow, the efficacy of CCS in any particular location could be compromised. Carelessly sequestered carbon could easily end up becoming mobilized and belched back into the atmosphere.

Economically there is a problem as well. Coal is currently not competitive with other forms of energy in the marketplace. It is, as we have already seen, a dying industry. Requiring that coal plants capture and sequester their carbon will only make it more expensive and hasten the collapse of the industry. Unless, of course, government (that is, taxpayers like you and me) pays for it. In that case, we would be subsidizing dirty energy that still carries climate risk, rather than the cheaper, clean energy that can mitigate it, a true perversion of the economic incentive structure.

Finally, there is the more fundamental limitation that CCS is not even carbon neutral in the best of circumstances. Even if the 90 percent rate of sequestration estimated by FutureGen scientists is correct, and representative more generally of CCS, that would mean that 10 percent of the carbon dioxide would still escape to the atmosphere. CCS-equipped coal-fired power plants would continue to emit tens of millions of tons of carbon dioxide every year. Moreover, most of the carbon dioxide that is captured in CCS is placed into tapped oil wells for enhanced oil recovery. The oil that is recovered, when burned, yields several times as much carbon dioxide as was sequestered in the first place by CCS. So much for carbon-friendliness!

Despite all the talk these days about "clean coal technology," such technology—in the sense of coal-based energy that is free of polluting greenhouse gases—does not yet exist. Until data from experimental sites have been collected and studied, a process that would take years, it will be unclear how much carbon dioxide is actually being sequestered by CCS. It could be decades before the efficacy of true long-term carbon burial could be established. Yet, we have seen that even a decade of additional business-as-usual greenhouse gas emissions could commit us to catastrophic climate change. As Michael Barnard, chief strategist for TFIE Strategy, Inc., a think tank focused on clean energy solutions, aptly put it, "we're in a hole that we've created by shoveling carbon out of the ground and into the sky. The first thing to do is stop shoveling. All CCS does is take teaspoons out of massive scoops of carbon and puts them back in the hole."[16]

CCS is attractive to fossil fuel companies, as it provides them with a license to continue extracting and selling fossil fuels. It is anathema to climate activists, however, because its claim to carbon neutrality is dubious. CCS, unsurprisingly, has been at the very center of the policy debate surrounding the Green New Deal.

Readers may recall from Chapter 5 a letter signed by leading environmental organizations proposing a particular version of AOC's Green New Deal warning that the groups "will vigorously oppose any legislation that . . . promotes corporate schemes that place profits over community burdens and benefits, *including market-based mechanisms . . . such as carbon and emissions trading and offsets.*"[17] What was obscured by those ellipses was the additional inclusion in the blacklist of "carbon capture and storage" (as well as "nuclear power" and "waste-to-energy and biomass energy"). Such overly restrictive language appears to have kept a number of prominent mainstream environmental organizations, including the Sierra Club, the Audubon Society, and the Environmental Defense Fund from signing the letter.[18]

One prominent group that *did* sign the letter, however, was the Sunrise Movement, the youth-led activist group that came to prominence in late 2018. It was in the news in particular over its efforts to pressure House Majority Leader Nancy Pelosi (D-CA) into creating a committee to draft a Green New Deal. Sunrise demanded that any potential plan must fund "massive investment in the drawdown and capture of greenhouse gases," which would seem to conflict with the restrictive language about carbon capture in the letter they had signed. But Sunrise now omits "capture," speaking only of the "drawdown of greenhouse gases," which would seem to indicate support for natural drawdown via reforestation and regenerative agriculture, but, by omission, not CCS.[19]

James Temple, senior editor for energy at the centrist *MIT Technology Review*, took issue with the environmentalists' letter in a piece he penned titled "Let's Keep the Green New Deal Grounded in Science." Temple argued that the sort of "rapid and aggressive action" the letter claims is necessary to avert the dangerous warming of 1.5°C (2.7°F) is likely incompatible with policies that take key options like carbon capture off the table.[20]

What has emerged here is a battle between climate progressives and climate moderates on the role of industry and market-driven mechanisms. And while my assessment of the science and economics leads me to side with climate moderates on the merit of climate pricing, for reasons outlined previously, I tend to side with the progressives on the dubious merit of CCS schemes for all the reasons discussed above (with the possible exception of currently difficult-to-decarbonize sectors like cement production).

GEOENGINEERING, OR "WHAT COULD POSSIBLY GO WRONG?"

So, if "clean coal" and natural gas "bridge fuels" aren't the solution, is there some other way we can engineer our way out of the climate crisis? Perhaps we should consider *geoengineering*—schemes that employ global-scale technological intervention with the planet in the hope of offsetting the warming effects of carbon pollution.

Many of these proposed schemes sound like they're taken right out of science fiction. And as with science fiction films, bad things tend to happen when we start tampering with Mother Nature. We might not get a planet run by apes, giant fire-breathing dinosaurs, or institutionalized cannibalism, but we could get worse droughts, more rapid ice-sheet melt, or any number of unpleasant surprises. When it comes to a system we don't understand perfectly, the principle of unintended consequences reigns supreme. If we screw up this planet with botched geoengineering attempts, there is no "do over." And, as they say, "there is no planet B."

Consider, for example, proposals to shoot reflective particulates—sulfate aerosols—into the stable upper part of the atmosphere known as the stratosphere, where they would reside for years. This human-produced effect would mimic the way volcanic eruptions cool the planet. An explosive tropical volcanic eruption can put enough reflective sulfate particles into the stratosphere to cool the planet for a while. (The Mount Pinatubo eruption of 1991 in the Philippines, for example, cooled the planet by 0.6°C [1°F] for about fifteen months.)[21]

This scheme has the *advantage* of being feasible. It would use custom-designed cannons to fire substantial amounts of sulfate aerosols into the stratosphere, easily as much as was released during the Pinatubo eruption. Doing the math, all it would take is a Pinatubo-size injection of particles every few years to offset the current warming effect of carbon emissions. It would also be relatively cheap to do (compared to other means of mitigation).[22]

The scheme has the distinct *disadvantage*, however, of potential major adverse climate side effects. First of all, we would get a very different climate from the one we're used to. The spatial pattern of the geoengineering-induced cooling isn't the mirror image of the pattern of greenhouse gas warming. That's because the physics is different. In the former case, we're reducing the incident sunlight, while in the latter case, we're blocking the escape of heat energy from Earth's surface. Those effects have very different spatial patterns. On average, the globe may not warm under the sulfate aerosol plan, but some regions would cool while others warmed. Indeed, some regions would likely end up warming even faster than they would have without the geoengineering. We could conceivably end up, for example, accelerating the destabilization of the West Antarctic or Greenland ice sheet and speeding up global sea-level rise. Climate model simulations indicate that the continents would potentially get drier, worsening droughts.[23]

There are other potentially nasty environmental side effects as well. It was, after all, the production of sulfur dioxide and the resulting sulfate aerosols in the lower atmosphere from coal-fired power plants that gave us the acid rain problem in the 1960s and 1970s, prior to passage of the clean air acts. The sulfate particles from geoengineering would be higher up—in the stratosphere—but they would ultimately still make it down to the surface, where they would acidify rivers and lakes. And then there's the "ozone hole." Though it has mostly recovered, there are enough ozone-depleting chemicals still in the stratosphere that, with the extra kick they would get from the injected sulfate aerosols, we would likely see continued destruction of the protective ozone layer.

As with any "cover-up" approach to climate change that doesn't deal with the root cause of the problem (continued carbon emissions), carbon dioxide would continue to build up in both the atmosphere and the ocean. The problem of ocean acidification, sometimes called "global warming's evil twin," would continue to get worse, further threatening the world's coral reefs and calcareous sea life such as shellfish and mollusks and wreaking havoc on ocean food chains.

Sulfate aerosol geoengineering is a Faustian bargain: it would require us to continue to inject sulfate aerosols into the stratosphere while carbon dioxide continued to accumulate in the atmosphere. Were there a major war, a plague, an asteroid collision, or anything else that might interfere with the regular required schedule of sulfate injections, the cooling effect would disappear within a few years. We would experience decades' worth of greenhouse warming in a matter of years, giving new meaning to the concept of "abrupt climate change."

One of the cruelest ironies of all with this prospective technofix is that it would likely render *less* viable one of the most important and safest of climate solutions: solar power. The sulfate aerosols would reduce the amount of sunlight reaching Earth's surface that is available to produce solar energy, making the already tough challenge of weaning ourselves off the fossil fuels at the root of the climate-change problem even more difficult.

Another widely discussed geoengineering scheme is ocean iron fertilization. Over much of the world's oceans, iron is the primary limiting nutrient for algae, or phytoplankton, which take up carbon dioxide when they photosynthesize. It is therefore possible to generate phytoplankton blooms by sprinkling iron dust into the ocean, which in turn metabolizes carbon dioxide. When the phytoplankton die, they tend to sink to the ocean bottom, burying their carbon with them.

One of the advantages of ocean iron fertilization is that it is solving the problem at its source, taking carbon out of the atmosphere. That means it also prevents the worsening of ocean acidification. It's an example of what is termed "negative emissions technology"—it actually takes carbon *out of* the atmosphere. The idea is appealing enough that a number of companies tried to commercialize the scheme more than a decade ago. One company even sold carbon credits, promising to bury a ton of carbon dioxide for only $5, a bargain for any organization or company seeking to lower its carbon footprint.

Subsequent experiments, however, showed that the scheme doesn't really work. Iron fertilization leads to more vigorous cycling of carbon in the upper ocean, but no apparent increase in deep carbon burial, which means no permanent removal of atmospheric carbon. To make matters worse, studies showed that it could actually favor harmful "red tide" algae blooms that create oceanic dead zones. Lacking evidence of efficacy, and with growing concern about unintended consequences, support for iron fertilization geoengineering has dissipated.[24]

Sticking with this theme, though, might there be other negative emissions technology that could be implemented safely and cost effectively? Trees do it, after all. They take carbon out of the atmosphere as they photosynthesize, and they store it in their trunks, branches, and leaves. Then they bury carbon in the ground, in their roots, and in the leaf and branch litter that falls and gets deposited onto the forest floor and buried in the soil.

Perhaps we can learn from the trees. Maybe even improve upon them. Trees, after all, don't do a perfect carbon burial job. Like us, they respire—putting carbon dioxide back into the atmosphere. And when they die and decompose, some of their carbon escapes back to the atmosphere. It's part of the long-term balance of the terrestrial carbon cycle.

We might try to make a more perfect (from a climate standpoint) "tree"—a tree that takes carbon out of the air more efficiently than regular trees and doesn't give any of it back to the atmosphere. Rather than dying and decomposing, synthetic trees (with "leaves" treated with sodium carbonate) could turn the carbon they extract from the atmosphere into baking soda, which can be buried for the long term. Such a scheme has not only been suggested by scientists, but its viability has already been demonstrated through proof-of-concept trials. It is calculated that an array of ten million synthetic trees around the world could take up a significant chunk, perhaps as much as 10 percent, of our current carbon emissions.[25] But this so-called direct air capture would be difficult and expensive to do, perhaps costing more than $500 per ton of carbon removed. A related approach that has been suggested recently, which involves atmospheric CO_2 removal through the artificial enhancement of weathering by rocks, might be less expensive—somewhere in the range of $50 to $200 per ton of carbon. But its proponents concede that it could remove, at the very most, only about two billion tons of carbon dioxide per year, a veritable drop in the bucket compared with current carbon emissions.[26]

These limitations mean that at present, it is far easier and cheaper to prevent the buildup of carbon dioxide in the atmosphere in the first place, by limiting fossil fuel burning. But the cost of this direct air capture could be brought down substantially with additional research and through the economies of scale of mass production. And if, after doing everything possible to reduce our carbon emissions, we still find ourselves headed toward catastrophic warming, we might need a stopgap solution.

Of all of the geoengineering schemes, direct air capture seems the safest and most efficacious. Unlike CCS, which continues our reliance on fossil fuels, this form of carbon burial could, along with natural reforestation (discussed later), be an important component of broader efforts to "draw down" carbon from the atmosphere, a strategy that arguably belongs in any comprehensive climate abatement program. But since we're only talking about 10 percent, at most, of current carbon emissions, it is obvious this cannot be a primary strategy for mitigation.

People have suggested many other schemes, from putting reflective mirrors in space to seeding low clouds over the oceans. All of them are fraught with political and ethical complications. For one, who gets to set the global thermostat? For low-lying island nations, current carbon dioxide levels are already too high—their people are already threatened with the loss of their land and their rich cultural heritages by the several feet of sea-level rise that is likely baked in. While the industrial world debates whether we can still avoid dangerous warming of 1.5 or 2°C (2.7 or 3.6°F), dangerous warming is already here for many. Some might want to set the thermostat at a lower temperature than others. Who gets to make the decision?

One could easily imagine a whole new form of global conflict wherein rogue states employ geoengineering to control the climate in a way that is optimal for themselves. A climate model simulation might show, for example, that sulfate aerosol injection could relieve the drought that plagues a particular nation. Yet, it would do so at the expense of causing a drought elsewhere. The perpetual conflict in the Middle East has arguably always been fundamentally about access to scarce freshwater resources.[27] Would geoengineering provide yet another weapon to fuel this ongoing battle?

A fundamental problem with geoengineering is that it presents what is known as a moral hazard, namely, a scenario in which one party (e.g., the fossil fuel industry) promotes actions that are risky for another party (e.g., the rest of us), but seemingly advantageous to itself. Geoengineering provides a potential crutch for beneficiaries of our continued dependence on fossil fuels. Why threaten our economy with draconian regulations on carbon when we have a cheap alternative? The two main problems with that argument are that (1) climate change poses a far greater threat to our economy than decarbonization, and (2) geoengineering is hardly cheap—it comes with great potential harm.

But despite the caveats, disadvantages, and risks, geoengineering has proven to be appealing to fossil fuel interests and those advocating for them.[28] They can have their cake and eat it too, claiming to support a putative climate "solution," but one that poses no threat to the fossil fuel business model. A 2019 report on geoengineering by the Center for International Environmental Law (CIEL) explains how "the most heavily promoted strategies for carbon dioxide removal and solar radiation modification depend on the continued production and combustion of carbon-intensive fuels for their viability." CIEL noted that "the hypothetical promise of future geoengineering is already being used by major fossil fuel producers to justify the continued production and use of oil, gas, and coal for decades to come."[29]

Geoengineering also appeals to free-market conservatives, as it plays to the notion that market-driven technological innovation can solve any problem without governmental intervention or regulation. A price on carbon, or incentives for renewable energy? Too difficult and risky! Engaging in a massive, uncontrolled experiment in a desperate effort to somehow offset the effects of global warming? Perfect!

It is thoroughly unsurprising, for example, that someone with as much skin in the carbon game as Rex Tillerson, former CEO of the world's largest fossil fuel company, ExxonMobil, has argued that climate change is "just an engineering problem."[30] Nor is it surprising that some of the now familiar inactivist players, such as Bjorn Lomborg and the Breakthrough Institute, have promoted geoengineering as a primary means of climate mitigation.[31]

Perhaps more eye-opening, though, is the fact that business magnates like former Microsoft CEO Bill Gates have embraced the concept. Writing in *Fortune*, journalist Marc Gunther reported that "Gates has been convinced that the risk of global warming is worse than most people think. He can see that the world's governments have failed to curb the emissions caused by burning coal, oil, and natural gas. . . . So the Microsoft billionaire and philanthropist has stepped into the breach to become the world's leading funder of research into geoengineering—deliberate, large-scale interventions in the earth's climate system intended to prevent climate change and its repercussions."

Gates gave millions of dollars to two climate scientists, David Keith of Harvard University and Ken Caldeira of Stanford University, to perform research and engage in experimentation with geoengineering. That includes relatively safe direct air capture but also potentially harmful stratospheric sulfate aerosol injection.[32] Perhaps relevant, in their *Guardian* commentary "The Fossil Fuel Industry's Invisible Colonization of Academia," Benjamin Franta and Geoffrey Supran singled out these two centers of geoengineering research—Stanford and Harvard—as exemplars of how "corporate capture of academic research by the fossil fuel industry is an elephant in the room and a threat to tackling climate change."[33]

Harvard's Keith has "done as much as any single researcher to push the touchy topic of geoengineering toward the scientific mainstream," according to James Temple of *Technology Review*.[34] Keith is affiliated with the Breakthrough Institute and a signatory of the "Ecomodernist Manifesto," a techno-optimist, pseudo-environmentalist polemic that *Guardian* columnist George Monbiot characterized as "generalisations, . . . ignorance of history, . . . unexplored prejudices . . . an astonishing lack of depth," and a "worldview that is, paradoxically, nothing if not old-fashioned."[35] Keith helps lead a for-profit venture financed by Bill Gates to implement geoengineering and is currently planning to do real-world experimentation testing the viability of sulfate aerosol stratospheric injection.[36]

Keith spearheaded a 2019 study of the ostensible impact of sulfate aerosol geoengineering on the global climate, which included a modeling experiment to simulate the effects.[37] He took to Twitter to promote his team's findings, claiming he and coauthors had demonstrated that "no region is made worse off" by solar geoengineering. Other leading climate scientists contested that claim. Chris Colose, a climate researcher at the NASA Goddard Institute for Space Studies, pointed out that the modeling experiment is a bit of a bait-and-switch: "They don't actually put aerosols in the atmosphere. They turn down the Sun to mimic geoengineering. You might think that is relatively unimportant . . . [but] controlling the Sun is effectively a perfect knob. We know almost precisely how a reduction in solar flux will project onto the energy balance of a planet. Aerosol-climate interactions are much more complex." Colose went on to point out the numerous other ways in which the modeling experiment they had done was a gross idealization of actual real-world implementation of their geoengineering scheme, emphasizing a number of the well-established flaws and caveats that we encountered earlier in our discussion of sulfate aerosol geoengineering.[38]

Ken Caldeira, the other of the two Gates-funded geoengineering scientists (who has now left his position at Stanford to work directly for Bill Gates), later weighed in, asserting, "The evidence is that solar geoengineering would be expected to reduce climate damage."[39] Again, many leading climate scientists begged to differ. Climate researcher Daniel Swain from the University of California at Los Angeles weighed in that he finds it "strange" that the regional details of climate model simulations are "taken pretty literally" in these idealized geoengineering experiments, "but are subject to huge caveats otherwise," adding that while there's "lots of evidence" that sulfate aerosol geoengineering would indeed reduce the global average temperature, "that's not all that matters!"[40] Jon Foley, executive director of Project Drawdown, added that relying on such idealized experiments is "a big gamble, especially when models have a hard time" reproducing detailed temperature patterns.[41] Matthew Huber, a leading climate researcher at Purdue University, expressed two concerns: whether humans could properly administer the highly structured geoengineering protocol required, and whether the models are reliable enough to capture some of the potential surprises that might be in store.[42]

One gets the distinct feeling that scientists like Keith and Caldeira suffer from some degree of hubris when it comes to leaping from the results of their highly idealized modeling experiments to sweeping conclusions about the real world. One also gets the feeling that their attitude toward real-world geoengineering potentially crosses the line from dispassionate inquiry to advocacy. As a scientist, that's okay as long as you're up front about it. I've argued as much in the *New York Times*.[43] But both of them seem uncomfortable acknowledging that they're engaged in advocacy. I can speak to this directly. Keith and Caldeira each responded rather defensively to a tweet of mine in which I stated that many "geoengineering *advocates* . . . see geoengineering as an excuse for continued business-as-usual burning of fossil fuels" (emphasis added).[44] At the time, I was bemused by the fact that they thought the tweet was directed at them (it wasn't), and I wondered aloud whether they *do* indeed consider themselves to be "geoengineering advocates." Each equivocated, drawing a distinction between advocacy for research and advocacy for implementation.[45] I would argue that their words and actions blur any such distinctions.

Finally, let's discuss the role here of climate *doomism*, a topic we will explore in depth in the next chapter. Geoengineering advocates have increasingly found common cause with climate-change doomsayers—those who believe that the situation is now so dire that truly desperate action is required, or that we're beyond the point where any effective action is possible.

Such misguided framing was beautifully captured in a December 2019 *Washington Post* op-ed, "Climate Politics Is a Dead End. So the World Could Turn to This Desperate Final Gambit."[46] In it the author, Francisco Toro, a Venezuelan political commentator, promotes a bleak climate policy outlook, articulating the view of some climate activists "that only a drastic push toward net-zero carbon emissions can save the world. But . . . the politics to achieve this don't exist." As an example he cites "the events of the past decade, including the failure of the climate conference in Madrid [COP25 in 2019]."

Toro then uses this defeatist narrative to justify the implementation of potentially dangerous geoengineering schemes ("Yes, a geoengineered future may be scary. But unchecked climate change is absolutely terrifying. And attempts to prevent it aren't working"). No inactivist polemic would be complete without deflection and a free pass for polluting interests ("Climate activists typically blame the failure to cut emissions on greedy corporations and crooked politicians. . . . The regrettable reality is that people around the world demand cheap energy"). He misleadingly invokes the "Yellow Vest" protests as evidence that people will "punish leaders who threaten their access to it."

This commentary exemplifies how climate doomism is being exploited to support dangerous technofixes that might be favored by polluters but could leave us worse off. It demonstrates the deep hypocrisy of polluting interests and the inactivists doing their bidding, who first sabotage climate negotiations like those in Madrid, and then proclaim that the *failure* of those negotiations is grounds for their proposed "solution" (geoengineering technofixes).[47]

The fundamental problem with geoengineering, in the end, is that tinkering with a complex system we don't fully understand entails monumental risk. Geoengineering expert Alan Robock of Rutgers University believes that geoengineering is too risky to ever try. "Should we trust the only planet known to have intelligent life to this complicated technical system?" Robock wondered. "We don't know what we don't know."[48] The CIEL report discussed earlier notes "the stark contrast between the . . . narrative that geoengineering is a morally necessary adjunct to dramatic climate action" and the reality that geoengineering is "simply a way of avoiding or reducing the need for true systemic change, even as converging science and technologies demonstrate that shift is both urgently needed and increasingly feasible." It highlights, furthermore, "the growing incoherence of advocating for reliance on speculative and risky geoengineering technologies in the face of mounting evidence that addressing the climate crisis is less about technology than about political will."[49]

GREENING THE PLANET

We've seen that one type of geoengineering that has been proposed— direct air capture—mimics what trees do naturally by capturing carbon through photosynthesis, storing it in their trunks and limbs, and burying it in their roots and branches and leaf litter. So why not just engage in the massive planting of trees—that is, large-scale reforestation of the vast regions of the planet that have been deforested (or *afforestation*—foresting regions that were previously something else). Such efforts could be supplemented by land use and agricultural practices that sequester additional carbon in soils.

What is appealing about this particular negative emissions option is that it's a "no regrets" path forward. After all, by planting trees we can get better-functioning ecosystems; maintain and even increase biodiversity; improve the quality of our soils, air, and water; and better insulate ourselves from the damaging impacts of climate change. Could efforts to "green the planet" make a major dent in our carbon emissions? Or mitigate them altogether? It's certainly proven to be convenient for some, who, to deflect attention away from the subject of what polluters should be doing, present "tree planting" as the solution and treat it as evidence of bold action on climate. Hence Donald Trump's "politically safe new climate plan" (promoted originally by some of his Republican congressional colleagues) of supporting efforts to plant hundreds of millions of new trees.[50] Is there actually merit to the suggestion?

Let's take a look at the prospects for reforestation and afforestation. One study claimed that an additional 0.9 billion hectares of the planet's surface is available for this purpose. That translates to billions of new trees that collectively could capture just over 200 billion tons of carbon dioxide over the next couple of decades.[51] That's a rate of carbon sequestration of roughly 11 billion tons of carbon dioxide per year. Other scientists have questioned the assumptions of the study and argued for much lower levels of potential carbon sequestration. In fact, the most recent IPCC report (2019) estimated that roughly only 60 billion tons of carbon dioxide could be sequestered through reforestation by the end of the century, which translates to less than 1 billion tons of carbon dioxide per year.[52] Nonetheless, let us, for the sake of argument, accept the much higher 11 billion number.

Regenerative agriculture based on recycling farm waste and using composted materials from other sources, combined with land use practices that enhance soil carbon sequestration, could potentially bury somewhere in the range of 3.5 to 11 billion tons of carbon dioxide emissions per year. Let us once again take the very optimistic upper limit of 11 billion tons per year.

Adding together these contributions gives us 22 billion tons of carbon dioxide per year. That sounds like quite a bit, but we are currently generating the equivalent of roughly 55 billion tons per year of carbon dioxide through fossil fuel burning and other human activities.[53] That means that even if we accepted estimates from the very upper limits of the uncertainty range, the combined effect of reforestation and agriculture and land use practices would at most only slow the buildup of carbon dioxide in the atmosphere by a factor of 44 percent. In other words, atmospheric carbon dioxide levels would continue to rise, just at a rate that is roughly half as fast.

That estimate, of course, is overly optimistic. We cannot ignore the massive demands on available land of 7.7 billion (and growing) people competing for space for settlement, agriculture, and livestock. When real-world economic constraints are taken into account, the actual land area available for reforestation may be only about 30 percent of the technically available land area assumed in the recent study.[54]

Climate change itself, furthermore, is likely to diminish the ability of forests to sequester carbon. The bushfires in the summer of 2019/2020 doubled Australia's total carbon emissions in the year that followed and were likely to cause a 1 to 2 percent increase in global carbon dioxide concentrations.[55] And Australia is not the only place that is burning. Wildfires taking place from the Amazon to the Arctic are releasing billions of tons of carbon dioxide a year.[56] A study reported in 2020 in the journal *Nature* demonstrated that the peak carbon uptake by tropical forests occurred during the 1990s and has declined ever since as a result of logging, farming, and the effects of climate change. The authors found that the Amazon could go from a sink (a net absorber of carbon) to a source (a net producer of carbon) within the next decade, which is decades ahead of schedule based on former climate model predictions.[57]

Such findings underscore one of the potential pitfalls of relying upon reforestation as a primary means of climate mitigation (or, for that matter, as the basis for carbon offsets or credits). Any carbon that is sequestered could easily be lost, perhaps in rapid bursts, because of forest burning. Ironically, the problem becomes worse as the planet continues to warm and conditions become more conducive to massive forest burning.

Moreover, as with geoengineering, there are potential unintended consequences. The coauthor of a recent government report on forest carbon burial told the BBC that "we would be crazy to undertake the massive scale of planting being considered if we did not also consider the wider effects upon the environment including impacts on wildlife, benefits in terms of reducing flood risks and effects on water quality, improvements to recreation and so on." The report noted that careless tree planting, ironically, could actually lead to increased carbon emissions. As the BBC noted, "carpeting upland pastures with trees would reduce the UK's ability to produce meat—which may lead to increasing imports from places that produce beef by felling rainforests."[58]

Finally, no discussion of natural carbon drawdown is complete without addressing proposals for using biomass for energy followed by the capture and sequestration of any carbon dioxide produced. This is known as "bioenergy with carbon capture and storage," or BECCS. The IPCC has emphasized this technology in its scenarios for stabilizing carbon dioxide concentrations that assume zero total effective emissions within a matter of decades. The IPCC does this by relying upon the presumption that BECCS can actually yield negative carbon emissions, which would offset some residual fossil fuel burning and other carbon-generating practices to achieve the needed zero net emissions.

How could this work? Readers may recall from the previous chapter Michael Moore's false claim, in his film *Planet of the Humans*, that "biomass releases 50 percent more carbon dioxide than coal and more than three times as much as natural gas." In reality, biofuels (neglecting the fossil fuel energy that might be used in processing and transportation) are carbon neutral, having taken as much carbon dioxide out of the atmosphere when they were plant matter as they release when they're burned. They are therefore far more carbon-friendly than fossil fuels, yielding energy with little or no carbon pollution. In fact, they can—in a sense—be made even *more carbon-friendly than renewables*, providing energy and drawing down carbon from the atmosphere *at the same time*.

This might seem like it violates some law of physics, but it doesn't. The idea is that you burn the biofuels to get energy as you would coal or natural gas. The process, as we have explained, is carbon neutral to start. Now, if you capture the carbon dioxide and bury it, then you're doing even better than carbon neutrality—you're actually drawing down carbon that came from the atmosphere and capturing and burying it. Of course, all of the concerns we encountered previously with carbon sequestration in the context of coal or natural gas apply here as well—namely, you have to be able to bury it efficiently, safely, and effectively permanently, and that's not easy to do. Moreover, as we already saw with CCS, capture is unlikely to be complete, so some of the carbon does make it back into the atmosphere.

As alluded to earlier, negative emissions technologies—and particularly BECCS—are assumed in the various IPCC emissions scenarios or "pathways," including those that allow us to stay below critical warming thresholds such as 1.5 or 2.0°C (2.7 or 3.6°F). Given the fact that BECCS has not yet been demonstrated to be commercially viable at the scale assumed in these scenarios, the IPCC could rightly be criticized for, in essence, "kicking the can down the road"—putting forth scenarios that allow substantial near-term carbon emissions and still avert dangerous planetary warming only by assuming massive negative emissions in future decades using currently unproven technology. What if that technology does not emerge? The "Faustian bargain" again rears its head.[59]

THE NUCLEAR OPTION

All reasonable options should be on the table as we debate how to rapidly decarbonize our economy while continuing to meet society's demand for energy. There is no easy solution, and there are important and worthy debates to be had in the policy arena as to how we accomplish this challenging task.

There is a good-faith argument to be made, for example, that nuclear energy should be part of the solution, and I have colleagues whom I deeply respect who are bullish on the role it might play as part of a comprehensive plan to tackle climate change. I myself remain skeptical that nuclear energy should play a central role in the required clean, green energy transition. Let me explain why.

There are a number of major obstacles, first of all, to safe, plentiful nuclear power. There is the risk of nuclear proliferation, and the danger that fissile materials and weapons-applicable technology could make it into the hands of hostile nations with militaristic intentions or terrorists. There is the challenge of safe long-term disposal of radioactive waste. And there are some profound examples of the acute environmental and human threat posed by nuclear power, most recently highlighted, for example, by the Fukushima Daiichi nuclear disaster north of Tokyo in March 2011.

Hitting closer to home—for me, literally—was the historic Three Mile Island nuclear disaster of March 1979. It took place in my home state of Pennsylvania on a long, narrow island in the Susquehanna River near Harrisburg, less than a hundred miles southeast of the Happy Valley, in which I currently reside. I'm reminded of the incident—a partial meltdown that led to the release of harmful radiation—every time I fly into the Harrisburg Airport over the plant's eerily iconic cooling towers. (The plant is now closed but not yet decommissioned.)

No means of energy production is without environmental risk, but nuclear power carries with it unique dangers. As noted by Robert Jay Lifton and Naomi Oreskes in a 2019 *Boston Globe* op-ed, improvements in design cannot eliminate the possibility of deadly meltdowns.[60] Nuclear power plants will always be vulnerable to natural hazards such as earthquakes, volcanoes, or tsunamis (like the one that triggered the Fukushima meltdown), or technical failure and human errors (like the ones responsible for Three Mile Island).

Climate change itself, ironically, increases the risk. As Lifton and Oreskes pointed out, extreme droughts have led to reactors being shut down as the surrounding waters become too warm to provide the cooling necessary to convey heat from the reactor core to the steam turbines and remove surplus heat from the steam circuit.[61] Some of my own research has shown that climate change is leading to less reliable flow for the very river—the Susquehanna—that supplied the Three Mile Island nuclear plant with needed cooling water.[62] A similar threat looms for many other active plants.

Some have argued in favor of a role for small modular reactors (SMRs), which, as the name implies, are considerably smaller than the massive reactors in Fukushima or Three Mile Island. They also require less up-front capital, and arguably they allow for better security of nuclear materials. Energy experts, however, have raised serious concerns about SMRs, including "locating sites for multiple reactors, finding water to cool these reactors, and the higher cost of electricity generation."[63] SMRs, in short, are not an obvious nuclear power "magic bullet."

Still others argue that the answer is so-called "next generation" or "generation IV" nuclear power plants, such as molten-salt reactors that automatically cool down when they get too hot, or very-high-temperature reactors (VHTR), which could be coupled to a neighboring hydrogen production facility for significantly reduced cost.[64] But as University of California, Berkeley, energy expert Dan Kammen noted, it "could easily take the advanced nuclear projects 30 years to get through regulatory review, fix the unexpected problems that crop up . . . and prove that they can compete." In the meantime, we could see a breakthrough in other technologies, such as electric storage and fusion. Kammen added that while "ultimately on a planet with 10 billion people, some amount of large, convenient, affordable, safe baseload power—like we get from nuclear fission, or fusion—would be just hugely beneficial," there are "other competitors in view on the straight solar side that 10 years ago sounded like science fiction—space-based solar, transparent solar films on every window. That world works, too."[65]

Some would argue that our energy choices amount to balancing different risks. True, nuclear energy has risks, they acknowledge, but they are worth it in the balance. They would say that though nuclear accidents are acute, they are rare. And while the damage can be fatal and long-lasting, it is regionally localized. Compare that to the risks posed by climate change, which are pervasive, global, and slowly but steadily growing. If we are forced into a choice between one risk or the other, a reasonable argument could be made that there's a significant role to be played by nuclear energy. The problem with this argument is that it buys into the fallacy that nuclear power is necessary for us to decarbonize our economy. Although it may well make sense to continue with the operation of *existing* nuclear power plants until they are retired (after twenty to forty years, their typical lifetime), given that the embodied carbon emissions associated with their construction is a "sunken" carbon cost, it makes little sense to build new ones.

As we have already seen, electrification of the various energy sectors in conjunction with decarbonization of the grid can already be achieved using renewables such as residential rooftop solar and solar plants, onshore and offshore wind farms, wave energy, geothermal energy, and hydroelectric and tidal energy. Researchers have shown how these existing renewable energy technologies could be scaled up to meet 80 percent of global energy demand by 2030 and 100 percent by 2050. To those who argue that nuclear is a cheaper option, the numbers indicate otherwise. As Lifton and Oreskes noted, the average nuclear power generating cost is about $100 per megawatt-hour, compared with $50 for solar and $30 to $40 for onshore wind. Renewable energy costs are now competitive with fossil fuels—even with the incentives that are currently skewed against them—and much lower than for nuclear.[66]

So if the math and logic don't obviously favor a nuclear solution, why do advocates fight so fiercely for it? For some, no doubt, it's a matter of principle. As I mentioned earlier, I have colleagues whom I respect deeply who are convinced that nuclear energy is critical to solving the climate crisis.[67] But for many, alas, it appears to be all about ideology and political tribalism. "Hippie punching"—establishing one's conservative bona fides by opposing perceived leftist environmentalists—has become de rigueur, as a common target for attack serves to unite conservatives in the climate arena. Consider, for example, the attacks on global-warming icon and conservative punching bag Al Gore. My friend Bob Inglis, a former Republican congressman from South Carolina, has said, "In my first six years in Congress from 1993 to 1999, I had said that climate change was hooey. I hadn't looked into the science. All I knew was that Al Gore was for it, and therefore I was against it."[68]

Support for nuclear energy has become a shibboleth for conservatives in the climate policy arena. It's easy to understand why. It was the left, after all, that protested nuclear power in the 1970s. While I was growing up in Massachusetts, and protests of the Seabrook nuclear plant were taking place in nearby New Hampshire, it was all granola-crunching tree-huggers, scruffy college students and aging flowerchildren.

"The enemy of my enemy is my friend" might not be a very satisfying explanation for the unusual amount of support for nuclear energy among conservatives, but it's difficult otherwise to explain it. Solar *should be* the preferred solution for conservatives: it can be deployed locally, and if installed privately it can help liberate users from dependency on overly regulated centralized utilities. Meanwhile, nuclear power plants require huge up-front capital investments and are not viable without governmental subsides, so they are hardly the free-market solution conservatives purport to favor.[69] Bob Inglis is of course famous as a conservative climate crusader. He is all about free-market solutions to the climate crisis. He also happens to have a nuanced view of the role of nuclear energy as a climate solution: "It used to be convenient for us as conservatives to blame enviros for why we're not building nuclear power plants," he told a reporter, "but if we update our rhetoric to the actual facts, what we find is it's more a question of economics."[70]

Inglis is the exception to the rule. Conservatives (and "conservative liberals" such as CNN commentator Fareed Zakaria) love big fixes like nuclear energy and geoengineering.[71] What do these "solutions" have in common? They divert resources and attention away from the more obvious solution—renewable energy. Indeed, a cynic might wonder whether some who staunchly advocate for these options are more interested in dampening enthusiasm for a renewable energy revolution than in actually solving the climate problem. The Breakthrough Institute promotes both nuclear energy and geoengineering. So do the "ecomodernists."[72] Former Democratic presidential candidate Andrew Yang promoted nuclear energy and geoengineering as well, as he sought, during his campaign, to thread the needle of maintaining credibility on climate while courting conservative Democrats.[73]

"ADAPTATION" AND "RESILIENCE"

The last refuge of the false solutionists is the language of "adaptation" and "resilience." That is not to say that both aren't important—they are. We have no choice but to adapt to those climate-change impacts that are now inevitable, and we need to establish greater resilience in the face of the heightened climate risk that already exists. The Global Commission on Adaptation, for example, has recommended pursuing five key areas of climate-change adaptation over the next decade: early warning systems, climate-resilient infrastructure, altered agricultural practices, protection of coastal mangrove ecosystems, and more resilient water resource management.[74]

But much as exclusive focus on individual action has been used in a deflection campaign to undermine systemic change, exclusive focus on adaptation and resilience has become a favored tactic of inactivists. It's another way of sounding like one is taking proactive steps to address the climate crisis while enabling business-as-usual burning of fossil fuels and the continued profits that go with it.

We see this language in the messaging of Republicans who are still trying to navigate a course between flat-out denial and indefinite delay. Consider Republican senator Marco Rubio of Florida, the state arguably most on the frontlines of climate-change impacts. In August 2018, he wrote an opinion piece in *USA Today* citing innovation and adaptation as the key to combating climate change.[75] In it, he insisted that the impacts of sea-level rise could be managed through restoration of the Everglades. Costly projects aimed at slowing the encroachment by the ocean of Florida's coastline might buy some of the wealthier communities some time, but, as a colleague and I responded in a commentary, without the ability to move to higher ground, coastal populations will become increasingly vulnerable to frequent flooding and toxic floodwaters, and coastal tourism and industry will suffer.[76]

There is no way to engineer our way out of sea-level rise. If we continue to emit carbon, warm the oceans, and melt the ice sheets, the oceans will ultimately prevail in this battle between humans and nature.

In early 2019, after Democrats had taken back the House of Representatives, there was both good news and bad news when it came to the Republican stance on climate. The good news was that Republicans were no longer contesting the basic scientific evidence—they'd finally, it seems, given hard denial a rest. The bad news was that they were still promoting inaction, only this time dressed up in the language of "innovation," "conservation," and "adaptation."[77]

The Republican approach to climate change resembled a person trying to fix a leak in his ceiling with buckets, towels, and mops, but no mention of repair or a handyman. As the *Washington Post*'s Steven Mufson reported in early 2020, "the GOP is still hammering out details, but some critics say the new Republican approach to climate change looks a lot like the old one. In addition to [proposals to plant] trees, senior Republicans are said to be considering tax breaks for research, curbs on plastic waste and big federally funded infrastructure projects in the name of *adaptation* or *resilience*. The already well-worn buzzword '*innovation*' will be their rallying cry, and natural gas, despite its carbon emissions, will be embraced" (emphasis added).[78] What's missing here? Any discussion of carbon emissions, fossil fuels, or renewable energy.

A week later, Republicans on the House Energy and Commerce Committee, Greg Walden of Oregon, Fred Upton of Michigan, and John Shimkus of Illinois, wrote an op-ed in which they acknowledged that "climate change is real," adding that they "are focused on solutions."[79] Their commentary predictably emphasized the belief that "America's approach for tackling climate change should be built upon the principles of *innovation, conservation,* and *adaptation*" (emphasis added). They promoted the usual conservative favorites of carbon capture and nuclear power. And where they mentioned renewable energy, they emphasized research and innovation with regard to clean energy technologies, batteries, and storage. There was no discussion about actual *deployment* of renewable energy or market mechanisms—such as incentives for renewables or a price on carbon—that might level the playing field and enable the rapid transition away from fossil fuels necessary to avoid a crisis.

This phenomenon is not unique to the United States. In Australia, with the massive shift in public sentiment that took place in the aftermath of the historic bushfires of summer 2019/2020, there was a grudging acceptance by conservatives of the climate threat.[80] Former deputy prime minister and National Party leader Barnaby Joyce, who had previously gone to great lengths to deny climate change, conceded in a *60 Minutes* special on the Australian bushfires (in which we were both co-panelists) not only that "the climate is changing," but that the bushfires were a consequence of climate change.[81] Even climate-change-denying columnist Andrew Bolt of the Murdoch-owned *Herald Sun* has now admitted the reality of human-caused climate change.[82]

Unfortunately, in spite of this grudging acceptance of the problem, there is no will in the current Australian government to do anything about it other than promoting "adaptation" and "resilience."[83] Such framing has been front and center in the messaging of Australia's fossil-fuel-industry-coddling prime minister, Scott

Morrison. When it comes to the record heat and drought that Australia has experienced, the collapse of major river systems (such as the Murray-Darling) that provide critical freshwater resources, the death spiral of the Great Barrier Reef, catastrophic flooding events, and unprecedented, widespread, intense, fast-spreading bushfires, the solution, Morrison seems to think, is simply to "build . . . resilience for the future."[84] That policy was satirically summarized as "get fucken used to it" in a mock governmental public service announcement produced by Juice Media of Melbourne that went viral in early February 2020.[85]

It's important once again to recognize that resilience does play a role. There is no doubt that the communities, individuals, and brave firefighters who battled the devastating Australian bushfires displayed remarkable resilience, courage, and fortitude, not only in fighting the fires but in dealing with the resulting death, loss, and destruction. But the political discourse of "resilience" does them— and indeed, everyone else—a disservice. In emphasizing "adaptation" and "resilience," Morrison was engaged in a rhetorical, rather than substantive, response, both to the immediate crisis of the bushfires and to the longer-term underlying crisis of human-caused climate change. The community-wide anger that resulted was therefore understandable.

The Morrison government had neglected a previous request by fire chiefs that would have funded a fleet of water-bombing aircraft— precisely the sort of equipment needed in the face of worsening firestorms.[86] When it came to action, all the Morrison government could muster were hasty, reactive announcements of government funding initiatives to deal with the bushfire crisis after it was already underway.[87]

Those actions amounted to political spin aimed at distracting the public from the serious conversation that is needed, not only about the underlying cause of the unprecedented extreme weather disasters, but about the need to decarbonize our economies. For Australia, dangerous climate change has arguably *already* arrived at roughly 1°C (1.8°F) of warming, and dramatic reductions in carbon emissions are necessary to avoid double that much warming. Yet, in the wake of the epic bushfires, Morrison announced a $2 billion plan to promote natural gas while his coalition partners were busy advocating for new coal-fired power stations. They also wanted to open new export-oriented coal basins.[88]

In the rhetoric that fossil-fuel-promoting politicians typically use in the aftermath of climate-change-fueled disasters, we encounter another form of deflection. Talk of reducing carbon emissions, blocking new fossil fuel infrastructure, and embracing renewable energy remain off limits. Instead, those defending the fossil fuel hegemony display a softer form of denial. Don't worry about mitigation and decarbonization, we'll just adapt to the "new normal." Perhaps we'll evolve to develop gills and fins. And fireproof skin. The onslaught of damaging extreme weather events in Australia and around the rest of the world reminds us that there are limits to adaptation and resilience in a rapidly warming world. There is no amount of resilience or adaptation that will be adequate if we fail to get off fossil fuels.

REAL CLIMATE SOLUTIONS . . .

A viable path forward on climate, as we have seen, involves a combination of energy efficiency, electrification, and decarbonization of the grid through an array of complementary renewable energy sources. The problem is that fossil fuel interests lose out in that scenario, and so they have used their immense wealth and influence to stymie any efforts to move in that direction. These interests, and those advocating for them, have attempted to deflect attention from these real climate solutions, promoting in their place ostensible alternatives. Their favored options include supposedly climate-friendly forms of fossil fuel burning, uncontrolled planetary-scale manipulation of the climate, and reliance on technologies such as massive reforestation and nuclear power whose viability as true climate solutions is dubious. Their other favored option is to engage in hollow rhetoric about "adaptation" and "resilience" that neglects the fundamental source of the problem—the burning of fossil fuels.

There are false prophets who promote these non-solutions. They come with progressive-sounding names, like the Breakthrough Institute or the "ecomodernists." But don't be fooled—what they're peddling is business-as-usual dressed up as progress. And don't fall for their crocodile tears over "divisiveness" whenever someone attempts to call out bad-faith efforts to promote false solutions and deflect attention from real ones.

Consider the plaintive lament by ecomodernist Breakthrough Institute affiliate Matthew Nisbet, who wrote that "those specializing in the dark arts of social media 'engagement' have used these platforms to hack our brains, training our focus on conservatives and the evildoings of the fossil fuel industry while the end times loom."[89] Nisbet asks us to accept an alternative reality in which social media, rather than having been exploited by denialists and inactivists, has somehow been gamed *against* them by some shadowy band of environmental activist "dark arts" practitioners. This outrageous claim is perhaps unsurprising coming from Nisbet, given that he authored a heavily criticized, un-peer-reviewed report some years ago that others have characterized as employing highly questionable accounting to level the rather absurd claim that green groups have outspent fossil fuel interests in the climate propaganda wars.[90]

But what about Nisbet's claim that climate activists might be governed by the fear that "end times loom"? Here he's at least partly right, but for the wrong reason. The false prophets have been successful, at least in part, in convincing some climate activists that desperate measures—like geoengineering—might be called for. Desperate times, after all, call for desperate measures, and there is a growing contingent within the climate movement that buys into a narrative of doom-and-gloom and desperation, a narrative that can, ironically, lead them down the very path of inaction that inactivists have laid out for them. It is the final front in the new climate war, a front that we explore in the next chapter.

The Truth Is Bad Enough

The only thing we have to fear is . . . fear itself—nameless, unreasoning, unjustified terror which paralyzes needed efforts to convert retreat into advance.
—FRANKLIN DELANO ROOSEVELT

The word "catastrophe" is not permitted as long as there is danger of catastrophe turning to doom.
—CHRISTA WOLF

AN OBJECTIVE ASSESSMENT OF THE SCIENTIFIC EVIDENCE IS adequate to motivate immediate and concerted action on climate. There is no need to overstate it. Exaggeration of the climate threat by purveyors of doom—we'll call them "doomists"—is unhelpful at best. Indeed, doomism today arguably poses a greater threat to climate action than outright denial. For if catastrophic warming of the planet were truly inevitable and there were no agency on our part in averting it, why should we do anything? Doomism potentially leads us down the same path of inaction as outright denial of the threat. Exaggerated claims and hyperbole, moreover, play into efforts by deniers and delayers to discredit the science, posing further obstacles to action.

DANGER IS HERE

There is no one well-defined threshold that defines dangerous human interference with our climate. There is no cliff that we fall off at

1.5°C (2.7°F) warming or 2°C (3.6°F) warming. A far better analogy is that we're walking out onto a minefield, and the farther we go, the greater the risk. Conversely, the sooner we cease our forward lurch, the better off we are.

Dangerous climate change has in fact already arrived for many: for Puerto Rico, which was devastated by an unprecedented Category 5 hurricane with Maria in September 2017; for low-lying island nations like Tuvalu and coastal cities like Miami and Venice, which are already facing inundation by rising seas; for the Amazon, which has seen massive forest burning and climate-change-induced drought; for the Arctic, too, which has seen unprecedented wildfires in recent years; and for California, which has experienced unprecedented death and destruction from wildfires that now occur year-round. And those are just a few examples. The United States, Canada, Europe, and Japan have collectively witnessed unusually persistent, damaging weather extremes in recent years. Africa has been subject to drought, floods, and plagues of locusts. Australia has witnessed virtually every possible form of weather and climate disaster in recent years. And the list goes on.

We often hear that climate change is a "threat multiplier" when it comes to conflict, national security, and defense, for it heightens the competition that already exists over critical resources—food, water, space. But that framing applies equally to other domains, including human health. As I was writing this paragraph in the isolation of my sabbatical residence in Sydney in mid-March 2020, overlooking a serene Pacific Ocean that took my mind off the ever-worsening coronavirus (COVID-19) pandemic rapidly spreading outside the confines of my apartment, I couldn't help but think about the lessons the crisis might offer us. Our infrastructure is already burdened by climate-related challenges. Australia hadn't yet recovered from the catastrophic weather disasters of the 2019/2020 austral summer. Along comes yet another assault on its basic societal infrastructure. Soon any capacity to cope and adapt is exceeded. I was forced to cut my sabbatical short and head back to the States. Things were even worse there.

So yes, it's fair to say that dangerous climate change has already arrived and it's simply a matter, at this point, of how bad we're willing to let it get. While climate-change deniers, delayers, and deflectors love to point to scientific uncertainty as justification for inaction, uncertainty is not our friend here. It is cause to take even more concerted action. We already know that projections historically have been too optimistic about the rates of ice-sheet collapse and sea-level rise.[1] They also appear to be underestimating the incidence and severity of extreme weather events.[2] The consequences of doing nothing grow by the day. The time to act is now.

Recognizing that dangerous climate change is here already is, in an odd way, empowering. For there is no "danger" target to worry about missing. It is too late to prevent harmful impacts—they're already here. *But how much* additional danger we encounter is largely up to us. There is agency in the actions we take. The latest science tells us that, to a good approximation, how much the surface of the planet warms is a function of how much carbon we've burned up until that point. It is our decision-making henceforth that will determine how much additional warming and climate change we get (with some important exceptions we'll discuss later).

It is for this reason that a "carbon budget" is a meaningful notion. We can only burn a finite amount of carbon to avoid 1.5°C warming. And if we exceed that budget, which seems quite possible at this point, there is still a budget for avoiding 2°C warming. Every bit of additional carbon we burn makes things worse. But conversely, every bit of carbon we *avoid* burning prevents additional damage. There is both urgency and agency.

There is a role for voicing concern. It is important to recognize the risks of unmitigated climate change, including the potential for unpleasant surprises. We must consider worst-case scenarios when assessing our vulnerability, particularly given the fact that we have historically underestimated the rate and magnitude of key climate-change impacts. It is appropriate to criticize those who downplay the threat.

But there is *also* a danger in overstating the threat in a way that presents the problem as unsolvable, feeding into a sense of doom, inevitability, and hopelessness. Some seem to think that people need to be shocked and frightened to get them to engage with climate change. But research shows that the most motivating emotions are worry, interest, and hope.[3] Importantly, fear does not motivate, and appealing to it is often counterproductive, as it tends to distance people from the problem, leading them to disengage from, doubt, or even dismiss it.

Max Boycoff of the University of Colorado is a recognized leader in the study of climate messaging. He has argued that "if there isn't some semblance of hope or ways people can change the current state of affairs, people feel less motivated to try to address the problems." Boycoff has a T-shirt (inspired by the work of climate communication expert Ed Maibach of George Mason University) that reads: "It's real; it's us; experts agree; it's bad; there's hope."[4] Note once again the carefully calibrated balance of urgency ("it's bad") and agency ("there's hope").

DOOMISM

On one hand, inactivists—as we have seen—attempt to downplay the threat of climate change, or even argue that it will be "good for us." Consider Bjorn Lomborg, who, as you'll recall, glibly writes off the displacement of nearly half a billion people by sea-level rise as "less than 6% of the world's population."[5] Or consider the pleadings of Trump's former EPA administrator (and Koch brothers lackey) Scott Pruitt, who infamously claimed that climate change would help "humans flourish."[6] And there's tone-deaf Murdoch media minion Andrew Bolt, who, in the wake of the devastating climate-change-fueled Australian bushfires of the summer of 2019/2020, insisted on the front page of the Melbourne *Herald Sun* that "warming is good for us."[7]

But if the inactivists tend to *understate* the threat from climate change, there is a segment of the climate activist community that not

TOO LATE

STILL TOO
SOON
TOO SOON
TOO SOON

CLIMATE ACTION
TIME

(Tom Toles)

only *overstates* it, but displays a distinct appetite for all-out doomism—portraying climate change not just as a threat that requires urgent response, but as an essentially lost cause, a hopeless fight. From the standpoint of climate action, that's problematic on several levels.[8] First, it provides a useful wedge for inactivists to employ as they attempt to divide climate advocates by raising the very emotional question of whether it is too late to act.

Doomism is a form of "crypto-denialism," or, if you like, "climate nihilism." The boundary between what constitutes denialism and what constitutes nihilism is fuzzy. As clean-tech author Ketan Joshi put it, "Doomism is the new denialism. Doomism is the new fossil fuel profit protectionism. Helplessness is the new message."[9] So it has been stoked by inactivists, primarily because it breeds disengagement.

This is hardly the first time it's been used in that way. In his 2011 book *Winston's War*, British historian Max Hastings made a compelling case that doomist framing was employed rather effectively by isolationists opposed to US involvement in World War II.[10] Hastings described how those opposing US involvement in the war transitioned rapidly from the argument that "our involvement isn't necessary" to the argument that "it's too late for our involvement to make a difference." The parallels with climate inactivism are compelling, and indeed, rather chilling. And the metaphor is worth extending, because it is arguable that what is needed to combat the climate crisis is in fact a World War II–like mobilization effort.

Climate doomism can be paralyzing. As one observer noted, "[climate] doomism has been used as a tool to turn people off action and to pervert election results."[11] That makes it a potentially useful tool for polluting interests looking to forestall or delay action. With many on the political right already opposed to meaningful climate action for ideological and tribal reasons, doomism provides a means for co-opting those on the left. It's a brilliant strategy for building a truly *bipartisan* coalition for inaction.

It is easy to understand why climate advocates have become somewhat disillusioned. In the space of a few years, we saw the United States go from playing a leading role in international climate negotiations to being the sole nation to renege on its commitment to the 2015 Paris Agreement. It is in this environment that doomism has flourished. Indeed, a September 2019 CBS News poll found that 26 percent of those who don't feel climate change should be addressed cite the belief that there is "nothing we can do about it," a larger percentage than those citing the belief that "it's not happening."[12] Doomism, it seems, now trumps denialism as a cause for inaction.

Doomist thinking has become widespread today even among ostensible environmental advocates. Consider in this vein the words of Morgan Phillips, codirector of The Glacier Trust, a not-for-profit organization that aims "to help communities at altitude adapt to and mitigate climate change."[13] Responding on Facebook to my June 2019 *USA Today* op-ed on the importance of systemic climate solutions, Phillips wrote, "You can't save the climate. . . . [T]he political, cultural and technological change required is impossible now. . . . We're very likely in the midst of a mass extinction event. . . . [I]t looks to me to be far too late to avoid runaway warming now."[14]

There is *no* scientific support whatsoever for such a claim. The state-of-the-art climate model simulations used, for example, in the IPCC's Fifth Assessment Report (2014) provide no support at all for a runaway warming scenario at even 4° or 5°C (7.2° or 9°F), let alone 3°C (5.4°F), which is where current policies (i.e., "business-as-usual") are now likely taking us as we slowly begin to decarbonize the economy.[15] As for "mass extinction," the most comprehensive study to date, published in April 2020 in the premier journal *Nature*, found that less than 2 percent of species assemblages will undergo collapse (what the authors call "abrupt ecological disruption") from climate change if we keep planetary warming below 2°C (3.6°F). The number rises to 15 percent if warming reaches 4°C. That is certainly very troubling, but it doesn't constitute a "mass extinction" event of the sort that is evident in the geological record.[16]

Now look where these false prophecies of doom lead Phillips. He continued: "There isn't a bottomless pit of resources available to spend on responses . . . to climate and ecological breakdown. Trade offs [*sic*] need to be made, we have to ask whether we want to spend billions on spurious 'green tech' silver bullets, or billions on disaster risk reduction in the global south." To summarize his argument: (1) there's nothing we can do to prevent catastrophic, "runaway" climate change, and (2) efforts to act will somehow siphon away critical resources from helping people adapt to the inevitable coming apocalypse. So doomism literally undermines his support for climate mitigation.

The flames of doomism are being fanned by polluting interests who don't want to see us change. We must fight back every bit as fiercely as we fight outright climate-change denial. Unsurprisingly, trolls and bots are being used to promote doom and inevitability. Doomist messaging has become omnipresent in my own Twitter feed. Let's consider a couple of particularly salient examples.

Canadian prime minister Justin Trudeau and his administration, as we have seen, have been targeted by trollbots over their implementation of carbon pricing. In response to a fairly anodyne tweet by Trudeau about Canadian governmental priorities, a Twitter user

quoted a previous statement by me that "meeting our Paris obligations alone doesn't get us to where we need to be . . ."[17] An account
named "DarleneLily," with a 66 percent Bot Sentinel trollbot score,
replied, *"There's no way you can control the planets temperature.* You
can't stop other countries from polluting and using up their own
natural resources. Truth the world is overpopulated. And *you can't
stop the Supreme deity. The world is ending"* (emphasis added).[18] It's
perfectly disabling doomist messaging.

A link I posted to my June 2019 *USA Today* op-ed touting the
importance of systemic solutions and the dangers of only emphasizing individual behavior triggered doomist troll-like responses.[19]
One Twitter user tweeted, "All-out war on climate change made
sense only as long as it was winnable. *Once you accept that we've
lost it, other kinds of action take on greater meaning . . ."* (emphasis added).[20] A few days later, the same person tweeted, "Carbon
tax caused yellow vest protests in France. Who can afford to buy
a new electric car. #MagicalThinking Too little, too late!"[21] Note
the combination of doomist thinking with an effort to undermine
agency—the Twitter user both discredited the role of electrification of transportation and invoked the "Yellow Vest" canard that
inactivists so often employ to throw damp water on carbon pricing.
That's some pretty sophisticated and savvy inactivist messaging.

Doomist social-media messaging is in fact often combined with
"both-siderism": that is, there is no hope because both major parties
in the United States are equally bad on climate (readers will recall
that Russia used this trope to suppress enthusiasm for Democratic
candidate Hillary Clinton in the 2016 election). In response to my
tweet of a link to Paul Krugman's *New York Times* op-ed titled
"The Party That Ruined the Planet: Republican Climate Denial Is
Even Scarier Than Trumpism," one user tweeted back, "Obama was
literally bragging this year about oil exports at an all time [*sic*]
high after his presidency. This is absurd. *Climate denial is absolutely
bipartisan.* Frankly, *establishment Democrats are worse because they
say it's real and still pursue policies that will kill us all"* (emphasis
added).[22] There are countless other examples in my Twitter feed

from the past few years of doomist messaging being used to suppress climate activism.

MESSENGERS OF DOOM

The problem can be as simple as the headline that is chosen by an editor. Consider, for example, the recent *Nature* study cited earlier showing that ecosystem collapse can be avoided by limiting warming to 2°C. On Twitter, the Pulitzer Prize–winning *Inside Climate News* instead said, "A new study warns that climate change will soon lead to massive ecosystem collapse as key species go extinct."[23] Note how agency, unintentionally perhaps, is stolen by not properly contextualizing the claim—namely, by not acknowledging that such a scenario can be avoided through concerted action. I coyly suggested a rewrite of their tag line: "Comprehensive new *Nature* Study Shows that Massive Ecosystem Collapse Can be Averted if Warming is Limited to 2C—Which is Still Possible."[24] The award-winning editor Bruce Boyes of the Australian *KM Magazine* concurred, explaining that "the reporting can spin science findings into the negative, with headlines that disengage rather than engage."[25] Another observer commented, "Turn away from climate doom and catastrophism, and suddenly a better future seems very possible."[26]

The New Yorker might as well be the member newsletter of America's liberal elite. Get a featured article there and you achieve the equivalent, with the progressive intelligentsia, of appearing on the cover of *Rolling Stone*. That's where Jonathan Franzen—known largely as a fiction writer—found himself in September 2019 with one of the most breathtakingly doomist diatribes that has ever graced a magazine's pages. In an article titled "What If We Stopped Pretending? The Climate Apocalypse Is Coming. To Prepare for It, We Need to Admit That We Can't Prevent It," Franzen gave inactivists one of the greatest gifts they've received in years.[27]

The reviews were decidedly negative. Ula Chrobak of *Popular Science* summarized Franzen's thesis thusly: "He's claiming that those advocating for climate action are practically delusional, and that renewable energy projects and high speed trains are futile efforts to stop a planet 'spinning out of control.'"[28] *Climate Nexus* executive director Jeff Nesbit explained that "this sort of 'climate doomism' is as much a trap as 'personal sacrifice' is. Both are clever narrative plots by forces opposed to any real action on climate."[29] Science journalist John Upton opined, "It's hard to imagine major outlets publishing essays declaring efforts to reduce poverty hopeless. Or telling cancer patients to just give up. Yet this Climate Doomist trope flourishes—penned, best I can tell, exclusively by older, comfy white men."[30] End Climate Silence founder Genevieve Guenther, too, was decidedly unimpressed: "This piece is completely incoherent: the apocalypse cannot be stopped due to 'human nature' (so says the white man) but we can endure it. . . . Jonathan Franzen has no particular authority on climate, and the NYer shouldn't run trash." And Project Drawdown executive director Jon Foley described the article as "a shallow, poorly researched, self-indulgent piece. Probably one of the worst climate pieces I've ever read outside the denier's camp."[31]

The fundamental problem with the article is that it attempts to build a case for doom on a flimsy foundation of distorted science. I can speak to this directly, because I was contacted by the *New Yorker*'s fact-checkers to evaluate a passage in an earlier draft of the article. The passage read, "To project the rise in the global mean temperature, scientists rely on complicated atmospheric modeling. They take a host of variables and run them through supercomputers to generate, say, ten thousand different simulations for the coming century, in order to make a 'best' prediction of the rise in temperature. What then gets reported in the media isn't the likeliest rise in temperature. It's the lowest temperature that shows up in ninety-three percent of all scenarios. When a scientist predicts a rise of two degrees Celsius, she's merely naming a number about which she's very confident: the rise will be at least two degrees. The likeliest rise is far higher."

I told the fact-checker: "This doesn't look correct to me. When scientists generate an ensemble (spread) of temperature projections, the quantity that is generally communicated is the *average* or *median* warming. There is roughly an equal likelihood that the true value is either less than or greater than that value. And most scientists do their best to communicate the spread itself, i.e. the uncertainty range, and not just the middle value."

Even after Franzen had been informed of his error, he ended up keeping the incorrect statement that "the rise will be *at least* two degrees" (albeit changing 'The likeliest rise is far higher" to 'The rise might, in fact, be higher"). The final wording still falsely implied that the model averages preferentially underestimate the warming, despite my having communicated to the fact-checker that there's an equal likelihood that they underestimate or overestimate the warming. The uncorrected error conveniently supported Franzen's doomist narrative. Alas, I had only been allowed to see this one passage.

The whole article, it turned out, was riddled with basic science errors. *Business Insider* summarized experts' assessment of his piece thusly: "Scientists blast Jonathan Franzen's 'climate doomist' opinion column as 'the worst piece on climate change.'"[32] The critical problem is one we've already encountered and discussed. Franzen argued that we will fail to limit warming to below 2°C. That in itself is not objectively defensible—it is certainly still within our ability to avert 2°C warming given rapid decarbonization efforts. But more problematically, he invoked the strawman that we will then fall off a climate cliff, with supposed runaway feedback loops that kick in, rendering mitigation efforts useless. To quote Franzen directly: "In the long run, it probably makes no difference how badly we overshoot two degrees; once the point of no return is passed, the world will become self-transforming." We've already seen that there is no objective scientific support for such runaway warming scenarios. Yet they form the entire basis for Franzen's false prophecy of doom.[33]

Franzen's feelings were apparently hurt by the overwhelmingly negative response to his article. In fact, he has blamed it—or at least online critiques of his brand of doomist prophesizing—for the lack of progress on climate. In an interview with *The Guardian* he complained of the "Twitter rage" against him, arguing that "online rage is stopping us tackling the climate crisis."[34] He insisted that the "messenger was being attacked even if the facts of the message were not being challenged." While I'm sympathetic to his concern about online rage, which—as I have noted myself—can be counterproductive to action, the critiques of his commentary were in fact, as detailed above, grounded in his fundamental misrepresentations of climate science.

One of the more baleful aspects of doomism is the way it endorses intergenerational inequity—that is to say, its total dismissiveness when it comes to the interests of future generations. Rupert Read is an academic from the University of East Anglia in the United Kingdom and a self-avowed spokesperson for Extinction Rebellion. He's also a messenger of doom. After Read delivered a particularly fatalistic public lecture, climate scientist Tamsin Edwards blasted him: "I am shocked at this talk. Please stop telling children they may not grow up due to climate change. It is WRONG . . ."[35] It certainly is.

There is something especially disturbing when middle-aged men scold teenage girls fighting for a livable future. It's even worse when other middle-aged men stand by and applaud. Perhaps I'm taking just a bit of poetic license here, but that's essentially what happened with provocateur and author Roy Scranton and *Vox* climate pundit David Roberts in an episode I'll now recount.

Scranton is the ultimate doomist. In 2018 he literally wrote a book titled *We're Doomed*.[36] He snidely criticizes youth climate activists, dismissing their efforts as "Pure Disney." Though he has since deleted his Twitter account, back in December 2018 he took to the social media outlet to castigate youth climate activists as unwitting tools: "Enlisting children to carry the message of catastrophic climate change is at the same time a *reprehensible* abdication of responsibility and an embarrassing display of sentimentality and magical thinking. Pure Disney logic" (emphasis added).[37]

Scranton uses "Disney" so often, in fact, I'm surprised he doesn't have to pay them royalties. He invoked the multinational entertainment conglomerate's name once again to dismiss the writing of environmental author and 350.org founder Bill McKibben.[38] Scranton's flippant language suggests he thinks this is all somehow funny. But in fact, it's dead serious, and others aren't laughing. In response to his reprehensible attack on the youth climate movement and its de facto leader Greta Thunberg, youth climate activist Alexandria Villaseñor retorted, "Greta sparked a movement that has thousands of youth learning about climate change and realizing they have power. What have you done @RoyScranton? besides tell us we're doomed . . ."[39]

What I found especially disappointing in this particular affair was the reaction of *Vox*'s David Roberts, a pundit whose views about environmental matters are often insightful. Roberts weighed in on a piece Scranton had written for the *Los Angeles Review of Books* that dismissed Bill McKibben and others for their efforts to present a viable path forward on climate.[40] Roberts glibly endorsed Scranton's doomist take: "I like this piece from @RoyScranton & agree that the forced hortatory uplift at the end of climate books/articles is always the worst part." He then contemptuously scorned those who right-fully push back on such doomism, saying it was "fascinating" to him "to watch how fiercely, even angrily," people responded to Scranton's piece.[41] It appears that Roberts has since deleted this tweet. I don't blame him.

People *should* be angry at anyone engaged in self-righteous and self-serving (yeah—doom porn *sells*!) propagandizing at the expense of our children and grandchildren's future. As a scientist who studies the projections and numbers, let me affirmatively state, for the re-cord, that Scranton—and Roberts and Read and Franzen and other doomist men—are dead wrong. Our demise is only assured if we fol-low their lead and surrender. If your midlife crisis has caused you to give up on the future, then step aside. Get out of the way. But please don't obstruct others stepping forward to do battle.

DOCTOR DOOM

Guy McPherson, a retired ecology professor from Arizona, is argu-ably the scientific leader of the doomism movement, a cult figure of sorts. McPherson, like other doomists, argues that we have already triggered irreversible vicious cycles (for example, the massive release of frozen methane) that will render the planet lifeless in a matter of years. There's nothing we can do about it. What he calls "exponential climate change" will render human beings and all other species extinct within ten years owing to supposed runaway warming—something for which there is, as we have already seen, no shred of scientific evidence. But, if you like, mark December 2026 on your doomsday calendar—that's when McPherson said we will meet our collective demise.[42] (In the wake of the COVID-19 crisis in early 2020, McPherson provisionally moved his doomsday estimate all the way up to November 1, 2020. So if you're reading this, you can breathe a sigh of relief now!)[43]

According to science journalist Scott Johnson, "McPherson is a photo-negative of the self-proclaimed 'climate skeptics' who reject the conclusions of climate science. He may be advocating the opposite conclusion, but he argues his case in the same way. The skeptics often quote snippets of science that, on full examination, don't actually support their claims, and this is McPherson's modus operandi. . . . Both malign the IPCC as 'political' and therefore not objective. And both will cite nearly any claim that supports their views, regardless of source—putting evidence-free opinions on par with scientific research."[44]

McPherson is prolific, writing books, doing countless lectures, and appearing in online videos where he trumpets his message of imminent doom. He counsels us to grieve for our demise and find solace in "love," ending each of his videos the same way: "At the end of extinction, only love remains." His message has spread like a virus through environmentally aware regions of the Internet, with copycats writing pieces like "Are We Heading Toward Extinction? The Earth's Species—Plants, Animals and Humans, Alike—Are Facing Imminent Demise. How We Got Here, and How to Cope" (this from the progressive *Huffington Post*).[45] Greenpeace cofounder Rex Weyler has even echoed McPherson's doomsaying of imminent extinction in a commentary posted on Greenpeace's website.[46]

This sort of framing, again, plays right into the hands of the forces of delay and inaction. It is readily used to suppress activism and reduce enthusiasm for action. If we're doomed, then why expend time and effort pushing for action on climate? Such efforts are curiously reminiscent of the way Russia sought to suppress democratic turnout in the 2016 presidential election by convincing enough Democratic voters that there was essentially no difference between Hillary Clinton and Donald Trump. If your vote doesn't matter, then why bother?

Russia used online social media campaigns to drive a wedge between supporters of Bernie Sanders, who promoted hard-core climate policies (such as the outright banning of fracking), and the ultimate Democratic nominee, Hillary Clinton, who favored more "centrist" climate policies.[47] It was a deeply cynical and indeed sinister campaign on Russia's part, for, as we know, Russia is opposed to international climate action and has used social media campaigns to promote climate contrarianism.[48] Russia clearly didn't support Sanders's aggressive climate stance. But its leaders understood that he was a spoiler, and so they ran a massive social media campaign to exaggerate and exploit perceived climate policy differences between him and Clinton. The objective was to make Clinton unacceptable to Sanders supporters and not worth voting for—to convince a large enough number of progressives to simply sit out the election. And they did, helping hand the presidency to Russia's preferred candidate: climate-change denier and fossil fuel stooge Donald Trump.[49]

The Trump connection is an interesting one. McPherson frequently does interviews on a webcast network called American Freedom Radio that features on its page a virtual smorgasbord of right-wing conspiracy theories.[50] McPherson posted a commentary on his website supporting Donald Trump in the 2016 presidential election. Quoting from it directly: *"Donald Trump is another manifestation of the cleansing fire. . . . [H]e has secured my vote* to quicken the demise—sparking the flame. If you are one of those folks going through an earlier stage of grief and still finding it hard to accept our fate . . . it's becoming more and more obvious that the jig is up and time is short. In context, my goals for today include being kind to someone, smiling at a stranger, and *calling a few friends to convince them to vote for The Donald*" (emphasis added).[51]

So with McPherson and other doomists we find ourselves in a very odd corner of the universe where right meets left and doom meets denial. Whether climate change is a hoax (as Donald Trump would have us think) or beyond our control (as McPherson insists), there is no reason to cut carbon emissions. It doesn't matter how we get there. To the inactivist agenda, only the destination matters.

Doomists will attack upon the mere suggestion that disaster can still be averted. As one observer noted, "I've seen people post links to apocalyptic films scorning climate activists for even trying to avert catastrophe."[52] I've experienced this personally. In early September 2019, I appeared on Ali Velshi's MSNBC show to discuss the findings of the IPCC's new "Special Report on the Ocean and Cryosphere in a Changing Climate." I was subsequently berated on Twitter by an individual describing himself as a New Green Deal–supporting "Ecotopian Berniecrat." He was upset that I had cited the report's apparently insufficiently doomist prognosis of five to six feet of sea-level rise by the end of the century under business-as-usual fossil fuel burning.[53]

Why would I and other leading climate scientists be lying to understate the climate threat? Climate doomists, like climate denialists, often subscribe to conspiracy theories about scientists. But in the doomist version, the scientists aren't conspiring to promote a massive hoax. Instead, they are engaged in a massive cover-up to hide how bad climate change really is. Scott Johnson noted that "the skeptics dismiss science they don't like by saying that climate researchers lie to keep the grant money coming," while doomists insist that "scientists are downplaying risks because they're too cowardly to speak the truth and flout our corporate overlords."[54] Commenting on the recent trend, my climate scientist colleague Eric Steig perhaps put it best when he asked, "Where did this 'climate scientists are lying to us—telling us it isn't so bad—because of grant money' come from? Is this a real 'movement,' or just a bunch of Russian bots?," adding, "I miss the days of arguing with climate deniers."[55]

Taking the conspiracy theory to its absurd limit, if climate scientists are lying to maintain their employment, then only unemployed climate scientists can be trusted. That was literally the argument made by one purveyor of climate doom whose Twitter account no longer exists: "I suggest reading Guy McPherson who is unemployed and so tells the truth (working academics are funded by big biz and can't—crowd control), Sam Carana who posts under a pseudonym and Peter Wadhams who is also unemployed—ice expert from Cambridge. All are in the imminent camp." (Actually, McPherson and Wadhams are both professors emeriti.) Note that "imminent" means "doomist."[56]

We've already learned about McPherson. What about these other two individuals? As noted by Dana Nuccitelli in *The Guardian*, Peter Wadhams predicted back in 2012 that we would see an ice-free Arctic by the summer of 2016.[57] It is 2020 and we are nowhere close to that point. Like McPherson, Wadhams insists that Arctic warming will lead to massive releases of trapped methane and abrupt resulting warming. Sam Carana isn't even a real person—it's a pseudonym—so we know nothing about his actual qualifications. What we *do* know is that, as Scott Johnson has stated, he "posts a great deal of strange and unscientific claims" about . . . you guessed it, Arctic methane.[58]

Why do the doomists seem to be inordinately obsessed with Arctic warming and methane? We know that methane is a very potent greenhouse gas. And some of the best-known natural examples of catastrophic past warming events appear to have involved substantial releases of methane trapped either in permafrost or in the so-called methane hydrate along the sea floor. For example, warming of roughly 14°C (25°F) occurred at the end of the Permian period 250 million years ago, resulting in one of the greatest mass extinction events in Earth's history: 90 percent of all life was wiped out. At the boundary of the Paleocene and Eocene epochs (what is known as the Paleocene-Eocene Thermal Maximum, or PETM) roughly 56 million years ago, Earth experienced warming of as much as 7°C (13°F), with, again, widespread extinction.[59]

So if you're looking for a dramatic, doomsday-like climate-change scenario, it's very tempting to look toward methane. More specifically, you might focus on mechanisms whereby warming of the Arctic releases massive amounts of methane previously frozen in the permafrost, leading to more warming, more melting ice, more methane release, and a runaway warming scenario. The problem is that, aside from the questionable claims of a handful of contrarian scientists, there's simply no evidence that the projected warming could lead to such an event. Authoritative reviews of the scientific literature on the topic reveal "no evidence that methane will run out of control and initiate any sudden, catastrophic effects."[60]

That hasn't stopped the methane catastrophists from looking for any scrap of data that might support their narrative. Back in September 2019, they were hyping a momentary spike recorded by one isolated methane measurement station in Barrow, Alaska. At the time, I explained that this was almost certainly an isolated blip, perhaps reflecting contamination of the site—and that there was no evidence it was part of a larger pattern or trend.[61] Sure enough, methane levels at that site subsequently returned to normal. At least one media outlet that had uncritically reported the putative methane spike issued a correction, noting that the data had not been "validated," were impacted by "local pollution," and may be "subject to change."[62]

There's another important point to be made here. Although there has been a global uptick in methane, as we noted in the previous chapter the evidence suggests it's coming from natural gas extraction and not natural sources such as melting permafrost.[63] The doomists thus have it completely backward here. Rather than it being out of our hands, with appropriate policies governing natural gas extraction and fugitive methane emissions we can likely prevent the continued buildup of methane in the atmosphere. There is *agency* on our part.

While doomism itself might be dismissed as a rather fringe movement, there is some evidence of "seepage" of doomist conspiracy-mongering into the mainstream climate discourse. Consider, for example, an exchange that took place between climate experts back in January 2020. It started with Kevin Anderson, a climate scientist who has been critical of the mainstream climate science community for what he perceives to be complacency and a lack of urgency in the face of a crisis.

Anderson is no doomist, but he's at the far end of the aggressiveness scale within the climate science community. He has, for example, publicly chastised scientists who continue to use air travel, going so far as to travel on a container ship to a scientific conference to make his point (long before this sort of thing became fashionable à la Greta). Going further, he has argued that even scientists engaged in fieldwork in remote locations should only travel this way: "People have gone to the Amazon for years without flying."[64] Let's leave aside any discussion of how presumptuous it is to tell scientists engaged in laborious and logistically challenging fieldwork that they must take several additional weeks out of their schedules to travel by boat to remote locations. Anderson obviously buys heavily into the "personal action" framing of climate solutions. But he has also blamed his fellow scientists for a failure of systemic action, which leads us back to our story.

In January 2020, Anderson criticized a report by the United Kingdom's Committee on Climate Change (CCC), an "independent, statutory body established under the Climate Change Act 2008 . . . to advise the UK Government . . . on emissions targets and report to Parliament on progress made in reducing greenhouse gas emissions and preparing for climate change."[65] Anderson plaintively asked, "Why is there so little critique of [the CCC's] 'net-zero' report by academics & the wider climate community? It is designed to fit with the current political & economic status quo, & in so doing proposes cuts in CO2 far smaller than those needed to meet our Paris 1.5–2°C commitments!"[66] Defending the CCC, one commenter pointed out that they "are mandated to fulfil the Paris Agreement but would most likely welcome more analysis of their work."[67]

It was then that Anderson leveled an accusation against the entire scientific community, responding, "I wish I had your confidence in the process of scrutiny. Fine to argue around the edges, but the overall framing is firmly set in a politically-dogmatic stone with academia & much of the climate community running scared of questioning this for fear of loss of funding, presitige [sic], etc."[68]

If that sounds like the sort of accusation we expect from climate deniers and doomists, it's because that's the sort of accusation we expect from climate deniers and doomists. The chief executive of the CCC, understandably perturbed by Anderson's attack, responded, "It's not a politically dogmatic stone. It's the UK's Climate Change Act, which we're obliged [to] follow."[69] UK climate scientist Tamsin Edwards objected to the collective smear against the climate science community, tweeting, "That's quite an accusation about academics. . . . On what basis do you make the claim that 'much' of the community are mendaciously or cynically silent to protect their own interests?"[70] Anderson's unsatisfyingly vague response? "Repeated discussion over many years with many academics (and others) who work speciifcally [sic] on mitigation."[71] A more likely explanation, in my view? Too much exposure to doomist rhetoric. Or perhaps its more civilized close cousin, *soft doomism*.

"DEEP ADAPTATION"

Doomism sometimes masquerades under a nom de plume. Consider what has come to be known as "Deep Adaptation," introduced and promoted by Jem Bendell, an academic from the University of Cumbria in the United Kingdom. In February 2019, Bendell published an article that *Vice* characterized as "The Climate Change Paper So Depressing It's Sending People to Therapy."[72] But it is not an academic article in the usual sense. It was rejected by scientific journals, and Bendell ultimately self-published it on his website.[73] That means it lacks the rigor of a peer-reviewed scientific article. It has nonetheless been viewed far more than any typical peer-reviewed scientific article—by one estimate, more than 100,000 people have read it.

Although Bendell's article is, at least on the surface, less hard-core than the doomist "all life will end in a decade" messaging of Guy McPherson, Bendell nonetheless argues that near-term "climate-induced *societal collapse*" (a somewhat more murky concept) "is now inevitable in the near term," which he clarifies to mean "about a decade" (emphasis added).[74] Bendell bases this prognostication on the now all-too-familiar (but discredited) claims of a supposed Arctic "methane bomb" that will precipitate runaway warming, the collapse of agriculture, exponential increases in infectious disease, near-term societal collapse, and *possibly*—he at least seems to imply in places—human extinction. Bendell exaggerates both the projected climate change and its impacts.[75]

Equally problematic, his prescription for how we might address this looming threat involves no real mitigation. There's no mention of reducing carbon emissions, just some vague language about "restorative agriculture" and "resilience" and the insistence that we must "adapt" to the inevitable demise of civilization as we know it.

The BBC interviewed a number of scientists, asking them to comment on the merit of Bendell's assertions.[76] Among them was Myles Allen, professor of geosystem science at the University of Oxford, who asserted that "predictions of societal collapse in the next few years as a result of climate change seem very far-fetched." Allen noted, moreover, that "lots of people are using this kind of catastrophism to argue that there's no point in reducing emissions."

I too was quoted. I described the Bendell paper as "a perfect storm of misguidedness and wrongheadedness," since "it is wrong on the science and its impacts." I said, "There is no credible evidence that we face 'inevitable near-term collapse,'" and I emphasized that Bendell's doomist framing was "disabling" and would "lead us down the very same path of inaction as outright climate change denial." I added that "fossil fuel interests love this framing."

And indeed they must, for it breeds disengagement from the climate battle. One alarmed reader of Bendell's article is quoted in *Vice* as saying, "We're fucked. . . . Climate change is going to fuck us over. . . . Should I just accept the deep adaptation paper and move to the Scottish countryside and wait out the apocalypse?'"[77] Another individual, quoted by the BBC, said that "a few months after reading the Deep Adaptation paper," he and his wife decided to sell their home and move out to the country. "When the crunch comes," he said, "there'll be a lot of people in a small area and it's going to be mayhem—and we'll be safer if we move further north because it's colder."[78] We have terms for such folks, like "doomsday preppers" and "survivalists."

If you take the most environmentally aware progressives, lead them to despair, and convince them to dissociate from civilization, they're not out there on the front lines participating in the political process, demonstrating and fighting for the needed systemic changes. Bendell's paper is a more powerful tool for disengagement than any article ever written by a climate-change denier.

SOFT DOOMISM

If outright doomism is generally too shrill to gain much currency in mainstream climate discourse, what we shall henceforth refer to as *soft doomism* has found its way to the very center of the conversation. Soft doomists don't quite argue for the inevitably of our demise as a species, but they typically imply that catastrophic impacts are now unavoidable and that reducing carbon emissions won't save us from disaster. It's doomism dressed up, you might argue, in more respectable clothing.

Soft doomists tend to use terms like "panic." "Time to Panic" was the headline on a 2019 *New York Times* op-ed by David Wallace-Wells, author of *The Uninhabitable Earth* (which I will discuss later).[79] According to Sheril Kirshenbaum, executive director of the nonprofit organization Science Debate and host of the National Public Radio podcast *Serving Up Science*, "stoking panic and fear creates a false narrative that can overwhelm readers, leading to inaction and hopelessness."[80]

"Panic" is a word that conjures images of people running screaming through the streets with their hands over their heads. It evokes irrational, desperate, rash behavior rather than considered, well-thought-out, deliberate action. The latter is helpful. The former is not. And it can lead us to very strange and uncomfortable places.

Let's concede that the "p" word is appropriate in some contexts. Consider, for example, Greta Thunberg. In her message to world leaders gathered at Davos, Switzerland, in January 2019 for the World Economic Forum, she chastised the crowd for having failed to act meaningfully on the climate crisis, telling them, "I want you to panic." In that context, it is reasonable to interpret her comments as suggesting that the attending politicians and opinion leaders deserve to feel the scorn of young people like herself calling for action. Indeed, her subsequent statement was "And then I want you to act."[81]

But unfocused and diffuse "panic" messaging can lead to counterproductive actions. As we have seen, it has led to support for potentially dangerous geoengineering schemes, which have been sold as a necessary last-ditch means of averting climate devastation. Read no further than the headline of the December 2019 *Washington Post* op-ed "Climate Politics Is a Dead End. So the World Could Turn to This Desperate Final Gambit."[82]

Soft doomism has become increasingly widespread. Its basic tenets have been adopted by groups like the aforementioned Extinction Rebellion, which takes the position that "we are facing an unprecedented global emergency. Life on Earth is in crisis. . . . [W]e have entered a period of abrupt climate breakdown, and we are in the midst of a mass extinction of our own making."[83] In mid-January 2020, a curious online article was making the rounds, ironically well-titled "Climate Fatalism."[84] While the article was unsigned, it was sponsored by an organization called the Freedom Lab, which describes itself as an "innovation hub" and "thinktank" that produces "actionable insights," which it shares "through regular publications and public events."[85]

The article embodies the ambivalence and internal contradictions that have come to characterize soft doomism. "Last year," it begins, "several alarming reports made it clear that immediate and radical action is needed to prevent disastrous levels of global warming." It's a promising start, acknowledging the problem and entreating the reader to action. However, in the very next line, the author writes, "Action is nowhere to be found and we are bound to hit the tipping points of global warming that will render any further action irrelevant." It's an abrupt turn toward doomism and futility that is made even more confusing by the sentence that follows, which warns of the threat of the very sort of fatalism that the article is promoting: "As this notion spreads, 2019 could see many of us falling prey to climate fatalism and a shift in political focus towards climate adaptation."

Despite the contradictions, the piece has an agenda. It concludes with a prescriptive statement masquerading as a predictive one: "We will see a shift from preventing climate change to adapting to (and battling) the effects. Much of this will entail engineering, to build dams and extreme-weather-proof buildings, for instance. It's likely that governments will shift funding from preventive measures to these kinds of adaptive solutions." The message is that climate change is bad—very bad, but we will fail to act to solve it, so we might as well just adapt, be more resilient, and, oh yeah—explore technofixes. We've heard this story before. It is the "non-solution solution" of the previous chapter.

Soft doomism in a sense plays the same role among progressives that soft denial plays among conservatives. That is to say, it is a form of doomist rhetoric that is tolerated in polite company. And unsurprisingly, some prominent progressive climate and environment pundits have engaged in its rhetoric. Consider again the otherwise generally insightful David Roberts of *Vox*. In late December 2019, Roberts tweeted, "We're not going to limit temp to 1.5C. The weird social pressure to continue pretending we can, or might, is weird to me. The situation is tragic. The people & institutions responsible deserve all the anger in the world. But it is what it is."[86]

Climate and energy policy pundit Jon Koomey chided Roberts: "Dave, please stop the defeatist pessimism. Not helpful, and probably not even right. We are able to do this, and given a sufficient shift in the politics, we will do it. But the longer we wait the more stranded assets there will be and the more costly it will be."[87] That comment, of course, precipitated its own feeding frenzy of doomist commentary. One individual wrote, "I know it is technically feasible. It is not socially and psychologically feasible."[88]

That comment, while misguided, usefully betrays an underlying point of confusion—a fallacy that is in fact commonly encountered in these sorts of discussions. The fallacy is conflating physics and politics. While the laws of physics are immutable, human behavior is not. And dismissiveness based on perceived political or psychological barriers to action can be self-reinforcing and self-defeating. Think World War II mobilization or the Apollo project. Had we decided a priori that winning the war or landing on the moon was impossible, these seemingly insurmountable challenges would never have been met. We have encountered compelling evidence that a clean energy revolution and climate stabilization are achievable with current technology. All we require are policies to incentivize the needed shift. That doesn't violate the Newtonian laws of motion, or the laws of thermodynamics. It only challenges us to think boldly. Scratch beneath the surface and we find that most soft doomism is premised not in the physical impossibility of limiting warming, but in a cynical, pessimistic belief that we lack the willpower to act. It's giving up before we have even tried. And once again, the inactivists are smiling all the way to the bank.

In this vein, let's talk about the so-called "Hothouse Earth" article mentioned in Chapter 5. It was published in August 2018, with Australian environmental scientist Will Steffen as the lead author.[89] In a sense, this article helped lay the groundwork for other doomist and soft doomist accounts like those by Scranton, Franzen, and Bendell. Like the Bendell paper, it went viral. Also like the Bendell paper, it wasn't peer-reviewed research, but simply a "perspective," more of an opinion piece than a scientific article. One important difference was that "Hothouse Earth" was published in the high-profile, prestigious *Proceedings of the National Academy of Sciences*, lending the imprimatur of the US National Academy of Sciences—the highest scientific authority in the land—to the study's findings.

Steffen, the principal author, is executive director of the Australian National University (ANU) Climate Change Institute. Readers may recall from Chapter 5 Steffen's unusually aggressive and prescriptive views on climate action: "*You have got to get away from the so-called neoliberal economics* . . . [and shift to something] more like wartime footing [to decarbonize society] at very fast rates" (emphasis added).[90]

The "Hothouse Earth" article makes similar claims to those we've encountered before among doomists and soft doomists—indeed, it is the likely inspiration for their thinking. But it is more nuanced and employs more caveats than other accounts, arguing that even if we keep warming under the oft-cited "dangerous" limit of 2°C, hypothesized amplifying feedbacks, such as "permafrost thawing" and "decomposition of ocean methane hydrates," *could* lead to climate change spiraling out of control. The article asserts that "even if the Paris Accord target of a 1.5°C to 2.0°C rise in temperature is met, we cannot exclude the risk that a cascade of feedbacks *could* push the Earth System irreversibly onto a 'Hothouse Earth' [4–5°C warming] pathway," with massive ice loss, sea-level rise, megadroughts, and other dire impacts.

Mainstream climate research, as already noted, doesn't support these claims—at least for the near term. Thus, rather than a summary of our current understanding, the "Hothouse Earth" article is "speculative" and more of an "interesting think piece," according to UK climate scientist Richard Betts.[91] Betts emphasized that there is "large uncertainty" in the "Hothouse Earth" authors' estimate of 2°C as the trigger point for cascading feedbacks, noting that it reflects "risk averse" assumptions, and that "even if the self-perpetuating changes do begin within a few decades, the process would take a long time to fully kick in—centuries or millennia."

The combination of the authority of a prestigious journal, high-profile authors, and dramatic claims nonetheless ensured that "Hothouse Earth" would get a huge amount of media attention—and naturally, all the nuance was lost in the media frenzy that ensued. A very similar follow-up commentary, coauthored by many of the principals of the earlier article, was published a year later in the prestigious journal *Nature*, triggering yet another round of publicity.[92] Collectively, the two reports were covered by hundreds of media outlets, including CNN, *Newsweek*, *The Guardian*, *National Geographic*, the BBC, the *Daily Mail*, the *Sydney Morning Herald*, the *New York Post*, and many others. With over-the-top headlines, like "Climate Change Driving Entire Planet to Dangerous 'Tipping Point'" (*National Geographic*), and "Scientists Warn Earth at Dire Risk of Becoming Hellish 'Hothouse'" (*New York Post*), the collective coverage suggested that we face imminent and unavoidable catastrophic climate change. It all played into a doomist narrative of helplessness—and, as we shall see later, fueled conservative efforts to caricature and discredit climate predictions.[93]

UNINHABITABLE EARTH?

There is one rendering of climate doomism that stands out above all others. It has been so influential that it deserves its own section. Albeit more nuanced than most of the doomist genre, "The Uninhabitable Earth," a July 2017 article by David Wallace-Wells that he later developed into a best-selling book, had a profound impact on the larger conversation about climate change.[94] The article, published in *New York Magazine*, predated "Hothouse Earth," Roy Scranton, Jonathan Franzen, Jem Bendell, and the rest of the lot. It was to climate doom porn what Shakespeare is to modern literature. It defined the genre, and its success generated considerable additional demand for more of the same. And make no mistake: climate doom porn *does* sell. "The Uninhabitable Earth" was the most read article in the history of *New York Magazine*.[95] Perhaps it's for the same reason people ride rollercoasters, engage in bungee jumping, or go skydiving—they sometimes just want to be scared out of their wits. Climate doom ostensibly gives them that same rush of adrenaline. Am I calling it a drug? I guess so. Am I calling its purveyors pushers? I guess, in a sense, I am.

It is perhaps redundant to say that an article entitled "The Uninhabitable Earth" presents an overly bleak view of our climate future. And the subtitle doubles down on the doom: "Famine, Economic Collapse, a Sun That Cooks Us: What Climate Change Could Wreak—Sooner Than You Think." But extraordinary claims, as Carl Sagan famously said, require extraordinary evidence. Does the article deliver?

I expressed my concern about the article initially in a Facebook post. "The evidence that climate change is a serious problem that we must contend with now," I wrote, "is overwhelming on its own. There is no need to overstate the evidence, particularly when it feeds a paralyzing narrative of doom and hopelessness. I'm afraid this latest article does that. That's too bad. The journalist is clearly a talented one, and this is somewhat of a lost opportunity to objectively inform the discourse over human-caused climate change."[96] I expanded on my critique in an op-ed I coauthored for the *Washington Post* warning against the threat of doomist thinking, using "The Uninhabitable Earth" as the central example.[97]

My fundamental point of contention will be familiar to readers by now because it reflects a recurrent problem: the overly pessimistic and bleak depiction of our prospects for averting catastrophic climate change based on overstatement of climate-change impacts. "Uninhabitable Earth" exaggerates, for example, the near-term threat of climate "feedbacks" involving the release of currently trapped methane. The scientific evidence, as we have already seen, doesn't support the notion of a game-changing, planet-melting methane bomb of the sort the article envisions.

The article incorrectly asserts that the planet is warming "more than twice as fast as scientists had thought." That statement was false. The study the article refers to simply showed that one particular satellite temperature dataset that had tended to show *less* warming than other datasets has now been brought in line with them after some problems were corrected for.[98] In fact, recent research (including work I was involved in) shows that past climate model simulations actually slightly *overpredicted* the warming during the first decade of the twenty-first century.[99] Once appropriate corrections are made, it turns out the models and observations are pretty much in line. While some climate-change impacts, like ice melt and sea-level rise, are indeed proceeding faster than the models predicted, the warming of the planet's surface is progressing pretty much as forecast. And that is plenty bad enough.

One could dismiss isolated mischaracterizations of the scientific evidence as innocent and innocuous oversights. But when there are many of them, and they all seem to point in the same direction—toward exaggerating the magnitude and pace of climate change—it suggests a cherry-picking of the evidence to support a particular narrative: a narrative of doom, in this case.

Even the story about the Svalbard seed vault that opens the article is, at best, misleading. Wallace-Wells begins his piece with "This past winter, a string of days 60 and 70 degrees warmer than normal baked the North Pole, melting the permafrost that encased Norway's Svalbard seed vault—a global food bank nicknamed 'Doomsday,' designed to ensure that our agriculture survives any catastrophe, and which appeared to have been flooded by climate change less than ten years after being built."

It's a nice story. But it's not true. I actually saw the vault a year after Wallace-Wells had written his piece, in October 2018, while attending a climate-change workshop on "Navigating Climate Risk" in Svalbard.[100] The vault was just fine. One of its founders explained that there really never was any flood. Rather, every year when the snow melts on the mountain, they get some water coming in at the top of the tunnel that leads to the seed vault. It's happened every year since it's been open, and they're working to address it.[101]

I'm just one scientist, and perhaps you might dismiss my concerns about the article as biased. After all, I was interviewed by Wallace-Wells at length and not mentioned or quoted. Perhaps there are sour grapes on my part?[102] Fortunately, you don't have to take my word for it. Climate Feedback is a climate-scientist-run website that evaluates the factual basis, reliability, and credibility of climate-themed articles that appear in the media based on evaluation by a panel of leading experts. Climate Feedback evaluated "Uninhabitable Earth."[103]

To be more specific, Climate Feedback had the article evaluated by fourteen climate scientists chosen for their expertise across the range of issues covered by the article (three more were added after the initial deadline, bringing the total to seventeen). The article earned an average scientific credibility score of –0.7 on a scale that goes from –2 (very low) to +2 (very high). A score of –0.7 puts it just above –1 (low). Climate Feedback provided the following summary: "Seventeen scientists analyzed the article and estimated its overall scientific credibility to be 'low.' A majority of reviewers tagged the article as: Alarmist, Imprecise/Unclear, Misleading."[104] It's one thing to be *alarmed*—and we should be given the evidence. It's something else to be *alarmist*—a term that implies an unfounded, potentially harmful exaggeration of risk or danger.

Some felt this critique was unfair. David Roberts, who, as we have seen, occasionally weighs in with pessimistic and doomist-sympathetic views of his own, dismissed the criticisms by scientists like myself as "off-base scientific niggling."[105] Does he have a point? Scientists, after all, are biased toward, well, the science. They might not, for example, appreciate the poetic license sometimes required for effective journalism. In November 2017, I participated in an event that was part of the New York University Arthur L. Carter Journalism Institute's Kavli Conversations on Science Communication. The host was Dan Fagan, professor of journalism and director of NYU's Science, Health and Environmental Reporting Program. The event was called "The 'Doomed Earth' Controversy" and billed as "The author of the controversial *New York Magazine* cover story about worst-case climate scenarios in conversation with a prominent critic."[106] Yes, that's Wallace-Wells and me, respectively. The discussion was moderated by Robert Lee Hotz, a science writer at the *Wall*

Street Journal and a Distinguished Writer in Residence at the NYU Journalism Institute.

After having listened to the roughly forty-five-minute discussion between the three of us (where there was actually more agreement than disagreement), host Dan Fagan took the floor and issued his verdict. He began by expressing his appreciation of a "great discussion" and went on to note that a journalist's "first obligation is to reflect reality." While he "salute[d] David for his piece because . . . all pieces of the bell curve . . . should be written about," he also criticized it. His main concern was that while it "had . . . boilerplate [language] . . . about likelihood, it felt . . . tossed in and it certainly wasn't part of the overall framing of the piece." The piece wasn't clear on "Is this happening in five years? Is this happening in a century?" and as a result it "violated some of the rules that I've been teaching." Namely, the article was "inadequately contextualized," though Fagan appreciated that Wallace-Wells was "operating from the frustration that many of us feel."

Wallace-Wells seemed to have taken the criticism to heart. In August 2018 he asked me to comment on the full-length book version of the article, also to be titled *The Uninhabitable Earth*. The way he described it gave me optimism: "The book is . . . in part a revision and expansion of [the] article," he said. It was "focused less on worst-case scenarios, and in part [was] a more essayistic meditation on what it will mean, for politics and culture etc., to live in a world transformed by climate change in the coming decades." He asked me to review the prologue in particular, which, as he put it, "frames the whole project." He told me to be "ruthless" in my assessment. I appreciated the opportunity and was happy to oblige. I read it over and reported back to him a few days later. I told him that "the science is solid," but that I had "a number of minor comments" (nine of them, to be specific) that I felt should be addressed. I outlined them for him.

Among my main points, I said that "the claim that 'few experts think we'll hit' the 2C target seems misleading. . . . Many experts have pointed out a viable path to 2C. . . . There are no physical obstacles to 2C stabilization. Only political ones—at this point." I also said, "The claim that none of the industrial nations are on track to meet their Paris commitments is questionable. Some analyses suggest that the U.S. is very much on target to do so . . . and China, the world's largest emitter (!) is on course to exceed its targets. That's the world's two largest emitters right there." Finally, I pointed out, "You say that scenarios exceeding 2C warming are shrouded, delicately, from view. By whom? Certainly not the scientific community. 'Business as Usual' warming scenarios of 4–5C are prominent in the IPCC reports, other scientific assessments, and many popular articles about climate change. If you mean that *journalists* (and the media) are shrouding these scenarios from view then [you] should say so."

The book came out in February 2019. I was disappointed to find that no substantive changes were made in the prologue in response to the points I had raised. As far as the rest of the book is concerned, while the sorts of blatant errors that marred the original article were largely gone, the pessimistic—and, at times, downright doomist—framing remained, as did exaggerated descriptions that fed the doomist narrative. Consider, for example, this passage:

> Some [climate feedbacks] work in [the] direction [of] moderating climate change. But many more point toward an acceleration of warming, should we trigger them. And just how these complicated, countervailing systems will interact—what effects will be exaggerated and what undermined by feedbacks—is unknown, which pulls a dark cloud of uncertainty over any effort to plan ahead for the climate future. We know what a best-case outcome for climate change looks like, however unrealistic, because it quite closely resembles the world as we live on it today. But we have not yet begun to contemplate those *cascades that may bring us to the infernal range of the bell curve* [emphasis added].[107]

The prose gives a reader the impression that there are all sorts of positive feedbacks that climate scientists haven't even "contemplated." And if "cascades that may bring us to the infernal range of the bell curve" isn't a doomist dog whistle for unjustified "runaway warming" scares, I don't know what is. This passage—and many others in the book—would lead readers to assume that we are completely flying blind with regard to climate change. It implies that climate projections are completely unreliable (reminiscent of the claims made by climate-change deniers). A reader would never suspect that, in fact, climate models (1) have done a remarkable job predicting the increase in global temperature over the past half century, and (2) show no evidence of the sort of "infernal cascade" Wallace-Wells asks us to fear.[108]

The publisher (Penguin Random House) features quotations from a variety of impressed reviewers on its webpage for the book. It's hardly surprising that most of these reviewers expressed alarm over what the book describes. One said "*The Uninhabitable Earth* hits you like a comet, with an overflow of insanely lyrical prose about our pending Armageddon." Another said, "*The Uninhabitable Earth* is the most terrifying book I have ever read." Yet another said "its mode is Old Testament" and called it a "white-knuckled tour through the cascading catastrophes that will soon engulf our warming planet."[109] This is climate doom porn. And, as I said before, climate doom porn sells. After its release on February 20, the book was on the *New York Times* Hardcover Nonfiction Best Sellers List for six weeks in a row.

If you still can't get enough of it, then have no fear, for the doom will be televised. HBO is turning *The Uninhabitable Earth* into a series. Well, sort of. According to Yessenia Funes of *Gizmodo*, it will "influence a fictional anthology series that examines what our future may look like as climate change progresses." The director, Adam McKay, "will help visualize the gloom and doom in all its horrible glory for the show's first episode." Funes doesn't hide her enthusiasm: "I am here for it. Let's freak everyone the hell out."[110] If you thought you just heard me groan, it's because you just heard me groan.

I was invited to appear with Wallace-Wells on the MSNBC *Morning Joe* program shortly after the publication of the book.[111] One of the show's hosts, Mika Brzezinski, opened with, "It could be a world of . . . mass extinctions and economic calamity. Our next guest argues that *fear* may be the only thing that saves us." Asked about how bad things are going to get, the first words out of Wallace-Wells's mouth were, "It looks pretty bleak." What ensued, however, was a more balanced and nuanced discussion about both the costs of inaction and the need to take action. I imagine that my participation in the segment helped steer the conversation in that direction.

During the commercial break, host Willie Geist turned to Wallace-Wells and said, "Isn't there any good news?" I joked, "I think that's what I'm here for." Then, back on the air, Geist turned to me and asked, in closing, "What's the good news you can tell people about climate change right now?" I pointed out that "there is urgency, as David has said, but there's also agency" (my first use of that framing). I went on to talk about how the conversation is now changing, with even some Republicans starting to come to the table. (We'll talk more about that in the next chapter.)

Wallace-Wells nonetheless continued with rather doomist language in his engagement with the public. A few days after the MSNBC segment, he did an interview with a reporter from *Vox*, Sean Illing. Illing titled his piece "It Is Absolutely Time to Panic About Climate Change: Author David Wallace-Wells on the Dystopian Hellscape That Awaits Us." Wallace-Wells told Illing, "As someone who was awakened from complacency into environmental advocacy through alarm, I see real value in fear."[112]

Wallace-Wells occasionally weighs in on Twitter with alarmist commentary that requires correction by climate scientists. For example, in September 2019 he tweeted that "the world could hit 1.5C—the target of all global climate action—as soon as 2021. It could hit 2C—'catastrophic warming' by 2025."[113] That's wrong. It's the result of an erroneous extrapolation of a claim someone had tweeted that "temperatures [are] up . . . 0.2°C just between 2011 and 2015."[114] No climate scientist would ever try to measure the warming trend based on a five-year period because of the huge amount of "noise" in measuring temperature differences from year to year. Things like El Niño and volcanic eruptions can skew short-term readings.

The true warming rate is about 0.2°C (~0.4°F) per *decade*. Since current warming stands at about 1.2°C (~2.2°F), it would at current rates take a decade and a half to reach 1.5°C (2.7°F) warming, and another two and a half decades to reach 2°C (3.6°F) warming. But even if we used the incorrect estimate of 0.2°C per five years, Wallace-Wells's math is still wrong. We wouldn't reach 1.5°C for the better part of a decade, and we wouldn't reach 2°C for another twenty years. So it's puzzling how Wallace-Wells came up with his numbers in the first place. What's clear is that it fits with a narrative of impending doom.

Climate scientists immediately corrected Wallace-Wells. Richard Betts noted, "Even if this extrapolation were correct (it isn't), a single year at 2C is not going to be 'catastrophic.' 2C above pre-industrial for decades would indeed bring profound & possibly self-reinforcing changes, but simply hitting 2C for the first time will not make it all kick off."[115] Eric Steig was more blunt: "This is the kind of thing that makes me want to say . . . 'leave the science communication to scientists' . . . it's utterly irresponsible and wrong."[116]

Wallace-Wells also continues to mischaracterize the progress that is being made on the policy front. In a December 2019 article in *New York Magazine*, referring to the Conference of the Parties in Madrid, he wrote, "It was, of course, the 25th COP, and judging by the only metric that matters—carbon emissions, which continue to rise—the conference followed 24 consecutive failures. Emissions set a new record in 2018, and are poised to set another again in 2019. Just three years since the signing of the Paris accords, no major industrial nation on Earth is on track to honor the commitments it made in Paris."[117] There are *all kinds of* wrong here.

First of all, he's just wrong. Emissions remained flat in 2019, with power-sector emissions actually dropping, and total emissions are poised to drop in 2020 (though in the latter case that's at least in part due to the COVID-19 pandemic). To quote the International Energy Agency (IEA), "Emissions trends for 2019 suggest clean energy transitions are underway, led by the power sector. Global power sector emissions declined by some 170 Mt [million metric tons], or 1.2%, with the biggest falls taking place in advanced economies where CO_2 emissions have dropped to levels not seen since the late 1980s (when electricity demand was one-third lower)."[118] We would like to be seeing them not just flattening but declining. However, it's wrong to claim they are rising or to ignore the transition that is clearly underway toward a renewable-energy-driven economy.

What about Wallace-Wells's assertion that no major industrial nations are on track to honor their 2015 Paris Agreement commitments? I challenged him on this very matter when reviewing the draft prologue of his book. He failed to make any changes, and he repeats the misleading claim here. China, the world's largest emitter, is on course to meet its Paris target early.[119] The United States may meet its obligations in spite of the Trump administration's policies.[120] While there are criticisms to be made about the limits of the Paris Agreement, and there are certainly countries that are failing to live up to their commitments, it's simply not the case that no major industrial nation is on track to honor its Paris obligations.

These errors and mischaracterizations aren't innocuous—they are in service of the doomist narrative Wallace-Wells continues to promote. He argues that the existing framework (the United Nations Framework Convention on Climate Change, or UNFCC, and the annual Conferences of the Parties) for global climate negotiations has failed us and should be abandoned. Instead, he insists, it should be replaced with something akin to an international version of the Green New Deal. He points to the perceived failure of the most recent climate negotiations in Madrid as motivation for this position.

This argument is misguided on several levels.[121] Not only does it engage in unhelpful despair-mongering, but it takes entirely the wrong message away from what transpired in December 2019 at the COP25 in Madrid. A small number of nations led by fossil-fuel-friendly regimes, including Australia, in essence conspired to sabotage the negotiations. Blaming the "COP model" and attributing blame broadly provides cover for, and enables, the relatively small number of bad state actors that are attempting to poison the well.

This current obstacle is a consequence of an unfavorable geo-political playing field that has allowed oligarchs and demagogues to rise to power in those countries in recent years. No alternative model for international climate cooperation is likely to circumvent that obstacle. Certainly not one based on, as Wallace-Wells is suggesting, a globalized version of the Green New Deal, which already carries ideological baggage and comes with so much opposition already baked in.

Wallace-Wells, moreover, by dismissing the entire history of efforts by the UNFCC and previous COPs, based on disappointment with COP25, is truly throwing the baby out with the bathwater. He is neglecting, for example, the highly successful COP21 Paris meeting in 2015, in which the nations of the world committed to substantial carbon emissions reductions. While those reductions don't alone solve the problem (they get us almost halfway to limiting warming below 2°C), and not every nation will meet its targets, the Paris Agreement was a monumental achievement. It put a framework in place for ratcheting up commitments as international negotiations proceed in subsequent COPs.[122]

Wallace-Wells, perhaps unsurprisingly, objected to these criticisms. He tweeted, "I haven't given up on the COP/UN model, but I don't think considering whether alternate approaches might be more effective is 'doomist.' We need to make progress wherever we can, and the European Green Deal (for instance) suggests at least one hopeful alternative (as I mention)."[123] Kalee Kreider is the head of communications at the National Geographic Society, former communications director for Al Gore, and a senior adviser to the United Nations Foundation. She took some offense to Wallace-Wells's dismissive comments about decades of climate policy efforts by the United Nations that she and so many others had contributed to. She replied to his tweet, sardonically, "*Cough*, the Paris Agreement was a US-China deal that then the rest of the world followed. *Cough*. That was how it got done" (emphasis added). In a subsequent tweet she linked to a November 2014 bilateral agreement between the United States and China, the world's two largest emitters, that laid the groundwork for the highly successful Paris international climate agreement.[124]

It is important that we hold our policymakers accountable for taking concerted action on climate, as activists like Greta Thunberg have done. But it's not constructive to dismiss the real progress that is being made, for it plays into the agenda of the inactivists, who have attempted to sabotage climate progress—including the 2019 COP25 negotiations. They would like nothing more than to see us throw up our hands in defeat and declare international climate negotiations dead.

I fear that defeatist rhetoric like Wallace-Wells's not only throws climate leaders who have spent their lives pushing for climate progress under the bus but also rewards the bad-faith efforts of inactivists. I suspect, moreover, that the attitude is contagious. Greta Thunberg not only follows Wallace-Wells on Twitter but retweets his often pessimistic missives.[125] In her January 2019 speech at the World Economic Forum in Davos, she declared that "pretty much nothing has been done" on climate change.[126] Not *enough* is being done, for sure, but to say that "nothing has been done" is simply false. It is dismissive of the actions that countries, states, cities, companies, and individuals are taking every day to help move us off fossil fuels, and it is dispiriting to the individuals who have worked so hard to improve the situation. It also neglects the hard data from the International Energy Agency demonstrating that we are indeed making progress toward decarbonizing the global economy.

In what can only be described as a case of journalistic whiplash, just days after his pessimistic December 16, 2019, *New York Magazine* article, Wallace-Wells published another piece in in the same magazine expressing a rather glowingly optimistic outlook. Titled "We're Getting a Clearer Picture of the Climate Future—and It's Not as Bad as It Once Looked," the piece, which came out on December 20, had the tag line: "Good News on Climate Change: Worst-Case Looks Unrealistic."[127] The basis of the article was an opinion piece that had just been published in *Nature*, the subtitle of which almost sounded like it was intended specifically for Wallace-Wells: "Stop Using the Worst-Case Scenario for Climate Warming as the Most Likely Outcome."[128] The piece didn't actually cast doubt on worst-case climate responses. It didn't in any way provide new evidence ruling out climate surprises or aggravating feedback mechanisms. It simply argued that the "business-as-usual" trajectory now points toward lower carbon emissions. Why? Because of the *policy progress* that is being made in decarbonizing our economy. The commentary, in short, challenged Wallace-Wells's basic thesis.

DOOMISM MEETS ALARMISM

The inactivists promote doomism for at least two different reasons. First, it leads to disengagement. It's another way to dampen enthusiasm among climate advocates and activists—simply convince them it's too late to do anything. But there's actually another reason that inactivists seek to promote doomism. To the extent that it can be portrayed as *alarmism*, it feeds a basic anti-environmental trope that has been a staple of inactivists for decades. As environmental author Alistair McIntosh succinctly put it, "by exceeding the consensus expert science whilst claiming to be based on it, [doomism] feeds denialists by discrediting real science . . . and it sets followers up for disillusion."[129]

Recall the attacks on Rachel Carson by industry groups back in the 1960s. She was denounced as "radical," "communist," "hysterical," "a fanatic defender of the cult of the balance of nature," and a mass murderer.[130] These slanders continue to this day: the fossil-fuel-funded Competitive Enterprise Institute currently claims that "millions of people around the world suffer the painful and often deadly effects of malaria because one person sounded *a false alarm* . . . that person is Rachel Carson" (emphasis added).[131]

Similar accusations were made against Paul Ehrlich, of *The Population Bomb* (1968) fame, whose early warnings of the impact of unrestricted resource depletion have ultimately proven prophetic;

against scientist and science communicator extraordinaire Carl Sagan; and against early climate messengers Stephen Schneider and James Hansen.[132] I myself am regularly dismissed as an "alarmist" by right-wing groups. Indeed, on the day that I wrote this paragraph I was called "the *most* staunch climate alarmist scientist" (emphasis added) in a commentary by CNS News, which is a project of the Media Research Center, a front group for fossil fuel interests and the right-wing Scaife family.[133]

For decades, "false alarm" and "alarmism" have been the rallying calls of conservative interest groups looking to discredit environmental concern—including climate change—as henny-pennyism. A favorite claim relates to the late great climate scientist and science communicator Stephen Schneider. In the early 1970s, when there was still some uncertainty about the relative impacts of warming from greenhouse gases and cooling from sulfur dioxide aerosol pollution, Schneider and coauthor S. Ichtiaque Rasool speculated that the latter effect might win out if sulfur emissions continued to accelerate. That didn't happen because the United States and other industrial nations passed clean air acts in response to the growing acid rain problem. These measures required sulfur dioxide to be "scrubbed" from smokestack emissions prior to entering the atmosphere.[134]

The fact that some scientists—like Schneider—were still wrestling with the competing effects of aerosol cooling and greenhouse warming in the early 1970s has nonetheless given rise to a widespread canard: the notion that "climate scientists were predicting an ice age in the 1970s." The implication is that if scientists so completely botched their predictions back then, why should we trust them now? The reality is (1) they didn't botch the predictions (they just couldn't predict the passage of the clean air acts), and (2) there was no scientific consensus about cooling in the 1970s, just a few scientists, like Schneider, speculating about that possibility.[135] But the notion of a discredited "1970s global cooling scare" has proven an enduring myth that denialists have continued to seize upon. During congressional testimony I gave in July 2006, for instance, climate-change-denying congresswoman Marsha Blackburn (R-TN) attempted to lecture me about how she "remembered" when she was growing up in the 1960s that climate scientists were worried about another ice age. She had obviously failed to study her denialist talking points closely enough, since the claim is supposed to be about the 1970s.[136]

It's hardly surprising that the forces of inaction would still be exploiting doomist narratives today. They can easily be caricatured as alarmism. "Prophets of doom" is the way Donald Trump described those who were advocating action on climate at the January 2020 World Economic Forum in Davos. Ideally, the accusation of alarmism is paired with the shopworn claim that climate scientists are promoting climate doom only to line their pockets with grant money.[137] There's nothing that fires up the conservative base more than right-wing pundits calling out "alarmist scientists who get . . . $89 billion in US government research money" by promoting doomist prophecies.[138] The doomists have made it all too easy for them.

Consider, for example, Jem Bendell's over-the-top "Deep Adaptation" article, which was the inspiration for Alistair McIntosh's warning about how doomism can feed denialism by playing into the agenda of the forces of anti-science.[139] McIntosh referred to a 1956 book, *When Prophecy Fails*, which uses the example of one particular doomsday cult to demonstrate the phenomenon. But a very specific example is at hand here.

Ronald Bailey, the author of *Global Warming and Other Eco Myths: How the Environmental Movement Uses False Science to Scare Us to Death*, reviewed Bendell's article for the libertarian magazine *Reason* in a piece titled "Good News! No Need to Have a Mental Breakdown over 'Climate Collapse.'"[140] In his commentary, Bailey invoked Paul Ehrlich, one of the inactivists' favored punching bags, to ridicule Bendell: "Ehrlich is still predicting an imminent ecological apocalypse and I suspect that Bendell will be doing the same thing in the year 2065." Bailey used Bendell's "concocted case for collapse fatalism" quite effectively to mock concern about climate change.

The "Hothouse Earth" article has also been used to caricature climate concern as an alarmist charade. The *Daily Caller*—which I've called "a Koch front group masquerading as a media outlet"—regularly features attacks on climate science and climate scientists.[141] "Scientists Issue 'Absurd' Doomsday Prediction," read its headline about "Hothouse Earth."[142] The *Caller*'s climate contrarian "energy editor," Michael Bastasch, proceeded to exploit the actual alarmist excesses of the article as an excuse to launch into boilerplate attacks on climate science (for example, "climate models have regularly over-predicted temperature rise"—no, as we have already seen, they haven't) and climate action (quoting, for example, climate contrarian Roger Pielke Sr., who said that an "absurd" emphasis on climate-change impacts "[harm] actual effective policies with reducing risks from extreme weather and other threats"[143]). Bastasch ends by warning of personal sacrifice, attempting to scare his conservative readers into thinking that the *true* threat is aggressive climate action and the dramatic lifestyle changes it will purportedly demand, which "means no fossil fuels . . . reducing consumption and a whole host of other activities." (The reality, of course, is that climate *inaction* is the greater threat to the economy and our way of life.)

Naturally, the Murdoch media is replete with "false-alarm" climate framing. Consider climate-change denier Miranda Devine, formerly of the Murdoch-owned Australian *Daily Telegraph*, *Sunday Telegraph*, and *Herald Sun*, who now pens columns for the Murdoch-owned *New York Post*. In the wake of the devastating Australian bushfires of the summer of 2019/2020, Devine wrote a column for the *Post* titled "Celebrities, Activists Using Australia Bushfire Crisis to Push Dangerous Climate Change Myth."[144] In the piece, she proceeded to dismiss the well-established linkages between climate change and the unprecedented wildfires based on the standard denialist canards. This included attributing the fires to "arson," "green groups," misguided "hazard protection," and "biodiversity" preservation policies. But her core message was summed up in this single sentence, "Whether or not you believe the most dire predictions of climate alarmists makes no difference. We can't dial down the Earth's temperature any more than we can lock up every teenage arsonist." Such a neat little package of denial, doomism, and deflection wrapped up and topped off with charges of alarmism. To err is Devine. To forgive is . . . well, difficult, in this case.

I have seen my own words misrepresented and weaponized by denialist media figures in an effort to portray the climate science community as doomist alarmists. A case in point involves the *Boston Globe*'s resident climate-change denier, Jeff Jacoby. His mischaracterizations of climate science have some scientists howling. They are so egregious that MIT climate scientist Kerry Emanuel, a Republican and political conservative, wrote a letter to the *Globe* in which he chastised Jacoby for presenting "a false choice between panic and the denial of risk." He went on to admonish the *Globe* for publishing a particular commentary that Jacoby had written: "Assessing and dealing with climate risk in an environment of highly uncertain science and expensive options is challenging enough without having to entertain the flippancy of your columnist."[145]

In a *Globe* column from March 15, 2020, "I'm Skeptical About Climate Alarmism, but I Take Coronavirus Fears Seriously," Jacoby quoted me in a way that implied that I myself had accused the climate science community of alarmism.[146] He wrote, "The horrors of pandemics have been documented and depicted often. Yet while climate activists have been forecasting world-ending doomsday scenarios since the 1960s, the apocalypse never seems to materialize." To support his claim, Jacoby then pointed to "facts" from the fossil-fuel-funded, climate-change-denying Competitive Enterprise Institute (rather than legitimate archival evidence). "Although climate is always in flux," he wrote, "unmitigated anthropogenic warming would doubtless lead to cataclysm. But human societies have a genius for mitigating and adapting their way out of existential threats. Which is why it's dangerous, as climatologist Michael Mann has written, to overstate the science of global warming 'in a way that presents the problem as unsolvable, and feeds a sense of doom, inevitability, and hopelessness.'" The source of this quote was my Facebook post criticizing David Wallace-Wells's doomist 2017 *New York Magazine* "Uninhabitable Earth" column.[147]

My actual position was, of course, very much the opposite of what Jacoby had implied. In a letter to the *Globe*, I responded,

> The truth is bad enough when it comes to the devastating impacts of climate change, which include unprecedented floods, heat waves, drought, and wildfires that are now unfolding around the world. . . .
>
> The evidence is clear that climate change is a serious challenge we must tackle now. There's no need to exaggerate it, particularly when it feeds a paralyzing narrative of doom and hopelessness.
>
> There is still time to avoid the worst outcomes, if we act boldly now, not out of fear, but out of confidence that the future is still largely in our hands. That sentiment hardly supports Jacoby's narrative of climate change as an overblown problem or one that lacks urgency.
>
> While we have only days to flatten the curve of the coronavirus, we've had years to flatten the curve of CO_2 emissions. Unfortunately, thanks in part to people like Jacoby, we're still currently on the climate pandemic path.[148]

A PATH FORWARD

It is important to communicate both the threat and the opportunity in the climate challenge. I learned this the hard way. For years my standard public lecture on climate change focused only on the science and the impacts, because I am a scientist. I would then pay lip service to "climate solutions," with the obligatory final slide depicting a montage of recycling efforts, wind turbines, solar panels, and the like. I was fortunate that my audiences were made up of thoughtful and sharing folks. And when they would linger afterward to talk with me, I heard the same thing over and over: "That was a great presentation. But it left me so *depressed*!"

My vanity led me to hear only the compliment and not the admonition that followed it. But the fact is that my presentation, by definition, was not *great*. It was *deficient*. I hadn't thought deeply about our predicament, and as a result I wasn't in a position to report on it responsibly. But I was inspired to do my due diligence and to inform myself about where we really stood, and what was truly necessary to avert catastrophe—to study the literature, crunch the numbers, and figure out how far down the climate-change highway we've gone and what exit ramps are still realistically available to us.

I can tell you that those who are paying attention are worried, as they should be, but there are also reasons for hope. The active engagement of many cities, states, and corporations, and the commitments of virtually every nation (with the United States currently a wildcard as this book goes to press), are very hopeful signs. The rapid movement in the global energy market toward cleaner options is another sign of hope. Experts are laying out pathways to avoid disastrous levels of climate change, and clearly expressing the urgency of action.[149] There is still time to avoid the worst outcomes if—to repeat myself—we act boldly now, not out of fear, but out of confidence that the future is largely in our hands.

What is the antidote to irrational, disabling, doom-and-gloom "futility messaging"? Motivating hope that is grounded in entirely legitimate and defensible reasons for cautious optimism that the worst can still be averted. Recognizing that some harm has already occurred, and that some additional harm is inevitable, provides some needed perspective. It's not a matter of whether we're "effed," after all. It's a matter of *how* "effed" we are.

Let us in this context revisit the two epigraphs that began this chapter, for they address the challenge we face. First is the famous Franklin D. Roosevelt quote: *"The only thing we have to fear is . . . fear itself—nameless, unreasoning, unjustified terror which paralyzes needed efforts to convert retreat into advance."* Roosevelt's famous admonition describes our climate predicament to a T; the surest path to catastrophic climate change is the false belief that it's too late to act.

Then there is the second quote, by the German literary critic, novelist, and essayist Christa Wolf: *"The word 'catastrophe' is not permitted as long as there is danger of catastrophe turning to doom."* It has become fashionable in the climate discourse to use terms like "catastrophe," "emergency," and even "extinction." We must not allow the policing of language to be used as wedge to divide us. But we cannot let words be used in a manner that robs us of agency. Once again it is important to convey both *urgency* and *agency* in talking about the challenge we face. Personally, I like to speak of the "climate crisis," as it embraces both elements (a "crisis," after all, is defined as "a time when a difficult or important decision must be made").

We do not face a scenario of near-term societal collapse or human extinction. The only assurance of such scenarios would be our abject failure to act. If there were not still a chance of prevailing in the climate battle, I would not be devoting my life to communicating the science and its implications to the public and policymakers. I know we can still avert catastrophe. And I speak with some authority on the matter. As a scientist who is still engaged in climate research, my views are informed by hard numbers and facts. In the final chapter of the book, we confront the remaining front in the new climate war—ourselves, our own self-doubt that we have it within ourselves as a species to meet the challenge at hand.

CHAPTER 9

Meeting the Challenge

The darkest hour is just before the dawn.
—Thomas Fuller

Hope is a good thing, maybe the best of things, and no good thing ever dies.
—Andy Dufresne (in *The Shawshank Redemption*)

Despite the challenges detailed in this book, I am cautiously optimistic—that is to say, neither Pollyannaish, nor dour, but objectively hopeful—about prospects for tackling the climate crisis in the years ahead. The reason for that optimism is a confluence of developments, a "perfect storm," if you will, of eye-opening events that are helping to prepare us for the task ahead. First, there have been a series of unprecedented, extreme weather disasters that have vivified the climate-change threat. Second, a global pandemic has now taught us key lessons about vulnerability and risk. And finally, we've seen the reawakening of environmental activism, and, in particular, a popular uprising by children across the world that has framed climate change as the defining challenge of our time.

The thesis of this book is that these developments—along with the collapse of plausible climate-change deniability—have provided us with an unprecedented opportunity for progress. The inactivists have been forced into retreat from "hard" climate denial to "softer" denial: downplaying, deflecting, dividing, delaying, and despair-mongering. These are the multiple fronts of the new climate war. Any plan for victory requires recognizing and defeating the tactics now being used by inactivists as they continue to wage war.

With immensely powerful vested interests aligned in defense of the fossil fuel status quo, it won't come without a fight. We will need the active participation of citizens everywhere aiding in the collective push forward. And we need to believe that it is possible. And it is. We can win the battle for our planet.

THE DENIAL DEATH SPIRAL

When *Washington Post* editorial cartoonist Tom Toles and I published our book *The Madhouse Effect* in the early fall of 2016, colleagues criticized us for writing a book about climate-change denial.[1] The age of denial, they said, was over. The discussion from here on out would be all about *solutions*.

But subsequent history did not cooperate. Climate-change denier Donald Trump was then elected leader of the world's most powerful country. During his administration we've seen the United States go from a leader in worldwide efforts to combat climate change to the only country threatening to withdraw from the 2015 Paris Agreement. We saw a veritable dismantling of fifty years' worth of environmental policy progress in the United States. The intransigence of the United States gave other polluters, such as China, an excuse to ease off on their own efforts. As a result, after flatlining for several years, and appearing to be poised to decline, carbon emissions rose for several years instead.

Something else happened around the same time. We witnessed unprecedented climate-change-fueled weather disasters in the United States and around the world. They came in the form of record floods, wildfires, heat waves, droughts, and superstorms. Damaging, deadly weather extremes drove home the fact that climate change is no longer theoretical and distant. It's here and now. The damaging impacts of climate change had arrived. We know the litany by now: Hurricane Maria in Puerto Rico; flooding in Houston and the Carolinas; wildfires in California; historic drought, flooding, and plagues of locusts in Africa; flooding, heat, drought, and bushfires in Australia. The list goes on. And on. And on.

To quote Groucho Marx, "Who ya gonna believe, me or your lying eyes?"[2] Denial simply isn't viable when people can see the unprecedented impacts playing out in real time on their television screens, their newspaper headlines, their social media feeds, and their backyards. And as a result, we are now seeing the last gasps of hard climate denial. We see it in the virtual disappearance of "false balance" in the mainstream media—the practice, widespread in the past, in which climate-change deniers were treated on a par with mainstream climate researchers when it came to journalistic climate coverage.[3]

Hard denial, today, is mostly confined to the media outposts of the fringe right, shoved to the edges of our discourse by a sliding "Overton window" driven toward reality by the stark facts on the ground. Climate denial operations are waning as fossil fuel interests and plutocrats reject their services in favor of the "kinder, gentler" forms of inactivism that make up the new climate war. The conservative Cato Institute, for example, closed up its climate-denial shop in 2019.[4]

The climate-denying Heartland Institute is increasingly ignored and unable to garner mainstream coverage.[5] Their 2019 "conference," held at the Trump International Hotel in Washington, DC, was reduced from the sprawling three days of its earlier incarnations to just a single-day affair. While it had attracted more than fifty sponsors in past years, it drew just sixteen in 2019—fifteen if you account for the fact that one was fake. Attendance was limited to a couple hundred attendees—predictably, given that the declining demographic of denialists is mostly older white men. Despite holding their "conference" at a Trump property and in Washington, DC, "no one from the Trump administration" was in attendance, a fact bemoaned by Heartland's "science director" (and convicted criminal) Jay Lehr.[6] Lehr insisted that this was "a huge loss" for the administration, since the conference would "reveal that neither science nor economics back up the climate scare." Heartland was forced to lay off staff in 2019.[7]

Even soft denial no longer seems to be getting the traction it once did. In June 2020, Michael Shellenberger, cofounder of the Breakthrough Institute, published a commentary titled "On Behalf of Environmentalists, I Apologize for the Climate Scare." Adopting the schtick of self-styled "Skeptical Environmentalist" Bjorn Lomborg, the piece engaged in the usual inactivist tropes of downplaying climate-change impacts and dismissing renewable energy, all out of alleged "concern" from an ostensibly reformed erstwhile "alarmist." The commentary was panned by the expert evaluators at Climate Feedback, who gave it an average credibility score of –1.2 (between "low" and "very low").[8] Shellenberger originally published the piece at *Forbes*, but they removed it within hours for violating their policies on self-promotion (he was essentially plugging his new book *Apocalypse Never: Why Environmental Alarmism Hurts Us All*). The commentary was subsequently republished by Murdoch's *Australian*. Shellenberger received coverage from the usual nexus of inactivism-promoting organizations and outlets (the Heartland Institute, Glenn Beck, Breitbart News, *Russia Today*, the *Daily Telegraph*, and the *Wall Street Journal*). But other than a critique by *The Guardian*, he got little mainstream coverage.[9]

Shortly thereafter, in mid-July, Bjorn Lomborg published his *own* book, *False Alarm*, once again offering up the same tired tropes. Nobel Prize–winning economist Joseph E. Stiglitz wrote a blistering review of the book for the *New York Times*, which ends thusly: "As a matter of policy, I typically decline to review books that deserve to be panned. . . . In the case of this book, though, I felt compelled to forgo this policy. Written with an aim to convert anyone worried about the dangers of climate change, Lomborg's work would be downright dangerous were it to succeed in persuading anyone that there was merit in its arguments. This book proves the aphorism that a little knowledge is dangerous. It's nominally about air pollution. It's really about mind pollution." There now seems to be little appetite for inactivist diatribes.

Republican communication experts recognize a sinking ship when they see one. Frank Luntz, the GOP messaging guru we encountered earlier, who coached climate-change-denying Republicans and fossil fuel interests on how to undermine public belief in human-caused climate change, has now flipped. In the summer of 2019 he testified to the US Senate's Special Committee on the Climate Crisis that "rising sea levels, melting ice caps, tornadoes, and hurricanes [are] more ferocious than ever. It is happening." He told the committee that he was "here before you to say that I was wrong in 2001"; now, he hoped to put "policies ahead of politics." He proceeded to advise the senators, based on wisdom derived from his polling and focus groups, on how best to frame the climate crisis to get buy-in from the electorate.[10]

Luntz is hardly alone. Douglas Heye, a former communications director at the Republican National Committee, warned of the threat to Republicans who continue to deny the climate crisis: "We're definitely sending a message to younger voters that we don't care about things that are very important to them. . . . This spells certain doom in the long term if there isn't a plan to admit reality and have legislative prescriptions for it."[11]

Republican policymakers seem to be getting the message, too. *Inside Climate News* noted that "an increasing number of Republican politicians have sought to distance themselves from climate denial." It cited the examples of House Minority Leader Kevin McCarthy of California, who recently "introduced a package of bills to promote carbon capture and sequestration technology," and Alaska senator Lisa Murkowski, who "has been attempting to lead a bipartisan effort to pass energy efficiency and technology investment."[12]

Even the fossil fuel industry has turned a corner, no longer denying that its product is warming the planet and changing the climate. In 2018, the cities of San Francisco and Oakland sued the oil companies BP, Chevron, ConocoPhillips, ExxonMobil, and Shell for the damages (due to sea-level rise) that they've caused, indirectly, through the extraction and sale of planet-warming fossil fuels. Citing the reports of the IPCC, a lawyer for Chevron, Theodore Boutrous Jr., assented unambiguously to the strength of the underlying science: "From Chevron's perspective, there is no debate about the science of climate change." The oil companies had admitted, in court, that, as *Grist* put it, "fossil fuels are the problem."[13]

You may have already guessed what came next. As *Grist* described it, Boutrous "twice read a quote from the IPCC that climate change is caused 'largely by economic and population growth.' Then, [he] added his interpretation. 'It doesn't say that it's the production and extraction that's driving the increase,' he said. 'It's the way people are living their lives.'" If you thought you heard a "ping" sound, that's because of the massive deflection we just witnessed.

If these proceedings were a bellwether, and I surely think they were, deniers have essentially thrown in the towel. When it comes to the war on the science—that is, the *old* climate war—the forces of denial have all but conceded defeat. But the new climate war—the war on *action*—is still actively being waged.

TIPPING POINTS—THE *GOOD* KIND

There is reason to be optimistic on the political side as well. The 2018 midterm elections in the United States resulted in a historic swing toward Democrats, ushering in prominent political "rock star" newcomers like Alexandria Ocasio-Cortez, who ran on a Green New Deal platform. Significantly, during the first climate-change hearing held by the House of Representatives' Science Committee under fresh new Democratic leadership, Republicans—seemingly aware of the dramatic shift in public perception—no longer sought to challenge the basic scientific evidence behind human-caused climate change. They instead argued for policy solutions consistent with their political ideology. We can argue over whether they are optimal solutions, but they go beyond the diversionary and deflective proposals we've seen from Republicans in the past, including mechanisms such as carbon pricing. There does now seem to be real political movement toward meaningful action on climate.

House Democrats put forward a bold climate plan in June 2020 that included incentives for renewables and support for carbon pricing.[14] Given an even modestly favorable shift in political winds, one could envision this passing the House and moving on to the Senate with a half dozen or more moderate conservatives crossing the aisle, joining with Senate Democrats to pass the bill within the next year or two. Indeed, it is a well-kept secret in Washington, DC, that many Republicans are quietly supportive of climate action but have been afraid to "come out of the closet" for fear of retribution from powerful ideological purists such as the Kochs and Mercers. *New York Times* columnist Justin Gillis met with one highly placed Republican operative who, requesting anonymity, acknowledged that "we are going to have to do a deal with the Democrats. We are waiting for the fever to cool."[15] I have also had amicable and productive anonymous meetings with prominent conservatives, including a well-known columnist for a Murdoch-owned Australian newspaper. That numerous Republican politicians and conservative opinion leaders *would* support climate action if they felt they were granted the license to do so by party power brokers adds to the notion that a climate-action tipping point could be looming in our near future.

This is not to say that it will be easy to pass climate legislation. Fossil fuel interests, ideologically driven plutocrats like Charles Koch, members of the Mercer and Scaife families, and the global Murdoch media empire are still doing all they can to muddy the waters and block progress. But, as we have seen, there are dramatic demographic shifts underway that favor action on climate. Frank Luntz's recent polling shows that Americans in general support carbon pricing by a four-to-one margin, and Republicans under the age of forty by an amazing six-to-one margin.[16] In short, climate denial is increasingly a liability, while the promise of climate action is an opportunity to win over younger voters.

History teaches us that social transitions are often not gradual but instead sudden and dramatic, and they don't even require a majority in support of change. A committed vocal minority can potentially push collective opinion past a "tipping point." A 2018 study suggested that "opinion of the majority [can] be tipped to that of the minority" once the latter reaches about 25 percent of the public.[17] We appear to have witnessed this phenomenon in action with the rather sudden, dramatic increase in support for marriage equality by Americans during the Obama years. According to Pew Research, public support for same-sex marriage rose from under 40 percent when Obama was elected to over 60 percent when he left office.[18]

Triggered by the horrific killing, captured on video, of a forty-six-year-old black man, George Floyd, by Minneapolis police, a similar tipping point on attitudes toward racial justice seems to have taken place in early summer 2020. One poll showed that the percentage of Americans who think that police are more likely to use excessive force against African Americans jumped from 33 percent to 57 percent. Public awareness and outrage led to massive demonstrations over the unjustified killing. Pollster Frank Luntz commented, "In my 35 years of polling, I've never seen opinion shift this fast or deeply. We are a different country today than just 30 days ago."[19]

It is not unreasonable to speculate that we might be close to such a tipping point on climate as well. According to a Pew Research poll in 2019, 67 percent of the public thinks we're doing too little to reduce the effects of climate change.[20] That, of course, doesn't mean that they prioritize it, or that they're actively pushing for action on climate. But another 2019 poll, conducted by CNN, found that "82 percent of registered voters who identified as Democrats or Democratic-leaning independents consider climate change a 'very important' top priority they'd like to see get the focus of a presidential candidate."[21] Let us account for the fact that roughly 80 percent of eligible citizens are registered, and that 40 percent of voters are Democrats and about 30 percent independent (which we'll conservatively assume split equally into 15 percent and 15 percent when it comes to which direction they lean).[22] That yields at least 36 percent of American citizens ($0.80 \times 0.55 \times 0.82$) who reasonably define the "issue public" for climate action—that is, the set of people who prioritize the issue. That percentage exceeds the 25 percent theoretical threshold required for generating a societal tipping point. It is comparable to the percentage of the American public that supported marriage equality at the beginning of the Obama era, just before that tipping point was reached.

In other words, there's reason to believe that we are currently primed for a marriage-equality-like tipping point with climate action. There is still opposition, but the opposing forces in this case—which include the world's most powerful industrial sector, fossil fuels—are considerably stronger and better funded than those that opposed marriage equality (the religious right). That means that the forward push to get us past the tipping point has to be all that much harder. Fortunately, the forces of progress appear to be aligning in a favorable manner: the visceral evidence of a climate crisis is now before us; we are seeing the demise of denial and the rise of climate activism, particularly from the children's climate movement; and we are learning critical lessons even now from another global crisis, the 2020 coronavirus pandemic.

One group of climate experts has in fact published a set of "concrete interventions to induce positive social tipping dynamics." They propose, as key ingredients, "removing fossil-fuel subsidies and incentivizing decentralized energy generation, building carbon-neutral cities, divesting from assets linked to fossil fuels, revealing the moral implications of fossil fuels, strengthening climate education and engagement, and disclosing greenhouse gas emissions information."[23] A lot of these basic ingredients indeed seem to be in place, or close to being in place.

First of all, as we have already seen, the fossil fuel industry is starting to "feel the heat." Oil-rich Saudi Arabia has "shifted its strategy in the era of decarbonization" by lowering the price of oil exports in a desperate attempt to maintain demand.[24] Coal, the most carbon-intense fossil fuel, is in a death spiral. The state of New York, for example, has retired its last coal-fired power plant.[25] The Canadian mining giant Teck Resources has withdrawn plans for its $20 billion tar sands project.[26] Natural gas is increasingly being recognized not as a "bridge to the future," but as a liability to local communities.[27]

And now, the banking and finance industry is rethinking its role in funding new fossil fuel infrastructure. The primary reason is what is known as *transition risk*. As we choose to decarbonize our economy, demand for fossil fuels will wane. That makes fossil fuel extraction, production, refining, and transport all bad investments. The finance and investment community increasingly fears a bursting of the so-called carbon bubble.

As *Guardian* correspondent Fiona Harvey explained, "investments amounting to trillions of dollars in fossil fuels—coal mines, oil wells, power stations, conventional vehicles—will lose their value when the world moves decisively to a low-carbon economy. Fossil fuel reserves and production facilities will become stranded assets, having absorbed capital but unable to be used to make a profit." Harvey also pointed out that "this carbon bubble has been estimated at between $1tn and $4tn, a large chunk of the global economy's balance sheet. . . . Investors with high exposure to fossil fuels in their portfolios will be hurt, as those companies and assets cease to be profitable." Especially worrying, "If the bubble bursts suddenly, as [experts suggest] it might, rather than gradually deflating over decades, then it could trigger a financial crisis."[28]

There is another reason investors are rethinking their fossil fuel investments, however. It is a generalized notion of *fiduciary responsibility*, which can be defined as "the legal and ethical requirement [of a financial adviser] to put your best interest before their own."[29] An expansive view of this responsibility would require that portfolio managers not make decisions that will mortgage the planet for their clients' children and grandchildren.

Under Australian law, such an expansive view of fiduciary responsibility already applies to pension (or so-called superannuation) fund managers.[30] And it turns out that this has broad international implications, because Australia is home to the world's third-largest net pension holdings, worth just under $2 trillion (a consequence also of Australian law, which requires employers to contribute at least 9 percent of a worker's salary to a superannuation fund[31]). That means that the decisions of Australian "superfund" managers substantially leverage global investment. If Australian superfund managers choose not to invest in fossil fuel companies, it will have reverberations for the fossil fuel industry writ large.

I participated in meetings with several groups of Australian superfund managers in Sydney and Melbourne during my sabbatical in Australia in early 2020. Repeatedly they told me that they now view their investment decision-making through the lens of their larger responsibilities to their clients—in particular, their responsibility not to laden them with risky long-term fossil fuel investments, and their responsibility not to invest in an industry that threatens future livelihood and livability. These audiences were as hungry for detailed facts, figures, and assessments of risk as any I've ever encountered. I left those meetings with the sense that "it may be banking & finance, rather than national governments, that precipitate a climate action tipping point."[32]

There is considerable evidence to support that conjecture. Investors are already taking preemptive actions. According to Axel Weber, the chairman of Swiss multinational investment bank UBS, the finance sector is on the verge of "a big change in market structure" because investors are increasingly demanding that the sector account for climate risk and embed a price on carbon in their portfolio decisions.[33] Mark Carney, governor of the Bank of England, said in early 2020 that because climate change could make fossil fuel financial assets worthless in the future, he is considering imposing a "penalty" capital charge on them.[34]

Insurance giant The Hartford, Sweden's central bank, and Black-Rock, the world's largest asset manager, have indicated they will stop insuring or investing in Alberta's carbon-intensive tar sands oil production.[35] BlackRock has gone even further, announcing it will no longer make investments that come with high environmental risks, including coal for power plants.[36] Goldman Sachs, Liberty Mutual, and the European Investment Bank—the largest international public bank in the world—are among the numerous banks and investment firms that are now pulling away from fossil fuel investments.[37] In the space of a few days in early July 2020, three multibillion-dollar oil and natural gas pipeline projects in the United States—Atlantic Coast, Dakota Access, and Keystone XL, were at least temporarily halted due to what the *Washington Post* characterized as "legal defeats and business decisions."[38] The carbon bubble sure appears ready to pop.

Younger investors, who are far more likely to prioritize action on climate, are playing a particularly vital role here. Consider the actions of twenty-four-year-old Mark McVeigh, an environmental scientist who works for the Brisbane City Council. McVeigh has sued his pension fund for failing to account for climate-change-related damages in its investment decisions. The case is currently working its way through the court system.[39]

While we're talking about the role of young folks, let us consider the impact of fossil fuel divestment, a college-student-led movement. I think back to my first semester at UC Berkeley in the fall of 1984. I had not been politically active in high school. My choice to matriculate to Berkeley had nothing to do with its legacy as a fount of political activism. It had nothing to do with the role it played in the protests of McCarthyism in the 1940s and 1950s, in the civil rights and free-speech movements in the 1960s, or in the Vietnam War protests of the late 1960s and early 1970s. As an aspiring young scientist, I was attracted to UC Berkeley because of its reputation as one of the leading institutions for scientific education and research.

The mid-1980s marked the "Reagan Revolution." Shortly after my arrival that fall, on the night that Ronald Reagan was elected to his second term as president, I watched the Berkeley College Republicans march triumphantly across campus. Complacency had replaced activism even at Berkeley. But activism wasn't dead. It was simply dormant. The anti-apartheid movement—opposing the South African government's brutal and violent policy of discrimination against nonwhites—however, was brewing.

It came to a full boil in 1985. The UC Regents had nearly $5 billion invested in the South African government, more than any other university in the country, helping prop up this system of discrimination. UC Berkeley students demanded the university divest of its holdings. When the Regents resisted, the students held increasingly large and well-publicized sit-ins and protests on famous Sproul

Plaza, the very place where Berkeley students before them had protested in decades past. The students were unrelenting. And in July 1986, under great pressure from the student body, the Regents finally agreed to divest of holdings in the apartheid government and companies doing business with them. That triggered a nationwide divestment movement, and by 1988, 155 institutions of higher learning had chosen to divest.[40] In 1990, five years after the protests had begun at Berkeley, South Africa initiated the dissolution of apartheid. Students at Berkeley—and all across the nation—had helped "change the world."[41] I was part of it.

In 2014, more than two decades later, Berkeley students would once again stage protests in Sproul Plaza. This time it was to demand that the UC Regents divest of fossil fuel holdings. The argument was twofold. First, fossil fuel companies, through the extraction and sale of their product, were causing dangerous planetary warming. Therefore, as with apartheid, there was an obvious moral argument to be made—that the university shouldn't be encouraging harmful activities with their investments. But there was another, more pragmatic reason the student protest made sense: simply put, fossil fuel companies are now bad, risky investments. Their main assets—known but as yet untapped fossil fuel reserves—must ultimately be left stranded.

Fossil fuel divestment has now spread across the country. More than a thousand college campuses and other institutions throughout the United States (accounting for more than $11 trillion in holdings) have divested of fossil fuel stocks.[42] The UC Regents are among them. In September 2019, roughly thirty-three years after their fateful decision to divest from the South African apartheid government, they announced they were divesting of fossil fuel holdings.[43] If past is indeed prologue, we might just speculate that *perhaps* we're just a few years from the bursting of the carbon bubble.

It has been said that "the stone age didn't end for want of stones."[44] Nor is the fossil fuel age ending for want of fossil fuels. It's ending because we recognize that the burning of fossil fuels poses a threat to a sustainable future. But it's also ending because something better has come along: renewable energy. As we have seen, even in the absence of widespread carbon pricing or adequate subsidies, renewable energy is surging owing to the fact that people are embracing clean sources of energy that are ever more competitive with dirty fossil fuel energy.

There is increasingly a sense of inevitably now in the clean energy revolution. The International Energy Agency, as we learned earlier, reported that "clean energy transitions are underway." The IEA attributed the fall in power-sector carbon emissions and the flattening of overall carbon emissions in 2019 to a combination of wind, solar, and other renewable energy sources. Clean energy collectively saved 130 Mt of carbon dioxide from being emitted that same year.[45] This global picture is encouraging.

What we see at the national level is no less promising. In the United States we've crossed a critical milestone. Renewable energy capacity has now reached 250 gigawatts (a gigawatt is a billion watts), amounting to 20 percent of total power generation, a consequence of growth in installed wind and solar voltaic capacity, enhanced energy storage, and an increase in electric vehicle sales.[46] Renewables, for the first time, outcompeted coal in power generation during the first quarter of 2020.[47] In Australia a similar story is underway. Tesla's big batteries are now outperforming fossil fuel generators on both performance and cost.[48] South Australia is now on its way to 100 percent renewable energy.[49] Similar success stories can be told around the world. We are ready to turn the corner. We are approaching a tipping point of the good kind.

THE *REAL* PANDEMIC

Opportunity can arise from tragedy. Such seemed to be the case with the COVID-19 outbreak of early 2020. Nature had afforded us a unique teaching moment. Watching the pandemic unfold, both the impacts and the response, was like watching a time lapse of the climate crisis.[50] Was this a climate-change practice run?

Though the climate crisis is playing out considerably more slowly than the pandemic, there is much to be learned about the former from the latter. These important lessons have to do with the role of science and fact-based discourse in decision-making; the dangers of ideologically driven denial, deflection, and doomism; the roles played by individual action and government policy; the threats posed by special interests hijacking our policy machinery; the fragility of our societal infrastructure; and the distinct challenges of satisfying the needs of nearly eight billion (and growing) people on a finite planet. Will we take away the right lessons?

What can we learn, for example, about the role of science? As with climate change, scientists had warned of the threat of a pandemic many years in advance.[51] They had designed theoretical models for just that scenario that proved essential for anticipating what would happen with the novel coronavirus. The initial spread occurred at an exponential rate, just as models predicted.[52] This meant we could anticipate that more and more people would become infected in the weeks and months ahead, which they did. We knew that the majority of those infected by COVID-19 would experience mild or no symptoms while remaining highly contagious, and we knew that for others, COVID-19 would create the need for emergency medical supports that are not available in sufficient supply.

A popular Internet meme is that "every disaster movie starts with the government ignoring a scientist." And the coronavirus provided some striking examples. Prime Minister Boris Johnson in the United Kingdom initially disregarded what the world's scientists were telling him and instead advocated for "herd immunity."—that is, simply letting the disease spread rampantly among the population, building collective resistance in the remaining population but needlessly sacrificing lives in the process.[53] This decision was based on what turned out to be a faulty analysis by his advisers.[54] Johnson then not only contracted COVID-19 himself but likely spread it to others through irresponsible personal behavior, becoming a poster child for the dangers of disregarding scientific predictions.[55]

The coronavirus outbreak also taught some important lessons about the cost of delay. The United States paid a terrible price by not acting quickly and decisively enough to avoid danger—more than 200,000 deaths at the time this book went to press. It is beginning to dawn on many that we are paying a similar price with the climate crisis. If we had acted decades ago, when a scientific consensus had been reached that we were warming the planet, carbon emissions could have been ramped down gently and much of the damage that we are now seeing could have been avoided. Now they must be lowered dramatically to avert ever more dangerous warming. With COVID-19, there is a two-week delay between intervention actions and changes to the rate of growth in transmissions and deaths. Both the United States and the United Kingdom were slow to take meaningful preventive measures. Whereas deaths had plateaued in most industrial countries by early April 2020, they continued to climb for these two countries.[56] For both climate change and coronavirus, taking appropriate action pays future dividends. Conversely, the slower we are to act, the higher the cost, as measured by both economic losses and deaths.

The parallels weren't lost on other observers. "By the time the true scale of the problem becomes clear, it's far too late," wrote Patrick Wyman in *Mother Jones*. "The disaster—a crisis of political legitimacy, a coronavirus pandemic, a climate catastrophe—doesn't so much break the system as show just how broken the system already was."[57] *The Guardian*'s Jonathan Watts weighed in, too, with a headline reading, "Delay Is Deadly: What Covid-19 Tells Us About Tackling the Climate Crisis."[58]

As with climate change, unwarranted doomism reared its head. Jem Bendell sought to connect the two phenomena explicitly, blaming the coronavirus on rising temperatures. Saijel Kishan at Bloomberg News reported, "Bendell is . . . willing to make the connection between coronavirus and climate change. He says that a warmer habitat may have caused the bats to alter their movements, putting them in contact with humans."[59] I know of no scientific evidence for that claim.

Lessons about the dangers of ideologically driven denial were of course in great abundance. The same individuals, groups, and organizations that have for years served as purveyors of climate-change denial were quick to attack and undermine public faith in the science of the coronavirus crisis. This strategy makes sense, given the common underlying ideology and politics. Climate-change denial serves the agenda of powerful corporations and the Trump administration. COVID-19 denial did the same, with corporate profits, near-term economic growth, and Trump's reelection prospects all threatened by large-scale lockdowns.

So we saw the standard denialist modus operandi in play. Russian trolls early on promoted disinformation and conspiracy theories.[60] Right-wing organizations pumped out anti-science propaganda. A dark-money-funded group called the Center for American Greatness published a commentary mocking the hockey-stick-like projections of coronavirus cases by epidemiologists, comparing them to the supposedly "widely refuted"—you guessed it—climate-change hockey-stick graph that my coauthors and I published more than two decades ago.[61] Even the subtitle of the article ("There's Still Time to Find a Balance Between Public Health and the Economy") cried false dilemma.

The usual denialist suspects were rounded up. Benny Peiser and Andrew Montford—two climate-change deniers—were given substantial real estate on the editorial pages of Rupert Murdoch's *Wall Street Journal* to insist that "scary" coronavirus projections were based on "bad data" and that we must not take "draconian measures" that might harm the economy.[62] As it was published on April 1, you could be forgiven for thinking it was an April Fool's joke. At that very moment, coronavirus cases in New York were surging toward their peak, as subsequent weeks would prove. The climate-change-denying Heartland Institute insisted that social-distancing measures should be lifted.[63] Online, meanwhile, a rogues' gallery of climate-change contrarians, including Judith Curry, Nic Lewis, Christopher Monckton, Anthony Watts, Marcel Crok, and William Briggs, all joined in on the frenzy.[64]

Trump himself emerged early on as a leading source of disinformation. As with climate change, he initially dismissed concerns about COVID-19 as a "hoax."[65] With both COVID-19 and climate change, "Trump . . . employed similar tactics—namely cherry-picking data, promoting outright falsehoods and using anecdotal experience in place of scientific data," reported *Energy and Environment News*.[66] And in both cases Trump depended upon agenda-driven anti-science contrarians to justify his course of inaction.[67] Writing for Pulitzer Prize–winning *Inside Climate News*, Katelyn Weisbrod described "6 Ways Trump's Denial of Science Has Delayed the Response to COVID-19 (and Climate Change)," with a subtitle noting that "Misinformation, Blame, Wishful Thinking and Making Up Facts are Favorite Techniques."[68]

Fearing a slowdown of the economy and threat to his reelection hopes, Trump repeatedly dismissed the public threat and discouraged people from taking the actions recommended by health experts, such as social distancing and mask-wearing. Jeff Mason wrote, in an article for Reuters, "Early on he said that the virus was under control and repeatedly compared it to the seasonal flu," and in late March "he argued the time was coming to reopen the U.S. economy, complaining that the cure was worse than the problem and setting a goal of economic rebirth by Easter on April 12." In early April, furthermore, Dr. Deborah Birx, leading the White House task force on the pandemic, told Americans they needed to "do better at social distancing." But, as Mason put it, "President Donald Trump didn't like the message."[69]

As time went on, and Trump's desperation with the lockdown grew, his anti-scientific and pseudoscientific response to the COVID-19 crisis itself constituted a mounting public health threat. There were his entirely unfounded and irresponsible suggestions that the virus could be cured by ultraviolet light or disinfectants. After having initially issued an emergency authorization in March 2020 for the use of two antimalarial medications, hydroxychloroquine and chloroquine, in response to pressure from Trump, the US Food and Drug Administration reversed that decision in June 2020, noting that the medications "were unlikely to be effective" for treating COVID-19, and that any potential benefits were outweighed by safety risks, including heart problems.[70]

Trump discouraged the use of face masks, a simple measure known to greatly reduce transmission of coronavirus. In June 2020, he held dangerous indoor political rallies in Tulsa, Oklahoma, and Phoenix, Arizona, that defied all public health measures (masks were not encouraged, and staff were even ordered to remove the social-distancing stickers on chairs in Tulsa). And he held a crowded "4th of July" event at Mount Rushmore that represented not only a public health threat but an environmental one as well, featuring a fireworks display that experts warned posed a severe fire hazard due to climate-change-fueled heat and drought conditions.[71]

Other conservatives aided and abetted Trump's efforts. At times, it would have been almost comical if it were not so dangerous. Indeed, the *Daily Show* was compelled to compile a "best of" reel it called the "Heroes of the Pandumic."[72] It featured assorted right-wing personalities, Republican talking heads, and politicians dismissing the threat of the virus. On Fox News, Sean Hannity complained that the "media mob" wanted people to think the pandemic was "an apocalypse," and Rush Limbaugh dismissed it as "hype," insisting that "the coronavirus is the common cold folks." Lou Dobbs on Fox warned, "The national left-wing media [is] playing up fears of the coronavirus." Commentator Tomi Lahren, also on Fox, mocked those who were concerned as crying, "The sky is falling because we have a few dozen cases," adding that she was "far more concerned with stepping on a used heroin needle."

The disdain for science and public health concern went on and on. Fox News personalities Jeanine Pirro, Dr. Marc Siegal, and Geraldo Rivera all dismissed coronavirus as no worse than the flu in what could readily be seen as a coordinated Fox News talking point. Other Fox personalities insisted they were not "afraid" of the virus, that it was "very difficult to contract," and that it was "milder than we thought." A Fox panel told viewers, "It's actually the safest time to fly."

Fox News and other right-wing media even resorted to orchestrated character attacks against the nation's top infectious disease expert, simply because he refused to act as a rubber stamp for Trump's most misguided coronavirus policy gambits. Media Matters described the phenomenon: "Dr. Anthony Fauci, director of the National Institute of Allergy and Infectious Diseases for the past 36 years, is a widely respected immunologist and major public face of the Trump administration's response to COVID-19. Despite his credibility established over decades as a public health official, right-wing media have begun to launch attacks against [him], blaming the medical expert for allegedly harming the economy and undermining President Donald Trump."[73] In what might sound all too familiar, the Trump administration even went so far as to circulate an opposition research document cherry-picking and misrepresenting Fauci's statements to try to discredit him as a scientist and as a messenger.[74]

Republican politicians followed suit, too. Trump's most loyal, fiercest bulldogs in Congress treated the pandemic like it was a joke. Congressman Devin Nunes (R-CA) told viewers to "just go out and go to a local restaurants." Matt Gaetz (R-FL) wore a gas mask on Capitol Hill to mock concern about coronavirus. When a reporter questioned James Inhofe (R-OK), the leading climate-change denier in the US Senate, about what precautions he was taking, Inhofe extended his arm and dismissively asked, "Wanna shake hands?" Eight governors—all Republicans—collectively ignored the words of Dr. Anthony Fauci, who had expressed concern about the lack of adequate lockdown.

Conservative coronavirus denial turned ever more deadly as a coordinated effort emerged among Republican politicians and talking heads to convince the elderly to "take one for the team." Texas lieutenant governor Dan Patrick said on Fox News that grandparents should be willing to die to save the economy for their grandchildren.[75] Conservatives doubled down on this talking point, with other leading personalities, like Fox News's Brit Hume, arguing that it was an "entirely reasonable viewpoint" for the elderly to risk their lives

to help the stock market.[76] One right-wing talk-show host took this progression to its logical extreme, insisting that "while death is sad for the living left behind, for the dying, it is merely a passage out of this physical body."[77]

Herein we see yet another remarkable parallel with climate-change inactivism: the transition over time from denial to false solutions, and then, eventually, to "it's actually good for us." This transition took more than a decade with climate inactivists; with the coronavirus deniers it happened in a matter of weeks.[78] Climate scientist Mike MacFerrin explained, "The right wing's instantaneous flip from 'it's a hoax' to 'let millions die in service to the "market"' is the same script they play with climate change, to a tee. They want you to do nothing."[79] And former CBS News anchor Dan Rather put it this way: "After years when we should have learned of the dangers of 'false equivalence' it baffles me that we are seeing a framing that pits the health of our citizens against some vague notion of getting back to work."[80] I noted, in turn, that it's "not unlike the false equivalence . . . that pits the health of our entire planet against some vague notion of economic prosperity."[81] The right-wing response to coronavirus was, indeed, a précis of the climate wars.

While it took years for the threat of climate change to crystallize, with the impacts of epic storms, floods, and wildfires, it took only weeks for the reality of the coronavirus to set in as people witnessed colleagues, friends, and loved ones contract the disease, and sadly, in some cases, perish from it. Under such circumstances, the consequences of denial and inaction became readily apparent to the average person on the street (or, more aptly, safely self-quarantined in their home).

The coronavirus pandemic thus provided an unexpected lesson on the perils of anti-science. As I told *Energy and Environment News*, the pandemic "exposes the dangers of denial in a much more dramatic fashion. We may look back at the coronavirus crisis as a critical moment where we were all afforded a terrifying view of the dangerous and deadly consequences of politically and ideology-driven science denial. We looked into the abyss, and I hope we collectively decide that we don't like what we saw."[82] Tweeted Steve Schmidt, former presidential campaign co-adviser for the late senator John McCain, "The injury done to America and the public good by Fox News and a bevy of personalities from Limbaugh to Ingraham . . . will be felt for many years in this country as we deal with the death and economic damage that didn't have to be."[83]

There were other key lessons to take away from the pandemic that had broad implications for the climate crisis. We were provided with more examples of the concept of a "threat multiplier"—that is, the compounding nature of multiple simultaneous threats. The damage already wrought by climate change in some places affected their ability to respond to the coronavirus threat. So extensive was the damage to Puerto Rico's health-care infrastructure after Hurricane Maria that vital equipment was lacking when coronavirus came along. A thirteen-year-old named Jaideliz Moreno Ventura was just one of the resulting casualties: she died because Vieques, where she lived, lacked the medical equipment to treat her.[84] Many others were similarly affected, and the tragedy was a legacy of the devastating, climate-change-fueled impacts of Hurricane Maria, along with the insufficient federal support for Puerto Rico under President Trump and his failure to send aid for hurricane recovery, including for critical public health infrastructure.[85]

The pandemic also crystallized the dual roles played by both individual action and government policy when it comes to dealing with a societal crisis. While containment required individuals to act responsibly by practicing social distancing, using masks, and following other advice regarding mitigative behavioral actions, it also required government action in the form of policies (like stay-at-home orders, restrictions on public gatherings, and so on) that would *incentivize* responsible behavior.

The coronavirus crisis, in fact, underscored the importance of government. The need for an organized and effective response to a crisis, after all, is one of the fundamental reasons we have governments in the first place. Crises, whether in the near term like COVID-19 or in the long term like climate change, remind us that government has an obligation to protect the welfare of its citizens by providing aid, organizing an appropriate crisis response, alleviating economic disruption, and maintaining a functioning social safety net.[86]

Citizens, in turn, have a responsibility to hold politicians accountable whenever government fails to uphold its end of the "social contract." In a democratic society, political action and individual action are inextricably linked. We need to deal with problems such as COVID-19 and climate change, and we need competent, science-driven leaders to do that. Consider the contrast between the United States and the United Kingdom, under Donald Trump and Boris Johnson, respectively, on the one hand—two politicians who dismissed the need for lockdown and social distancing—and, on the other hand, New Zealand and Germany, which saw limited impact under their respective leaders Jacinda Ardern and Angela Merkel, who instead embraced such measures.

As I'm writing, we don't yet know the outcome of the upcoming presidential election that will determine the fate of climate policy in the United States, and indeed the world, for years to come. But is seems plausible that voters will recognize the shortcomings of a president who had "received [his] first formal notification of the outbreak of the coronavirus in China" at the beginning of January 2020, including "a warning about the coronavirus—the first of many—in the President's Daily Brief," and "yet . . . took 70 days from that initial notification . . . to treat the coronavirus not as a distant threat or harmless flu strain . . . but as a lethal force . . . poised to kill tens of thousands of citizens."[87] It seems equally plausible that an administration that exploited the pandemic by stripping away environmental protections at the behest of big polluters, greenlighting the construction of controversial new fossil fuel infrastructure, and criminalizing climate protests while the public was distracted will see a reckoning come the election.[88]

The most important question of all, though, is this one: Can an event like the coronavirus crisis become a turning point, an opportunity to bring needed focus to an even greater crisis—the climate crisis? The climate crisis is, after all, the greatest long-term health threat we face. Even as we battled the pandemic, climate change continued to loom in the background. "Earth Scorched in the First 3 Months of 2020," reported *Mashable*.[89] In Australia, where I was residing in early 2020 when the COVID-19 epidemic was just beginning to unfold, Australians were still recovering from the calamitous bushfires of the summer of 2019/2020. Meanwhile, the Great Barrier Reef was beginning to suffer the third major bleaching event in five years, an unprecedented and foreboding development.[90]

The COVID-19 pandemic spoke to the fragility of our expanding, resource-hungry civilization and our reliance on massive but fragile infrastructure for food and water on a planet with finite resources. Some argued that this crisis might be sounding the death knell of resource-extractive neoliberalism.[91] I myself am not so sanguine.[92] But I do think it has generated a long-overdue discussion about the public good and environmental sustainability.

Some ecologists believe that our resource-hungry modern lifestyle—in particular, the destruction of rain forests and other natural ecosystems—may be an underlying factor favoring the sorts of pandemics we have just witnessed.[93] That raises some disturbing possibilities, but to appreciate them, we must take a brief scientific digression into the concept of *Gaia*, the ancient Greeks' personification of Earth herself.

Put forward by scientists Lynn Margulis and James Lovelock in the 1970s, the Gaia hypothesis says that life interacts with Earth's physical environment to form a synergistic and self-regulating system.[94] In other words, the Earth system in some sense behaves like an organism, with "homeostatic" regulatory mechanisms that maintain conditions that are habitable for life. Although the concept has often been taken out of context and misrepresented—for example, to depict Earth as a sentient entity—it is really just a heuristic device for describing a set of physical, chemical, and biological processes that yield stabilizing "feedback" mechanisms maintaining the planet within livable bounds. There is no consciousness or motive. It's simply the laws of physics, chemistry, and biology at work in a fascinating and fortuitous manner.

There is evidence that the hypothesis holds within the range of its assumptions. Earth's *carbon cycle*, which governs the amount of CO_2 greenhouse gas in the atmosphere, is heavily influenced by life on Earth. Photosynthetic organisms, such as cyanobacteria (blue-green algae) and plants, for example, take in CO_2 and produce oxygen, which is needed by animals like us. There is evidence that as the Sun has become brighter over Earth's lifetime of the past 4.5 billion years, the carbon cycle has intensified, decreasing atmospheric CO_2 levels and helping keep Earth from becoming inhospitably hot. A specific example is the famous Faint Young Sun Paradox—the surprising finding that Earth was habitable to basic lifeforms more than 3 billion years ago despite the fact that the Sun was 30 percent dimmer—which we encountered back in Chapter 1. Readers may recall that the great Carl Sagan proposed an explanation: namely, there must have been a considerably larger greenhouse effect at the time. (Incidentally, Sagan and Margulis were married for about seven years. I often wonder what other scientific synergies must have emerged in their daily dinner-table conversations.)

During the height of the COVID-19 crisis, air traffic, transportation, and industrial activity greatly diminished, and pollution, including carbon emissions, was reduced. I couldn't help but pose a rhetorical question.[95] Are pandemics such as coronavirus, metaphorically speaking, acting like Gaia's immune system, fighting back against a dangerous invader? Aren't *we*—through the damage we are inflicting on the planet, its forests, its ecosystems, and its oceans and lakes, actually the metaphorical *virus*?[96] I wasn't the only one asking such questions.[97] My question was intentionally provocative, and I was sensitive about even asking it, since such thinking can easily be misconstrued and abused for misanthropic and ecofascist purposes.[98]

Here's the point, though. Unlike microbes, human beings have agency. We can choose to behave like a virus that plagues our planet, or we can choose a different path. It's up to us. Our response to the coronavirus pandemic shows it's possible for us to change our ways when we must. The COVID crisis was acute and immediate, and the penalty of inaction was swift. Climate change may seem slower than coronavirus and farther away, but it is very much here, and it requires many of the same behavioral changes. In this case our commitment must be sustained rather than fleeting. We must flatten the curve—of carbon emissions—to get off the climate pandemic path.[99]

While the coronavirus pandemic was truly a tragedy, we must consider the opportunities it has brought along in its wake as we attempt to work our way back to normal life and governments implement economic stimulus plans to jump-start their economies. The pandemic has given us an opening to get off the path of climate distress and onto a healthier path. We must work even harder to decarbonize our economy and minimize our environmental footprint. There are clear side benefits to an economy that is less vulnerable to disruptions in the production and transport of fuel. Regardless of what else happens, the sun will still shine, and the wind will still blow. Renewable energy is both safer and more reliable than fossil fuels. We were already seeing the decoupling of our global economy from fossil fuels before the pandemic. (We had substantial economic growth in 2019 without a rise in carbon emissions.) Why not take this opportunity to accelerate the transition from fossil fuels to renewable energy?

The good news is that this seems to be happening, despite the Trump administration's best efforts to impede this transition by seeking to fast-track the further dismantling of climate and environmental protections.[100] *Inside Climate News* reported in July 2020 that two of the world's largest oil companies, Shell and BP, were lowering their outlooks for demand for their products and slashing the value of their assets by billions, saying the coronavirus pandemic could accelerate a shift to clean energy.[101] In early April 2020, a group of state officials from agencies such as the California Energy Commission, collectively representing more than 25 percent of total US power generation, announced a new coalition dedicated to 100 percent clean energy. In doing so, they explicitly acknowledged both the challenges and the opportunities for change in the wake of the pandemic.[102] New York State, the world's eleventh-largest economy, put forward a COVID-19 recovery plan centered on renewable energy.[103] ClimateWorks Australia had a stimulus-ready plan already in place for Australia to move toward net-zero carbon emissions.[104] It appears we may, indeed, be turning a corner. That's just one reason to be optimistic. There are others.

THE WISDOM OF CHILDREN

The Bible prophesied that "a little child shall lead them" (Isaiah 11:6). And such has been the case with climate action. Over the past few years, we have witnessed the rise to prominence of Greta Thunberg, a teenager from Sweden, who achieved by the age of sixteen an iconic global cultural status typically reserved for pop stars and Hollywood celebrities. She has been nominated for the Nobel Peace Prize and was featured on the cover of *Time* magazine. Thunberg has been diagnosed as having Asperger syndrome, but instead of seeing it as a liability, she calls it her "superpower."[105] Now seventeen, she possesses a remarkable ability to speak truth to power in strong, laser-focused, perfectly delivered language.

In 2018, at age fifteen, she began protesting outside the Swedish parliament to raise awareness about the threat of climate change. Her efforts garnered increasing levels of media attention. She went on to speak at the 2019 United Nations Climate Change Summit, to the British and European Parliaments, and, perhaps most famously, to the attendees of the 2019 World Economic Forum in Davos, where she chided the politicians and other influential individuals gathered there for their failure to address the existential challenge of our time, warning them "our house is on fire."

Thunberg's efforts have been infectious. She has sparked a global youth movement called "Fridays For Future," with literally millions of children around the world marching, striking, and protesting for climate action weekly. Kids in the United States wear T-shirts bearing her likeness. Adults are now mobilizing to support the movement, too. Inactivists have become so worried that they've even manufactured and promoted an "anti-Greta," a teenager who dismisses the climate crisis, in a desperate and feeble attempt at distraction and misdirection.[106]

They *should* be worried. In response to this popular uprising, the UK and Irish parliaments have now both declared a "climate emergency."[107] The majority of UK voters now support dramatic action to lower greenhouse gas emissions to nearly zero by 2050 regardless of cost.[108] There is clearly a sense of urgency. But there is also recognition of *agency*—a sense that action is possible, that our future is, to a great extent, still in our hands.

While Thunberg has garnered the lion's share of attention, there are other leaders of this movement. Among them is Alexandria Villaseñor, who, beginning in December 2018, at the age of fourteen, skipped school every Friday to protest against lack of climate action in front of United Nations Headquarters in New York City. She cofounded the US Youth Climate Strike and Earth Uprising youth climate activist groups. Then there's Jerome Foster, who as of 2020 was eighteen years old. An activist from Washington, DC, he is founder and editor in chief of *The Climate Reporter*. I joined Villaseñor and Foster in Easthampton, Long Island, in August 2019 in a panel event called "The Youth Climate Movement Could Save the Planet"—a sentiment with which I agree.[109] Afterward, the two even *inducted* me "officially" into the youth climate movement after I was able to demonstrate competency in Instagram technique.

It was a light moment, but the topic couldn't be more serious. The local paper, summarizing the discussion, said of the youth leaders, "Despite little meaningful movement to address a growing emergency, they have hope. Their generation, they said, is mobilizing to preserve a livable world."[110] These kids have helped accomplish what seemingly nobody else could. They've helped place climate change on the front page of the papers and at the center of our public discourse. They are the main reason I'm optimistic that we're finally going to win this battle.

In solidarity with these youths, a group of just under two dozen climate scientists, myself included, published a letter in *Science* magazine that was ultimately cosigned by thousands of other scientists around the world. The letter offered support to them for their efforts.[111] It read, in part, "The enormous grassroots mobilization of the youth climate movement . . . shows that young people understand the situation. We approve and support their demand for rapid and forceful action. We see it as our social, ethical, and scholarly responsibility to state in no uncertain terms: Only if humanity acts quickly and resolutely can we limit global warming . . . and preserve the . . . well-being of present and future generations. This is what the young people want to achieve. They deserve our respect and full support."

They deserve not only our respect and support but our protection as well.[112] We saw earlier, back in Chapter 4, how leaders of the youth climate movement like Greta Thunberg and Alexandria Villaseñor have been targeted by trolls and bots and even heads of state, including Donald Trump and Brazilian president Jair Bolsonaro.

The attacks on Thunberg reached fever pitch in the lead-up to the high-profile, high-stakes events of September 2019: the Global

Youth Climate Strike and the UN Climate Change Summit in New York City. Andrew Bolt, the Australian climate-change-denying propagandist at Murdoch's *Herald Sun*, attacked Thunberg, then sixteen, as "strange" and "disturbed."[113] Christopher Caldwell, a Senior Fellow and contributing editor for the right-wing Scaife-funded Claremont Institute, was granted space in the *New York Times* to attack her in a piece titled "The Problem with Greta Thunberg's Climate Activism: Her Radical Approach Is at Odds with Democracy."[114] Patrick Moore, chairman of the board of directors of the CO2 Coalition, a climate-change-denying Koch brothers front group that is the modern-day successor to the infamous George C. Marshall Institute we encountered back in Chapter 2, went so far as to tweet "Greta = Evil."[115]

The Eye of Sauron is focused upon these kids. The most powerful industry in the world, the fossil fuel industry, sees them as an existential threat and has them firmly in its sights. Consider the recent actions of the Organization of the Petroleum Exporting Countries (OPEC), a trillion-dollar international organization founded in 1960 by five petrostates—Iran, Iraq, Kuwait, Saudi Arabia, and Venezuela. It now consists of fourteen oil-exporting countries that own 80 percent of the world's proven oil reserves.

In July 2019, OPEC's secretary general, Mohammed Barkindo, referred to the youth climate movement as the "greatest threat" the fossil fuel industry faces. He expressed concern that the pressure being brought to bear on oil producers by the mass youth movement was "beginning to . . . dictate policies and corporate decisions, including investment in the industry." Barkindo acknowledged that even the children of OPEC officials were now "asking us about their future because . . . they see their peers on the streets campaigning against this industry."[116]

Unintimidated, members of the youth climate movement actually welcomed the comments "as a sign the oil industry is worried it may be losing the battle for public opinion." The criticism, as *The Guardian* characterized it, "highlights the growing reputational concerns of oil companies *as public protests intensify along with extreme weather*" (emphasis added).[117] Note, by the way, the acknowledgment here of the role played by a synergy of underlying factors—in this case both the youth climate movement and mounting weather disasters. It is indeed no single factor—but a convergence of them—that has led both to intensified attacks by inactivists and an unprecedented opportunity for change.

The kids are at the center of it all. And they are being attacked simply for fighting for their future. It is morally incumbent upon the rest of us to do more than just pat them on the back. Communication expert Max Boycoff expressed the worry, in a September 2019 op-ed, "that we adults, who got us into this mess, are not doing enough. . . . Adult utterances about 'legacies' and 'intergenerational' generally ring hollow when the scale of engagement and action pales in comparison to the scale of the ongoing challenge."[118]

The children have created an opportunity that didn't exist before—they've gained a foothold for the rest of us. It is time for us to take the opportunity we've been given as we prepare for battle—the battle to preserve a livable planet for our children and grandchildren.

THE FINAL BATTLE

Though they are on the run, the forces of climate-change denial and inaction haven't given up. Nor are they, as Malcolm Harris wrote in *New York Magazine*, "planning for a future without oil and gas." "These companies," Harris observed, after attending a fossil-fuel-industry planning meeting, "want the public to think of them as part of a climate solution. In reality, they're a problem trying to avoid being solved."[119]

Climate inactivists are now engaged in a rear-guard action as their defenses start to crumble under the weight of the evidence and in the face of a global insurgency for change. But let us also recognize that they are still in possession of a powerful arsenal as they wage the new climate war. It includes an array of powerful *D*s: disinformation, deceit, divisiveness, deflection, delay, despair-mongering, and doomism. The needed societal tipping point will not easily be reached as long as these immensely powerful vested interests remain aligned in defense of the fossil fuel status quo and in possession of these formidable weapons. It will only happen with the active participation of citizens everywhere aiding in the collective push forward.

It is the goal of this book to inform readers about what is taking place on this front and to enable people of all ages to join together in the battle for our planet. With that goal in mind, let's revisit the four-point battle plan outlined at the very start, reflecting now on everything we've learned:

Disregarding the Doomsayers: We have seen how harmful doomism can be. It is disabling and disempowering. And it is readily exploited by inactivists to convince even the most environmentally minded that there's no reason to turn out for elections, lobby for climate action, or in any other way work toward climate solutions. We must be blunt about the very real risks, threats, and challenges that climate change *already* presents to us. But just as we must reject distortions of the science in service of denialism, so, too, must we reject misrepresentations of the science—including unsupported claims of runaway warming and unavoidable human extinction scenarios—that can be used to promote the putative inevitability of our demise.

Unfortunately, *doom sells*! That's why we've seen a rash of high-profile feature articles and best-selling books purveying what I call "climate doom porn"—writing that may tap into the adrenaline rush of fear but actually inhibit the impulse to take meaningful action on climate. It's why we see headlines with an overly doomist framing of what the latest scientific study shows (or at least plays up the worst possible scenarios).[120]

Feeding doomism is the notion that climate change is just too big a problem for us to solve. Especially pernicious in this regard is the dismissal of climate change as a "wicked problem." While definitions vary, what's relevant here is how it is defined in common parlance. Wikipedia defines a wicked problem as "a problem that is difficult *or impossible to solve* because of incomplete, contradictory, and changing requirements that are often difficult to recognize" (emphasis added).[121]

The idea that the climate problem is fundamentally unsolvable is itself deeply problematic. Jonathan Gilligan, a professor in earth and environmental science at Vanderbilt University, agrees, explaining, in a Twitter thread, "There are profound problems with the 'wicked problem' idea, that tend to produce a sense of helplessness because wicked problems are, by their definition, unsolvable."[122]

Others weighed in on how the "wicked problem" framing can constitute a form of soft denial. Paul Price, a policy researcher in Dublin City University's Energy and Climate Research Network, explained, "Social science use of 'wicked' & 'super-wicked' too often seems a form of 'implicatory denial,' a rhetorical fence to avoid physical reality."[123] Atmospheric scientist Peter Jacobs added, "There is almost literally no environmental problem that one couldn't successfully reclassify as 'wicked' at the outset if one wanted to, even topics where we've successfully mitigated much of the harm (ozone depletion, acid rain, etc.)."[124]

In any case, the "wicked problem" framing is convenient to polluting interests, which have worked hard to sabotage action on climate. And it's wrong. The truth is, if we took the disinformation campaign funded by the fossil fuel industry out of the equation, the climate problem would have been solved decades ago. The problem is not hopelessly complicated.[125]

Nevertheless, the forces of doomism and despair-mongering remain active, and we must call them out whenever they appear. In March 2020, as I was writing the final section of this book, social media was abuzz: Bernie Sanders had dropped out of the Democratic presidential primaries, leaving Joe Biden the presumptive nominee. Some Sanders supporters were particularly aggressive in insisting that this spelled climate Armageddon. A commenter tweeted at me, "If we don't reduce carbon output by 50% of 2018 by 2030 climate change becomes a run away [*sic*] process that cannot be stopped."[126] I responded, "That's false. . . . Climate-change deniers distort the science. Let's not resort to their tactics."[127] The commenter continued, "Biden's plan doesn't come close to accomplishing that. There is no reason to vote in this election because it's apocalypse either way."[128]

It was a perfectly toxic brew of misguided thinking, consisting of distortions of the science in the service of doomist inevitability and false equivalence—between a president who has done notable damage to international climate efforts and a candidate whom Politifact calls "a climate change pioneer."[129] The cherry on top is the overt and cynical nihilism—the notion that there is nothing we can do, so we might as well simply give up. It would be easy to dismiss this as a one-off comment. But in fact it is reflective of a hostile online atmosphere that has been fueled by bad state actors. Bernie Sanders had said just a month earlier, "In 2016, Russia used Internet propaganda to sow division in our country, and my understanding is that they are doing it again in 2020. Some of the ugly stuff on the Internet attributed to our campaign may well not be coming from real supporters."[130]

This sort of propaganda may be more harmful now than climate-change denial itself. It must be treated as every bit as much of a threat to climate action. Those who promote it should be called out in the strongest terms, for they threaten the future of this planet. When you encounter such doomist and nihilistic framing of the climate crisis, whether online or in conversations with friends, coworkers, or fellow churchgoers, call it out.

Don't forget, once again, to emphasize that there is both *urgency* and *agency*. The climate crisis is very real. But it is *not* unsolvable. And it's *not* too late to act. Every ounce of carbon we don't burn makes things better. There is still time to create a better future, and the greatest obstacle now in our way is doomism and defeatism. Journalists and the media have a tremendous responsibility here as well.

A Child Shall Lead Them: Back in 2017, I coauthored a children's book, *The Tantrum That Saved the World*, with children's book author and illustrator Megan Herbert.[131] It told the story of a girl, Sophia, who is frustrated by the animals and people—including a polar bear, a swarm of bees, a Pacific Islander, and others—who continue to show up at her door. They've been displaced from their homes by climate change. Sophia becomes increasingly frustrated by this disruptive activity and throws a tantrum. But she ultimately redirects her anger and frustration—and the tantrum itself—in an empowering way. She becomes the change she wishes to see in the world, starting a whole movement that demands accountability by the adults of the world to act on the climate crisis. Less than a year later, Greta Thunberg would rise to prominence and the youth climate movement would take the world by storm. Yes, life does indeed sometimes imitate art—in this case, in a most profound way.

The children speak with a moral clarity that is undeniable to all but the most jaded and cynical. It is a game-changer. But, as we've seen, that's what makes them such a threat to vested interests—the heads of petrostates and the fossil fuel industry itself. They have attacked the children because the children pose a serious threat to the industry's business-as-usual model. Fossil fuel interests rely on that model continuing for record profits.

Some colleagues of mine blithely dismiss the notion that we are, even if involuntarily on our part, in a "war" with powerful special interests looking to undermine climate action. Ironically, they are engaged in a form of denial themselves. The dismissiveness of soothing myths and appeasement didn't serve us well in World War II, and it won't serve us well here either. Especially when we are dealing with an enemy that doesn't observe the accepted rules of engagement. To carry the analogy one step further, the attacks on child climate activists most surely constitute a metaphorical violation of the Geneva Conventions. So yes, we are in a war—though not of our own choosing—and our children represent unacceptable collateral damage. That is why we must fight back—with knowledge, passion, and an unyielding demand for change.

This problem goes well beyond science, economics, policy, and politics. It's about our obligation to our children and grandchildren not to leave behind a degraded planet. It is impossible now not to be reminded of this threat whenever I have an opportunity to share the wonders of this planet with my wife and fourteen-year-old daughter. In December 2019, before I began my sabbatical in Australia, I traveled with my family to see the Great Barrier Reef. Within a month after our visit the third major bleaching event of the past five years, the most extensive yet, was underway. Some experts fear that the reef won't fully recover.[132]

It fills me with an odd sort of "survivors' guilt" to have seen the reef with my family just in the nick of time. The next stop on our vacation was no less sobering. We went to the famous Blue Mountains of New South Wales. Unfortunately, the majestic vistas were replaced by a thick veil of smoke from the unprecedented bushfires that were spreading out across the continent.

I feel some wistfulness about the fact that my daughter, when she grows up, may not be able to experience these same natural wonders with her children or grandchildren. It's appropriate to feel grief at times for what is lost. But grief about that which is wrongly presumed to be lost yet can still be saved—and which is used, under false pretenses, in the service of despair and defeatism—is pernicious and wrong. Since I have already used at least one *Lord of the Rings* metaphor in this book, you'll forgive me if I use another. I'm reminded of the Steward of Gondor, who wrongly presumes his son to be dead and his city to be lost, telling the townspeople to run for their lives, and his assistants to take his still-living son off to be burned. Fortunately, Gandalf whacks him upside the head with his staff before his orders can be carried out. Sometimes I feel that way about doomists who advocate surrender in the battle to avert catastrophic climate change.

Educate, Educate, Educate: As we have discussed, the battle to convince the public and policymakers of the reality and threat of climate change is *largely* over. The substantive remaining public debate is over how bad it will get and what we can do to mitigate it. So online, don't waste time engaging directly with climate-change-denying trolls and bots. And where appropriate, report them. Those who seem to be *victims* of disinformation rather than *promoters* of it deserve special consideration. Try to inform them. When a false claim appears to be gaining enough traction to move outside the denialist echo chamber and infect honest, well-meaning folks, it should be rebutted.

You have powerful tools at your fingertips. A personal favorite resource of mine is Skeptical Science (skepticalscience.com), which rebuts all the major climate-change-denier talking points and provides responses that you can link to online or via email. Inform yourself about the latest science so you are armed with knowledge and facts, and then be brave enough to refute misinformation and disinformation. Online there are Twitter accounts you can follow that provide up-to-date information about the science, impacts, and solutions. A few personal favorites of mine are @ClimateNexus, @TheDailyClimate, @InsideClimate, and @GuardianEco. Feel free to follow yours truly (@MichaelEMann), too, if you don't mind the occasional cat video!

Climate-change deniers constantly complain about language and framing. Don't fall for it. Don't make concessions to them. In sports parlance, they're trying to "work the refs." The classic example is the shedding of crocodile tears over use of the term "climate-change denier" itself. In point of fact, it's an appropriate, accepted term to describe those who reject the overwhelming evidence. The goal of the critics in this case is to coerce us into granting them the undeserved status of "skeptics," which actually *rewards* their denialism. Legitimate skepticism is, as we know, a good thing in science. It's how scientists are trained to think. Indiscriminate rejection of evidence based on flimsy, ideological arguments is not.

When we falsely label climate-change denialism as "skepticism," it legitimizes disinformation and muddies the climate communication waters. It makes concessions to those who have no interest at all in good-faith engagement, are unmovable in their views, and are intentionally trafficking in doubt and confusion. What is so pernicious is that at the same time it actually *hinders* efforts to convince and motivate the "confused middle"—those who are liable to throw up their hands in frustration when presented with the apparent predicament of a debate between two ostensibly legitimate camps.

But enough about climate-change deniers. They are increasingly a fringe element in today's public discourse, and our efforts to educate are best aimed at those in the confused middle. These folks accept the evidence but are unconvinced of the urgency of the problem and are unsure whether we should—or can—do anything about it.

My advice is to spend your time on those who are reachable, teachable, and movable.[133] They need assistance. As we have seen, far too many have fallen into climate despair, having been led astray by unscientific, doomist messaging, some of it promoted by the in-activists in a cynical effort to dispirit and divide climate activists. Others are victims of other types of climate misinformation. When you encounter, for example, the claim that it's too expensive to act, point out that the opposite is true. The impacts of climate change are already costing us far more than the solutions. And indeed, 100 percent green energy would likely pay for itself.[134]

Call out false solutions for what they are. We've seen that many of the proposed geoengineering schemes and technofixes that have been proposed are fraught with danger. Moreover, they are being used to take our eye off the ball—the need to decarbonize our society. Even some of the fiercest climate hawks are sometimes way off base here. Elon Musk, for instance, has suggested that nuclear bombs could be used to make Mars's atmosphere habitable. While such proposals seem almost amusingly flippant, they are dangerous—not because we might expose little green men to nuclear radiation, but because they offer false promise for a simple escape route, providing fodder for those who argue "we can just find another planet if we screw up this one."

Climate change is arguably the greatest threat we face, yet we speak so little about it. Silence breeds inaction. So look for opportunities to talk about climate change as you go about your day—that's the gateway to all of the solutions we've discussed. Unlike corona-virus, we cannot look forward to a literal vaccine for the planet. But in a metaphorical sense, knowledge *is* the vaccine for what currently ails us—denial, disinformation, deflection, delayism, doomism, you know the litany by now. We must vaccinate the public against the efforts by inactivists to thwart climate action, using knowledge and facts and clear, simple explanations that have authority behind them. That's empowering, because it means we can *all* contribute to the cure.

Changing the System Requires Systemic Change: Inactivists, as we have seen, have waged a campaign to convince you that climate change is *your* fault, and that any real solutions involve individual action and personal responsibility alone, rather than policies aimed at holding corporate polluters accountable and decarbonizing our economy. They have sought to *deflect* the conversation toward the car you drive, the food you eat, and the lifestyle you live.

And they want you arguing with your neighbor about who is the most carbon pure, dividing advocates so they cannot speak with a unified voice—a voice calling for change. The fossil fuel industry and the inactivists who do their bidding fear a sober conversation about the larger systemic changes that are needed and the incentives they will require. And it's for one simple reason: it means the end of their reign of power.

Make no mistake. Individual action is part of the solution. There are countless things we can do and ought to do to limit our personal carbon footprint—and indeed our total environmental impact. And there are many reasons for doing them: they make us healthier, save us money, make us feel better about ourselves, and set a good example for others to follow. But individual action can only get us so far.

We were recently afforded a cautionary tale about the limits of behavior change alone in tackling the climate crisis. The dramatic reduction in travel and consumption brought about by the global lockdown response to the coronavirus pandemic reduced global carbon emissions by only a very modest amount.[135] Referencing this fact, Glen Peters, research director on past, current, and future trends in energy use and greenhouse gas emissions at the Center for International Climate Research (CICERO), posed a question: "If such radical social change leads to (only) a 4% drop in global emissions, then how do we get a 100% drop by ~2050? Is #COVID19 just going to show how important technology is to solve the climate problem?"[136] It's a valid point.

The answer is that there *is* no path of escape from climate-change catastrophe that doesn't involve polices aimed at societal decarbonization. Arriving at those policies requires intergovernmental agreements, like those fostered by the United Nations Framework Convention on Climate Change (UNFCCC), that bring the countries of the world to the table to agree on critical targets. The 2015 Paris Agreement is an example. It didn't solve the problem, but it got us on the right path, a path toward limiting warming below dangerous levels. To quote *The Matrix*, "There's a difference between knowing the path and walking the path." So we must build on the initial progress in future agreements if we are to avert catastrophic planetary warming.

The commitments of individual nations to such global agreements can only, of course, be met when their governments are in a position to enforce them through domestic energy and climate policies that incentivize the needed shift away from fossil fuel burning and other sources of carbon pollution. We won't get those policies without politicians in office who are willing to do our bidding over the bidding of powerful polluters. That means that we must bring pressure to bear on politicians and polluting interests. We do that through the strength of our voices and the power of our votes. We must vote out politicians who serve as handmaidens for fossil fuel interests and elect those who will champion climate action. That brings us full circle, because we are now back to talking about the responsibility of individuals—but now, it's about the responsibility to vote and to use every other means we have to collectively influence policy.

Herein we have encountered a new challenge. Opposition to key policy measures is now coming not just from the right, as traditionally expected, but from the left, too. While a vast majority of liberal democrats (88 percent) support carbon pricing, there is a movement underway, as we have seen, among some progressive climate activists to oppose it. Their opposition is based on the perception that it violates principles of social justice (though there's no reason that needs to be the case), or that it buys into market economics and neoliberal politics.[137] Others insist that it can't pass because it's unpopular with voters (the opposite is actually true), or that it could too easily be reversed by a future government (which one could say of any policy that isn't codified as a constitutional amendment).[138]

Some climate opinion leaders are in denial of this development. In early April 2019, I complained that "the greatest trick the devil ever pulled was getting progressives to oppose carbon pricing." I was referring not to the majority of self-identifying progressives, but to the small number of progressive climate activists who now oppose such measures.[139]

The often vituperative pundit David Roberts defensively tweeted in response that "the number of progressives who outright oppose carbon pricing is tiny & utterly insignificant in US politics. Just another example of phantom leftists against when [sic] Reasonable People can define their own identities."[140] This argument ignores the most prominent progressive in modern American politics, Bernie Sanders, who, in response to direct questioning by the *Washington Post* in November 2019, indicated he didn't support carbon pricing.[141]

It's not just Sanders. Roberts was immediately contradicted by Twitter users who came out of the woodwork to demonstrate my very point.[142] One self-avowed Unitarian Universalist (a religion known for its progressive philosophical and political outlook[143]) responded to Roberts, "I'm an advocate for climate action thru [Citizens Climate Lobby] & other groups. Almost ALL progressive folks I encounter (friends, Twitter, EJ) reflexively oppose carbon price of all sorts. They generally retreat to 'just ban FFs' as more likely & better. There's a lot of work to do."[144]

This opposition to carbon pricing seems to be tied to a larger trend on the left against "establishment" politics. This development has been fueled at least in part by state-sponsored (Russian) trolls and bots looking to sow division in Democratic politics in an effort to elect fossil-fuel-friendly plutocrats like Donald Trump to power. That tactic was successful in the 2016 presidential election and was very much still in play during the 2020 election, as detailed by the *Washington Post* in a February 2020 article.[145]

The same witches' brew that helped bring Donald Trump to power in 2016—interference by malevolent state actors, cynicism, and outrage, including among some on the progressive far left—appears, as this book goes to press, to be a potent threat to climate action today.

Let's recognize, though, that while some of the outrage has been manufactured by bad actors who have magnified and then weaponized divisions, some legitimate underlying grievances have also played a role. Some environmental progressives profess a distrust of neoliberal economics. And why not? It's gotten us into this mess. Some prominent figures, such as Naomi Klein, have openly challenged the notion that environmental sustainability is compatible with an underlying neoliberal political framework built on market economics. It's entirely conceivable she's right.

Some progressives feel that current policies don't do enough to address basic societal injustices. At a time when we see the greatest income disparity in history, along with a rise in nativism and intolerance, surely they have a point. They argue that any plan to address climate change must address societal injustice, too. But I would argue that social justice is *intrinsic* to climate action. Environmental crises, including climate change, disproportionately impact those with the least wealth, the fewest resources, and the least resilience. So simply *acting* on the climate crisis is acting to alleviate social injustice. It's another compelling reason to institute the systemic changes necessary to avert the further warming of our planet.

Yes, we have other pressing problems to solve. And climate change is just one axis in the multidimensional problem that is environmental and societal sustainability. I don't purport to propose, in this book, the solution to all that ails us as a civilization. I do, however, offer what I see as a path forward on climate.

As we pass the milestone of the fiftieth anniversary of the very first Earth Day (April 22, 1970), I believe that we are at a critical juncture. Despite the obvious political challenges we currently face, we are witnessing an alignment of historical and political events—and acts of Mother Nature—that are awakening us to the reality of the climate crisis. We appear to be nearing the much-anticipated tipping point on climate action. In a piece titled "The Climate Crisis and the Case for Hope" published in September 2019, my friend Jeff Goodell, a writer for *Rolling Stone*, posited that "a decade or so from now, when the climate revolution is fully underway and Miami Beach real estate prices are in free-fall due to constant flooding, and internal combustion engines are as dead as CDs, people will look back on the fall of 2019 as the turning point in the climate crisis."[146] We can debate the precise date of the turning point. But I concur with Jeff's larger thesis.

It is *all* of the things we have talked about—behavioral change, incentivized by appropriate government policy, intergovernmental agreements, and technological innovation—that will lead us forward on climate. It is not any one of these things, but *all of them* working together, at this unique moment in history, that provides true reason for hope. To repeat one of the epigraphs that began this final chapter, "Hope is a good thing, maybe the best of things." Alone it won't solve this problem. But drawing upon it, we will.

Epilogue

THE ORIGINAL HARDCOVER EDITION OF *THE NEW CLIMATE WAR* went to press in late summer 2020, a time of great uncertainty, anxiety, and trepidation. We were living through an unprecedented global pandemic with no end yet in sight. An increasingly unhinged American president seemingly grew more desperate and dangerous by the day as he and his supporters feared their grip on power was slipping away. We were in the home stretch of what was arguably the most consequential election in modern US history. The future of our planet lay quite literally in the balance. The reelection of Donald Trump to a second term would have solidified efforts by polluters to dismantle federal climate policies, continued to weaken our diplomatic relationships with other countries, and threatened not just domestic but global action on climate change as well.

Yet I argued for cautious optimism in spite of all of this. I pointed to signs of a turning point on climate action—a tipping point "of the good kind," if you will. Though hardened climate cynics and doomers chided me as a purveyor of "hopium," several favorable factors supported my view.

Among them was the continuation of what I referred to previously as a "death spiral of denial" that was well underway, driven by climate catastrophes playing out in real time right before our eyes—an incessant procession of heat extremes, withering droughts, infernal wildfires, drenching floods, and catastrophic superstorms.

That hardly sounds like good news. But it rendered denial unviable. Climate denialists had become all but irrelevant, widely mocked for the carnival barkers they were seen to be, dismissed in serious policy circles, and relegated to the fringes of the right-wing fever swamp.

We also witnessed what I had presaged as a "modest shift in the political winds" that would soon create renewed opportunity for meaningful climate policy in the United States, with global reverberations. That faith proved justified. Albeit by the slimmest of margins, voters in the November 2020 election vested full control of our federal government—the presidency and both houses of Congress—in the Democratic Party, a party that, at the very least, acknowledged the climate crisis. There was even a fleeting glimmer of bipartisan support for climate action—a COVID-19 relief bill containing a healthy $35 billion in clean energy funding that passed with overwhelming support from Republicans and Democrats alike.[1]

I wondered if the pandemic, as tragic and deadly as it had proven to be, might have taught us some important lessons. There were lessons about the importance of listening to science, about the fragility of our existence on this planet and the need to find a way to live sustainably upon it, and about the opportunity we might have to change our mindsets as we began to put the COVID-19 crisis behind us and faced the even greater crisis—the climate crisis—that still lay in front of us. Would we come away having learned the right lessons?

Finally, I emphasized the importance of a global youth climate movement that has recentered the climate debate on fundamental issues of intergenerational and distributive justice. The reverberations are still being felt far and wide. There had been a dramatic movement of capital away from fossil fuels toward renewable energy. Some monumental setbacks had emerged for the fossil fuel industry. Consider the rebellion by ExxonMobil stockholders that led to the election of three climate advocates to its board of directors; a court decision that one of the largest oil companies, Royal Dutch Shell, must reduce its carbon emissions by roughly 50 percent within the next decade; and cancellation of the Keystone XL Pipeline (which my colleague James Hansen once warned would be "game over for the climate") after the Biden administration revoked a key permit for the project.[2] And that was just during one week in May 2021.

It's fair to say, overall, that the case for cautious optimism originally laid out in *The New Climate War* remains compelling, even if some serious challenges remain (not the least of which is a Republican-led authoritarian movement that threatens the very foundation of American democracy). Let's look more closely at what has happened and speculate about where we may be headed.

DENIAL DEATH SPIRAL

The march of extreme weather events continued on in 2020 and 2021. A record-breaking thirty Atlantic tropical storms were named in the fall of 2020 at the same time that late-season wildfires continued to blaze across the western United States. A devastating cold blast, which some scientists argued was symptomatic of a warming Arctic and slower, wobblier jet stream, overwhelmed the fossil-fuel-driven power grid in Texas, leading to a deadly power outage and hundreds of lives lost to extreme cold. Meanwhile, sustained winter drought in the western states portended an ominous summer fire season in 2021. And when summer came, it added an exclamation mark to this series of events. At a joint press conference with President Joe Biden in early July, as twin weather disasters played out in the United States and Europe (a deadly "heat dome" over the Pacific Northwest and unprecedented rainfall and flooding in Western Europe), German prime minister Angela Merkel urged dramatic action on climate.

Weather extremes played out across the Northern Hemisphere in the weeks and months that followed.[3] In the western United States, over a million acres caught fire, while in New York City, tropical afternoon thunderstorms turned subway stations into underground rivers and streets into lakes. Deadly floods continued to spread across Europe and Asia. Over two dozen people perished in Mumbai after nearly a foot of rain fell in just twenty-four hours. Tens of millions of people were impacted by devastating flooding in central China that caused twenty-five deaths and delivered a year's worth of rainfall in just three days. Meanwhile, more than two hundred fires burned in Siberia, scorching nearly six thousand square miles, an area larger than the state of Connecticut. These extreme weather disasters and their connection with climate change were for once highlighted in mainstream news coverage. Even Fox News covered it, grudgingly acknowledging that the "sweltering weather and historic drought" in the United States were "exacerbated by climate change."[4]

If it was *implausible* to deny the climate crisis in mid-2020, it was *impossible* in 2021. As devastating climate-fueled disasters continued to pile up, and as fossil-fuel-driven utilities were themselves feeling the crush of extreme weather events, some wondered aloud whether the long-awaited "turning point" on climate had finally arrived.[5] Gone were the congressional show trials so common just a few years earlier when congressional Republicans carried out assaults on the science and the scientists. Most telling was a virtual hearing held by the House of Representatives Homeland Security Subcommittee on Emergency Preparedness, Response and Recovery in early June. It was, for the first time I can recall, impossible to discern, from testimony alone, which witnesses had been invited by the Democrats and which had been invited by the Republicans.[6] I noted with pleasant surprise that Pamela Williams, the executive director of the BuildStrong Coalition (and the sole witness invited by Republican committee members), tied the increasingly damaging weather events to climate change. All the witnesses, Democratic and Republican, affirmed that climate change had reached crisis levels.[7]

Consider one James Taylor, the president of the Koch-funded Heartland Institute, a central cog in the denial machine. In August 2021, Taylor conceded that, "yes, the climate is changing," and that "humans may be playing some role."[8] He continued to insist, in a seemingly desperate effort to salvage some of the delayism agenda, that it was "not an emergency." But the goal posts had moved.

Even the uber-denialist *Wall Street Journal* editorial page seemingly gave up the denialism ghost, instead favoring "new climate war" messaging acknowledging the reality of human-caused climate change but downplaying the threat and the urgency of action.[9] Such a nontrivial concession on the part of a central player in the climate wars signals a substantial, favorable shift in climate politics.

WINDS OF CHANGE

It's fair to say that we have witnessed something a bit more than the merely "modest" favorable shift in political winds anticipated in *The New Climate War*. Having seized full control of the federal government, President Biden and congressional Democrats have made substantial progress on the policy front. The $35 billion green energy stimulus in the December 2020 COVID relief bill negotiated by president-elect Biden and passed on a bipartisan basis by Congress was just an appetizer for what was to come. Within a week of his inauguration, Biden announced the most wide-ranging package of executive actions in history to address the climate crisis, promoting climate-forward policies in every sector of the federal government. He appointed former US senator John Kerry—a climate hawk with enviable diplomatic bona fides, who helped negotiate the 2015 Paris Agreement as Barack Obama's secretary of state—as his special envoy on climate. While Kerry was tasked with reestablishing US climate leadership on the international stage, another climate hawk, former EPA administrator Gina McCarthy, headed up the domestic climate policy agenda as White House national climate adviser.[10]

With renewed US leadership, other countries returned to the negotiating table. China agreed to cooperate with the United States once again, as it had during the Obama years, in addressing the climate crisis with due urgency.[11] At a virtual climate summit he held on Earth Day, April 22, 2021, Biden announced that the United States would cut carbon emissions by more than 50 percent by 2030, a bold pledge that is commensurate with the effort to hold warming below the 1.5°C (2.7°F) danger level. Other nations followed suit. The United Kingdom and the European Union committed to cut carbon emissions by 68 percent and 55 percent, respectively, by 2030. China, the world's largest current emitter, but with a far briefer legacy of fossil fuel burning, has vowed to reach peak emissions by 2030 and achieve carbon neutrality by 2060.[12] Pledges, of course, are promises, not actions. And there remains an "implementation gap" between the commitments these countries have made and the policies they have put in place. That gap does appear to be closing though.

For Biden to make good on his pledge, Congress will have to act. They have now passed a trillion dollar bipartisan infrastructure deal that provides substantial support for climate-friendly measures, including a major investment in passenger rail, a massive expansion of electric vehicle charging stations, and resilience measures aimed particularly at disadvantaged and frontline communities. But that's not enough. As this book went to press, the key climate provisions were still in limbo—including those that would allow the Biden administration to make good on its pledge to cut carbon emissions in half over the next decade (such as those in a far more ambitious "Build Back Better" plan, which could in principle pass Congress through the reconciliation process, on a party-line Democratic vote). The current version of the bill includes a "clean energy standard," wherein utility companies are required to purchase an increasing fraction of electricity from clean sources, setting a goal of 80 percent carbon-free energy by 2030 and 100 percent by 2035. It's not perfect. There's language that supports risky strategies, including nuclear energy and carbon capture. I'm skeptical about these methods, given that safe options—renewable energy along with electrification and storage—can get us where we need to go.[13] But I also believe in not letting the perfect be the enemy of the good, and in the role of realpolitik when confronting the Herculean task of shepherding bold climate legislation through a fifty-fifty divided US Senate.

A *DECENT* RESET

There is both challenge and opportunity in how we choose to collectively respond to the COVID-19 pandemic. The World Economic Forum (WEF) has unpacked it for us: "The Covid-19 crisis, and the political, economic and social disruptions it has caused," it pointed out, "is fundamentally changing the traditional context for decision-making. The inconsistencies, inadequacies and contradictions of multiple systems—from health and financial to energy and education—are more exposed than ever amidst a global context of concern for lives, livelihoods and the planet. Leaders find themselves at a historic crossroads, managing short-term pressures against medium- and long-term uncertainties." The WEF argued that we have a "unique window of opportunity to shape the recovery" and to "build a new social contract that honours the dignity of every human being."[14] A name has even been given to this agenda: "the Great Reset."

As much as I share these hopes and aspirations, I am less sanguine about near-term prospects for a "great" reset. I hope that what we have gone through collectively as a species has instilled in us a sense of humbleness and a recognition that we must strive to live more sustainably on this planet. But that ultimate goal remains elusive, even in the wake of the pandemic. What we *can* expect, I believe, is a "decent" reset.

We're not quite building back "green." It's more like "olive"— somewhere between green and brown. Globally, for example, only 10 percent of the roughly $17 trillion global bailout funds have gone toward efforts to cut carbon emissions and address environmental degradation.[15] Although there is movement in the right direction and meaningful progress is being made, expansive "green new deal"–like legislation is nowhere on the horizon here in the United States. As I warned in the first edition of *The New Climate War*, only narrower, more targeted climate legislation, relying mostly on demand-side market mechanisms (for example, the "clean energy standard" in the 2021 infrastructure package), can likely become law in the current, closely divided US political environment.

Compromise and realpolitik emerge, too, in the actions the Biden administration has (or more aptly, *hasn't*) taken on the *supply side* of the problem. To the disappointment of many climate advocates, the administration's actions have been decidedly mixed when it comes to the approval of new fossil fuel infrastructure. Joel Clement, a whistleblower who resigned from the US Department of the Interior under Trump in protest of the administration's climate obstructionism, has gently critiqued the Biden administration as well. He noted, for example, that although the Biden administration has suspended controversial oil and gas leases in the Arctic National Wildlife Refuge (and recall that it also blocked the construction of the Keystone XL Pipeline), "oil and gas permitting at the Interior Department has continued apace, if not increased: Multiple new fossil energy projects have gotten the greenlight, and early messaging from Interior Secretary Deb Haaland suggests that the existing ban on new leasing may not stand."[16] Clement suggested that these actions reflected a compromise, on the part of the Biden administration, aimed at appeasing moderate oil-state Republicans, such as Senator Lisa Murkowski (R-AK), who are still beholden to fossil fuel interests, and that such a compromise may have been needed to get some of Biden's larger agenda items through Congress.

What about the lessons that COVID-19 offered us about the dangers—indeed, the deadliness—of science denial, with all its obvious implications for climate-change denialism? Hundreds of thousands of Americans had died, after all, because of a president, Donald Trump (along with enabling Republican politicians and right-wing news outlets), that downplayed the threat posed by the virus and actively worked to undermine public faith in the messaging from public health officials. Surely even MAGA-hat-wearing Trump supporters would come around as they witnessed friends, acquaintances, and family members suffer and succumb to the pandemic.

That was not to be, however. The shadowy, conspiracy-theory-promoting "QAnon" movement, in concert with malicious state actors such as Russia and conservative media outlets, continued to target political conservatives in the United States, spreading disinformation through social media to sow doubt and confusion with regard to the efficacy of social distancing, mask-wearing, and even the vaccine.[17] Refusing the vaccine has become a purity test for one's allegiance to Donald Trump. It is now apparently worth dying to "own the libs."[18]

So it is a tale of two nations: Blue State America, where the word of public health officials was heeded, vaccination rates remained high, and the virus was largely contained; and Red State America, where right-wing disinformation ran rampant, social distancing and mask-wearing were lacking, vaccination rates remained low, and COVID-19 cases continued to climb. The great Carl Sagan's dreaded prophecy has now been at least half realized.

In his 1996 book *The Demon-Haunted World*, Sagan warned, "I have a foreboding of an America in my children's or grandchildren's time—when the United States is a service and information economy; when nearly all the key manufacturing industries have slipped away to other countries; when awesome technological powers are in the hands of a very few, and no one representing the public interest can even grasp the issues; we slide, almost without noticing, back into superstition and darkness." Sagan added, "The dumbing down of America is most evident in the slow decay of substantive content in the enormously influential media . . . lowest common denominator programming, credulous presentations on pseudoscience and superstition, *but especially a kind of celebration of ignorance*." That "celebration of ignorance" has begun.

I've said that Carl Sagan predicted QAnon more than two decades ago.[19] Perhaps as much as half of America has been gaslighted into favoring conspiracy theories and fantasy over fact. And with QAnon, malign state actors, and right-wing media collaborating in an effort to cast doubt on the threat from climate change, it is clear that many conservatives will not be part of the sort of awakening envisioned in a *great* reset.[20] But a *decent* reset is still possible without them.

A BURSTING CARBON BUBBLE

Another reason for cautious optimism is the bursting of the "carbon bubble" that now appears to be underway. We can thank the youth climate movement and its leaders—Greta Thunberg and so many others—in substantial part for this development. Their incessant, urgent messaging to power brokers and opinion leaders seems to have finally broken through.

Consider what transpired with ExxonMobil, the world's largest publicly traded fossil fuel company, in May 2021. ExxonMobil shareholders went against the recommendation of the company and elected no less than three climate advocates (belonging to the activist hedge fund "Engine No. 1") to their board of directors. As CNN described it, "Upset with Exxon's financial performance and its foot-dragging on climate, the hedge fund sought to oust four directors at the company's annual shareholder meeting." With apologies to Meat Loaf, "three out of four ain't bad." CNN said it was "a major milestone in the battle against climate change."[21] In another major blow for the fossil fuel industry that same week, a Dutch court ordered Royal Dutch Shell, the world's fourth-largest fossil fuel company, to cut its greenhouse gas emissions by 45 percent (relative to 2019 levels) by 2030.[22]

These are hardly the only challenges the fossil fuel industry now faces. *The New Climate War* reviewed the dramatic movement of capital by investment and finance institutions away from the funding of new fossil fuel infrastructure. That trend has accelerated in the short time since the publication of the first edition. In 2020 alone, fossil fuel financing dropped by a whopping 9 percent, while investment in clean energy technology rose by the same amount.[23] Experts cited worries about stranded fossil fuel assets and a preference for "greener markets" on the part of investors.[24]

In May 2021, the rather conservative International Energy Agency (IEA), known for its somewhat bearish stance on the renewable energy transition, issued a shocking proclamation: that it is still possible to decarbonize our economy on the time frame necessary to avoid dangerous planetary warming, but there *can be no new investment in fossil fuel infrastructure*.[25] The pronouncement generated shock waves felt by international governments around the world. Just days later, the Group of Seven, or G7, an intergovernmental organization comprising the world's major industrial economies, announced that it would no longer support financing of new coal infrastructure.[26]

The IEA report made clear, however, that nations must go further—canceling investments in new natural gas and oil projects as well. Given that more than half of the pandemic economic recovery funds for G7 countries during 2020 and early 2021 were earmarked as aid to the fossil fuel industry, it's clear that there is considerable work still to be done to narrow the gap between aspiration and action.[27]

THE RIGHT PATH FORWARD

Toward the end of *The New Climate War*, I wrote, "We are at a critical juncture. Despite the obvious political challenges we currently face, we are witnessing an alignment of historical and political events—and acts of Mother Nature—that are awakening us to the reality of the climate crisis. We appear to be nearing the much-anticipated tipping point on climate action." I still believe that to be true.

The dangers posed by climate change have become obvious to the public. Political winds have shifted in a favorable direction. An unprecedented global pandemic has refocused attention on the sustainability of our civilization. A global youth climate movement has placed heightened pressure on decision-makers and power brokers. The carbon bubble is bursting. Never before has such an alignment of factors favored climate action.

Yet I'm concerned that one of the primary obstacles to action I cited in the book—the advocacy of false solutions—poses an even greater threat today. Just weeks after *The New Climate War* was published, another book—by none other than Microsoft founder and premier tech-lord Bill Gates (*How to Avoid a Climate Disaster*)—appeared on the scene.[28] While I welcome someone of Gates's profile using his considerable platform to raise awareness of the climate crisis, I worry that he has offered the wrong prescription for what ails us.

Gates, in his book, is overly dismissive of the role that renewable energy can play in decarbonizing our economy, despite leading experts in the field of energy economics having argued that we can meet 80 percent of energy demand by 2030, and 100 percent by 2050, with *existing* renewable energy technology. What is needed is political willpower, not a technological "miracle," as Gates has asserted. Gates wrote, "I don't have a solution to the politics of climate change." But the primary remaining obstacles are not fundamentally technological in nature at this point. They *are* political.

As they say, if your only tool is a hammer, everything looks like a nail. So it's not especially surprising that Gates advocates for a technocratic approach to addressing the climate crisis. But in downplaying the role that existing renewable energy technologies can play, he instead argues for very risky strategies, such as "geoengineering" and "next-generation" (i.e., not yet existing) nuclear energy. It's worth noting that Gates has personal investments in both of these technologies.[29] Gates also promotes technologies such as carbon capture that haven't been proven at scale, but provide a convenient argument for polluters and their enablers, who are looking to kick the carbon can down the road with the equivalent of modern-day indulgences (i.e., "We'll pollute now, and clean up later—trust us!").

That's simply not the right path forward.[30] As the *New Statesman*, a UK magazine, put it, "Bill Gates's faith in a technological fix for climate change is typical of privileged men who think they can swoop in and solve the problems others have spent decades trying to fix."[31] Gates was able to take advantage of that very privilege in achieving best-seller status and saturation coverage of his book, where he continued to promote his flawed technocratic framing of the climate crisis.[32] He received an assist from power brokers and influentials. Much to the chagrin of some climate advocates, former president Barack Obama promoted Gates's book shortly after it appeared, and Gates was given a prime speaking role at President Joe Biden's virtual Earth Day summit.[33]

The damage could be seen in the talking points now being adopted by the nascent Biden administration. John Kerry—Biden's special envoy on climate—falsely asserted during an interview in March 2021, to the dismay of more informed observers, that half of all carbon reductions between now and 2050 would have to come from technologies that don't yet exist. He thus implied that much of the reduction would have to come from the sorts of untested and risky technologies—next-gen nuclear, geoengineering, carbon capture—that Gates had promoted.[34]

Kerry said, "I am told by scientists that 50% of the reductions we have to make to get to net zero are going to come from technologies that we don't yet have. That's just a reality." No, John, that's a *falsehood*. Leading experts say that we can get there with safe, existing renewable energy technologies, conservation, storage, and smart grid technology.[35] It's possible that Kerry and his staff simply misinterpreted a recent IEA report. As I noted in a tweet, the IEA had argued that in 2050, 50 percent of *contemporaneous* reductions would likely come from what we currently consider "future tech" (after all, a lot of R&D and progress will take place during the intervening three decades!).[36] But this does not remotely support Kerry's assertion that 50 percent of the *cumulative reductions* through 2050 will come from future technology. The vast majority of those reductions, including—importantly—the 50 percent reductions needed over the next decade, will come from existing technology.

But it feels like there is a pile on now when it comes to this idea that we must rely on risky and/or unproven future tech in decarbonizing our economy. Intransigent politicians, such as Australia's Scott Morrison, are now using the promise of massive carbon capture decades down the road to justify a policy of near-term inaction, best embodied by the "gas-led" economic recovery he has promised.[37] Some of our leading institutions, including Harvard University and the US National Academy of Sciences, have lent their imprimatur to the geoengineering gambit.

Consider the academy's recent report (*Reflecting Sunlight*) on schemes aimed at reducing incoming solar radiation by, for example, shooting sulfate pollution into the stratosphere.[38] It is one thing to say that we should study and understand such geoengineering proposals. In the report, however, the authors go much further. They appear to justify not just their study but indeed their *implementation*, stating that "despite overwhelming evidence that the climate crisis is real and pressing, emissions of greenhouse gases continue to increase. . . . *The pandemic is thus providing frustrating confirmation of the fact that the world has made little progress in separating economic activity from carbon dioxide emissions* (emphasis added)."

In an op-ed I coauthored with Oxford University climate scientist Ray Pierrehumbert, we pointed out "a fundamental misconception" in the report's presumption that "we likely won't achieve the necessary decarbonization of our economy in time to avoid massive climate damages, so this technology might be needed." We noted that this is the "worst possible justification for developing solar geo-engineering technology," as it "is laden in moral hazard—providing, as it does, an excuse for fossil fuel interests and their advocates to continue with business as usual. Why reduce carbon pollution if there is a cheap workaround?"[39]

Other proposals, such as those involving massive carbon capture and sequestration, might not come with quite the same perils, but they still fall victim to the fundamental problem of moral hazard, namely, the promise of a future technofix that may or may not work in exchange for permission to continue to pollute now. Companies that have promised massive capture of fossil fuel emissions, such as Chevron, have recently conceded that they are unable to deliver on that promise.[40] And the only carbon capture coal-fired power plant in the United States shuttered its doors in early 2021, after failing to achieve viable, cost-effective capture of carbon pollution.[41] Nothing could be more foolish than to place all of our eggs in the geoengineering and carbon capture basket only to come up short a decade or two from now, with no recourse because we've already burned through our carbon budget.

We have just under a decade to halve global carbon emissions if we are to remain on course for keeping warming below the truly dangerous levels. We cannot afford any dead ends or wrong turns. The answer is simple: We've got to stop burning fossil fuels as soon as possible. We need policies and politicians that will make that happen. The Glasgow climate summit (COP26) of November 2021 didn't yield everything climate advocates would have liked, such as a ban on new fossil fuel infrastructure and an end to all subsidies for fossil fuels. But with renewed US leadership on climate, substantial progress was made. New commitments from major emitters, such as India, may potentially cut warming in half (under 2°C) relative to where we were headed (roughly 4°C) going into the 2016 Paris climate summit. There is of course still much more work that must be done, both in getting countries to rachet up their commitments to bring warming below the 1.5°C danger level, and, most important of all, getting governments to institute policies that can actually get us there.[42]

As individuals, we must use our voices, in every way possible, to demand that politicians act on our behalf rather than at the behest of polluters. We must vote in politicians who are committed to climate action, and vote out those who aren't. We must, in short, win the new climate war. Failure isn't an option. The future of our planet hangs in the balance. This is our time. Let's make it happen, readers.

Acknowledgments

I am grateful to the many individuals who have provided help and support over the years. First and foremost are my family: my wife, Lorraine; daughter, Megan; parents, Larry and Paula; brothers, Jay and Jonathan; and the rest of the Manns, Sonsteins, Finesods, and Santys.

I give special thanks to Bill Nye, "The Science Guy," for his friendship, leadership, inspiration, and support over the years.

I am indebted to all those who have inspired me, mentored me, and served as role models, including, but not limited to, Carl Sagan, Stephen Schneider, Jane Lubchenco, John Holdren, Paul Ehrlich, Donald Kennedy, Warren Washington, and Susan Joy Hassol. I thank leaders of the youth climate movement, including Greta Thunberg, Alexandria Villaseñor, Jerome Foster, and Jamie Margolin, for the inspiration they have provided.

I am greatly indebted to the various politicians on both sides of the political spectrum who have stood up against powerful interests to support and defend me and other scientists against politically motivated attacks and who have worked to advance the cause of an informed climate policy discourse. Among them are Sherwood Boehlert, Jerry Brown, Bob Casey Jr., Bill Clinton, Hillary Clinton,

Peter Garrett, Al Gore, Mark Herring, Bob Inglis, Jay Inslee, Edward Markey, Terry McAuliffe, John McCain, Christine Milne, Jim Moran, Alexandria Ocasio-Cortez, Harry Reid, Bernie Sanders, Arnold Schwarzenegger, Arlen Specter, Malcolm Turnbull, Henry Waxman, Sheldon Whitehouse, and their various staffs.

I thank my colleagues in the Departments of Meteorology and Geosciences and elsewhere at Penn State for the supportive environment they have helped foster. Among them are President Eric Barron, former president Graham Spanier, Dean Lee Kump, Earth and Environmental Systems Institute (EESI) director Sue Brantley, department head David Stensrud, and former department head Bill Brune, as well as the ever-helpful staffs in the Department of Meteorology, the College of Earth and Mineral Sciences, and EESI.

I also want to thank my agents, Jodi Solomon, Rachel Vogel, and Suzi Jamil; PublicAffairs editorial staff members Colleen Lawrie, Jeff Alexander, and Brynn Warriner; publicity team Brooke Parsons and Miguel Cervantes; and copyeditor Kathy Streckfus for all their hard work and support with this project.

I wish to thank the various other friends, supporters, and colleagues past and present for their assistance, collaboration, friendship, and inspiration over the years, including John Abraham, Kylie Ahern, John Albertson, Ken Alex, Richard Alley, Yoca Arditi-Rocha, Ed Begley Jr., Lew Blaustein, Doug Bostrom, Christian Botting, Ray Bradley, Sir Richard Branson, Jonathan Brockopp, Bob Bullard, James Byrne, Mike Cannon-Brookes, Elizabeth Carpino, Nick Carpino, Keya Chaterjee, Noam Chomsky, Kim Cobb, Ford Cochran, Michel Cochran, Julie Cole, John Collee, Jennifer Collins, Leila Conners, John Cook, Jason Cronk, Jen Cronk, Heidi Cullen, Hunter Cutting, Greg Dalton, Fred Damon, Kert Davies, Didier de Fontaine, Brendan Demelle, Andrew Dessler, Steve D'Hondt, Henry Diaz, Leonardo DiCaprio, Paulo D'Oderico, Pete Dominick, Finis Dunaway, Brian Dunning, Andrea Dutton, Bill Easterling, Matt England, Kerry Emanuel, Howie Epstein, Jenni Evans, Dan Fagan, Morgan Fairchild, Chris Field, Frances Fisher, Patrick Fitzgerald, Pete Fontaine, Josh Fox, Jennifer Francis, Al Franken, Peter Frumhoff, Jose Fuentes, Andra Garner, Peter Garrett,

Peter Gleick, Jeff Goodell, Amy Goodman, Nellie Gorbea, David Graves, David Grinspoon, Genevieve Guenther, David Halpern, Thom Hartmann, David Haslingden, Susan Joy Hassol, Katharine Hayhoe, Tony Haymet, Megan Herbert, Michele Hollis, Rob Honeycutt, Ben Horton, Lee Hotz, Malcolm Hughes, Jan Jarrett, Paul Johansen, Phil Jones, Jim Kasting, Bill Keene, Sheril Kirshenbaum, Barbara Kiser, Johanna Köb, Jonathan Koomey, Kalee Kreider, Paul Krugman, Lauren Kurtz, Deb Lawrence, Stephan Lewandowsky, Diccon Loxton, Scott Mandia, Joseph Marron, John Mashey, Roger McConchie, Andrea McGimsey, Bill McKibben, Pete Meyers, Sonya Miller, Chris Mooney, John Morales, Granger Morgan, Ellen Mosely-Thompson, Ray Najjar, Giordano Nanni, Jeff Nesbit, Phil Newell, Gerald North, Dana Nuccitelli, Miriam O'Brien, Michael Oppenheimer, Naomi Oreskes, Tim Osborn, Jonathan Overpeck, Lisa Oxboel, Rajendra Pachauri, Blair Palese, David Paradice, Rick Piltz, James Powell, Stefan Rahmstorf, Cliff Rechtschaffen, Hank Reichman, Ann Reid, Catherine Reilly, James Renwick, Tom Richard, David Ritter, Alan Robock, Lyndall Rowley, Mark Ruffalo, Scott Rutherford, Sasha Sagan, Jim Salinger, Barry Saltzman, Ben Santer, Julie Schmid, Gavin Schmidt, Steve Schneider, John Schwartz, Eugenie Scott, Joan Scott, Drew Shindell, Randy Showstack, Hank Shugart, David Silbert, Peter Sinclair, Dave Smith, Jodi Solomon, Richard Somerville, Amanda Staudt, Alex Steffen, Eric Steig, Byron Steinman, Nick Stokes, Sean Sublette, Larry Tanner, Jake Tapper, Lonnie Thompson, Sarah Thompson, Kim Tingley, Dave Titley, Tom Toles, Lawrence Torcello, Kevin Trenberth, Fred Treyz, Leah Tyrrell, Katy Tur, Ana Unruh-Cohen, Ali Velshi, Dave Verardo, David Vladeck, Nikki Vo, Bob Ward, Bud Ward, Bill Weir, Ray Weymann, Robert Wilcher, John B. Williams, Barbel Winkler, Christopher Wright.

Notes

INTRODUCTION

1. Neela Banerjee, Lisa Song, and David Hasemyer, "Exxon: The Road Not Taken. Exxon's Own Research Confirmed Fossil Fuels' Role in Global Warming Decades Ago," *Inside Climate News*, September 16, 2015, https://insideclimatenews.org/news/15092015/Exxons-own-research-confirmed-fossil-fuels-role-in-global-warming.

2. Naomi Oreskes and Erik M. Conway, *Merchants of Doubt: How a Handful of Scientists Obscured the Truth on Issues from Tobacco Smoke to Global Warming* (New York: Bloomsbury Press, 2010), 6.

3. David Hagmann, Emily Ho, and George Loewenstein, "Nudging Out Support for a Carbon Tax," *Nature Climate Change* 9, no. 6 (2019): 484–489.

4. M. E. Mann, R. S. Bradley, and M. K. Hughes, "Global-Scale Temperature Patterns and Climate Forcing over the Past Six Centuries," *Nature* 392 (1998): 779–787.

5. Michael E. Mann (@MichaelEMann), Twitter, April 5, 2020, 2:53 p.m., https://twitter.com/MichaelEMann/status/1246918911989334016.

CHAPTER 1: THE ARCHITECTS OF MISINFORMATION AND MISDIRECTION

1. Rachel Shteir, "Ibsen Wrote 'An Enemy of the People' in 1882. Trump Has Made It Popular Again," *New York Times*, March 9, 2018, www.nytimes.com/2018/03/09/theater/enemy-of-the-people-ibsen.html.

2. David Michaels, *Doubt Is Their Product* (New York: Oxford University Press, 2008).

3. See Mark Hertsgaard, "While Washington Slept," *Vanity Fair*, May 2006; "Hot Politics," PBS Frontline, April 3, 2006, www.pbs.org/wgbh/pages /frontline/hotpolitics/interviews/seitz.html; "Smoke, Mirrors and Hot Air: How ExxonMobil Uses Big Tobacco's Tactics to Manufacture Uncertainty on Climate Science," Union of Concerned Scientists, January 2007, www.ucsusa .org/sites/default/files/2019-09/exxon_report.pdf.

4. Rachel Carson, *Silent Spring* (Boston: Houghton Mifflin, 1962).

5. Quoted in Christopher J. Bosso, *Pesticides and Politics: The Life Cycle of a Public Issue* (Pittsburgh: University of Pittsburgh Press, 1987), 116.

6. Naomi Oreskes and Eric M. Conway, *Merchants of Doubt: How a Handful of Scientists Obscured the Truth on Issues from Tobacco Smoke to Global Warming* (New York: Bloomsbury Press, 2010).

7. "Rachel Carson's Dangerous Legacy," Competitive Enterprise Institute, SAFEChemicalPolicy.org, March 1, 2007, www.rachelwaswrong.org.

8. Clyde Haberman, "Rachel Carson, DDT and the Fight Against Malaria," *New York Times*, January 22, 2017.

9. Robin McKie, "Rachel Carson and the Legacy of Silent Spring," *The Guardian*, May 26, 2012.

10. "Henry I. Miller," Competitive Enterprise Institute, https://cei.org /adjunct-scholar/henry-i-miller; "Gregory Conko Returns to CEI as Senior Fellow," Competitive Enterprise Institute, https://cei.org/content/gregory-conko -returns-cei-senior-fellow.

11. Henry I. Miller and Gregory Conko, "Rachel Carson's Deadly Fantasies," *Forbes*, September 5, 2012, reprinted at Heartland Institute, www.heartland .org/_template-assets/documents/publications/rachel_carsons_deadly _fantasies_-_forbes.pdf.

12. For a description of a recent peer-reviewed article demonstrating the threat to bird populations associated with a class of pesticides known as neonicotinoids, see "Controversial Insecticides Shown to Threaten Survival of Wild Birds," *Science Daily*, September 12, 2019, www.sciencedaily.com /releases/2019/09/190912140456.htm.

13. Joseph Palca, "Get-the-Lead-Out Guru Challenged," *Science* 253 (1991): 842–844.

14. Benedict Carey, "Dr. Herbert Needleman, Who Saw Lead's Wider Harm to Children, Dies at 89," *New York Times*, July 27, 2017.

15. While the group no longer lists its funders, a screen capture from some twenty years ago shows present or past funding from Exxon as well as the Sarah Scaife Foundation. See "George C. Marshall Institute: Recent Funders," George C. Marshall Institute, on Internet Archive Wayback Machine, 14 captures from August 23, 2000, to December 17, 2004, http://web.archive.org /web/20000823170917/www.marshall.org/funders.htm.

16. Oreskes and Conway, *Merchants of Doubt*, 6–7, 249. Oreskes and Conway used the term "free-market fundamentalism" to describe blind faith in the

ability of the free market to solve any problem without the need for government intervention.

17. "Whatever Happened to Acid Rain?," *Distillations* radio program, hosted by Alexis Pedrick and Elisabeth Berry Drago, Science History Institute, May 22, 2018, www.sciencehistory.org/distillations/podcast/whatever-happened-to-acid-rain.

18. Osha Gray Davidson, "From Tobacco to Climate Change, 'Merchants of Doubt' Undermined the Science," *Grist*, April 17, 2010, https://grist.org/article/from-tobacco-to-climate-change-merchants-of-doubt-undermined-the-science/full; Orestes and Conway, *Merchants of Doubt*, 82.

19. William H. Brune, "The Ozone Story: A Model for Addressing Climate Change?," *Bulletin of the Atomic Scientists* 71, no. 1 (2015): 75–84.

20. See Andrew Kaczynski, Paul LeBlanc, and Nathan McDermott, "Senior Interior Official Denied There Was an Ozone Hole and Compared Undocumented Immigrants to Cancer," CNN, October 8, 2019, www.cnn.com/2019/10/08/politics/william-perry-pendley-blm-kfile/index.html.

21. Lee R. Kump, James F. Kasting, and Robert G. Crane, *The Earth System*, 2nd ed. (New York: Pearson, 2004).

22. Sasha Sagan's first book, *For Small Creatures Such as We: Rituals for Finding Meaning in Our Unlikely World*, was published in October 2019 by G. P. Putnam's Sons. The publisher describes it as "a luminous exploration of Earth's marvels that require no faith in order to be believed."

23. Keay Davidson, *Carl Sagan: A Life* (Hoboken, NJ: John Wiley and Sons, 1999).

24. Sagan coauthored a peer-reviewed article in the leading scientific journal *Science* on December 23, 1983, presenting the scientific case for nuclear winter based on computer modeling. The article has become known as "TTAPS," after the first letters of the various authors' last names. The full reference is R. P. Turco, O. B. Toon, T. P. Ackerman, J. B. Pollack, and Carl Sagan, "Nuclear Winter: Global Consequences of Multiple Nuclear Explosions," *Science* 222, no. 4630 (1983): 1283–1292.

25. In *Merchants of Doubt*, Oreskes and Conway detailed how cold war hawks distrusted the scientists questioning the efficacy and appropriateness of developing missile defense systems like the Reagan administration's Strategic Defense Initiative.

26. Oreskes and Conway, *Merchants of Doubt*, chap. 2.

CHAPTER 2: THE CLIMATE WARS

1. See Mark Bowen, *Censoring Science: Inside the Political Attack on Dr. James Hansen and the Truth of Global Warming* (New York: Dutton, 2007), 224–227.

2. Michael Mann, *The Hockey Stick and the Climate Wars: Dispatches from the Front Lines* (New York: Columbia University Press, 2013).

3. Neela Banerjee, Lisa Song, and David Hasemyer, "Exxon: The Road Not Taken. Exxon's Own Research Confirmed Fossil Fuels' Role in Global Warming Decades Ago," *Inside Climate News*, September 16, 2015, https://insideclimatenews.org/news/15092015/Exxons-own-research-confirmed -fossil-fuels-role-in-global-warming.

4. See Kyla Mandel, "Exxon Predicted in 1982 Exactly How High Global Carbon Emissions Would Be Today," *Climate Progress*, May 14, 2019, https://thinkprogress.org/exxon-predicted-high-carbon-emissions-954e514b0aa9. The internal Exxon report, dated November 12, 1982, with the subject line "CO_2 'Greenhouse' Effect," 82EAP 266, under the Exxon letterhead of M. B. Glaser, manager, Environmental Affairs Program, was obtained by *Inside Climate News* and is posted at http://insideclimatenews.org/sites/default /files/documents/1982%20Exxon%20Primer%20on%20CO2%20Greenhouse %20Effect.pdf.

5. Élan Young, "Coal Knew, Too: A Newly Unearthed Journal from 1966 Shows the Coal Industry, Like the Oil Industry, Was Long Aware of the Threat of Climate Change," *Huffington Post*, November 22, 2019, www.huffpost.com /entry/coal-industry-climate-change_n_5dd6bbebe4b0e29d7280984f.

6. Naomi Oreskes and Eric M. Conway, *Merchants of Doubt: How a Handful of Scientists Obscured the Truth on Issues from Tobacco Smoke to Global Warming* (New York: Bloomsbury Press, 2010), 186.

7. Jane Mayer, "'Kochland' Examines the Koch Brothers' Early, Crucial Role in Climate-Change Denial," *New Yorker*, August 13, 2019, www.newyorker .com/news/daily-comment/kochland-examines-how-the-koch-brothers-made -their-fortune-and-the-influence-it-bought.

8. According to Ross Gelbspan, Lindzen admitted in an interview to receiving (as of 1995) roughly $10,000 per year from fossil fuel indus-try consulting alone. Gelbspan noted that Lindzen "charges oil and coal interests $2,500 a day for his consulting services; his 1991 trip to testify before a Senate committee was paid for by Western Fuels, and a speech he wrote, entitled 'Global Warming: The Origin and Nature of Alleged Scien-tific Consensus,' was underwritten by OPEC." Ross Gelbspan, "The Heat Is On: The Warming of the World's Climate Sparks a Blaze of Denial," *Harp-er's*, December 1995, https://harpers.org/archive/1995/12/the-heat-is-on. Lindzen's official bio lists him as a member of the Science and Economic Advisory Council of the Annapolis Center for Science-Based Public Policy. "Richard S. Lindzen," Independent Institute, www.independent.org/aboutus /person_detail.asp?id=1215. According to the *DeSmog* blog, the Annapo-lis organization received funding from ExxonMobil. "Annapolis Center for Science-Based Public Policy," *DeSmog* (blog), n.d., www.desmogblog.com /annapolis-center-science-based-public-policy.

9. Mann, *The Hockey Stick and the Climate Wars*.

10. See Gelbspan, "The Heat Is On," 46–47; James Hoggan, with Richard Littlemore, *Climate Cover-Up: The Crusade to Deny Global Warming* (Vancou-ver: Greystone, 2009), 30, 80, 138–140, 156–157.

11. Keith Hammond, "Wingnuts in Sheep's Clothing," *Mother Jones*, December 4, 1997, www.motherjones.com/politics/1997/12/wingnuts-sheeps-clothing. See also J. Justin Lancaster, "The Cosmos Myth," OSS: Open Source Systems, Science, Solutions, updated July 6, 2006, http://ossfoundation.us/projects/environment/global-warming/myths/revelle-gore-singer-lindzen; "A Note About Roger Revelle, Justin Lancaster and Fred Singer," *Rabett Run* (blog), September 13, 2004, http://rabett.blogspot.com/2014/09/a-note-about-roger-revelle-julian.html (contains a comment by Justin Lancaster stating his views about these matters).

12. S. Marshall, M. E. Mann, R. Oglesby, and B. Saltzman, "A Comparison of the CCM1-Simulated Climates for Pre-Industrial and Present-Day CO_2 Levels," *Global and Planetary Change* 10 (1995): 163–180.

13. An excellent account of these events can be found in Chapter 6 of Oreskes and Conway, *Merchants of Doubt*.

14. See Intergovernmental Panel on Climate Change, *IPCC Second Assessment: Climate Change 1995*, available at IPCC, www.ipcc.ch/site/assets/uploads/2018/06/2nd-assessment-en.pdf.

15. Mann, *The Hockey Stick and the Climate Wars*, 181–183.

16. These events are summarized in Mann, *The Hockey Stick and the Climate Wars*.

17. Fred Pearce, "Climate Change Special: State of Denial," *New Scientist*, November 2006.

18. "Kyoto Protocol to the United Nations Framework Convention on Climate Change," Chapter 27, section 7.a, December 11, 1997, United Nations Treaty Collection, https://treaties.un.org/pages/ViewDetails.aspx?src=TREATY&mtdsg_no=XXVII-7-a&chapter=27.

19. See Daniel Engber, "The Grandfather of Alt-Science," *FiveThirtyEight*, October 12, 2017, https://fivethirtyeight.com/features/the-grandfather-of-alt-science.

20. "Skepticism About Skeptics," *Scientific American*, August 23, 2006.

21. M. E. Mann, R. S. Bradley, and M. K. Hughes, "Global-Scale Temperature Patterns and Climate Forcing over the Past Six Centuries," *Nature* 392 (1998): 779–787.

22. See "The Environment: A Cleaner, Safer, Healthier America," Luntz Research Companies, reproduced at SourceWatch, www.sourcewatch.org/images/4/45/LuntzResearch.Memo.pdf.

23. J. T. Houghton, Y. Ding, D. J. Griggs, M. Noguer, P. J. van der Linden, X. Dai, K. Maskell, and C. A. Johnson, "Climate Change 2001: The Scientific Basis," Contribution of Working Group I to the Third Assessment Report of the Intergovernmental Panel on Climate Change, 2001, www.ipcc.ch/site/assets/uploads/2018/03/WGI_TAR_full_report.pdf.

24. Mann, *The Hockey Stick and the Climate Wars*.

25. In 2012, a team of seventy-eight leading paleoclimate scientists representing the PAGES 2k Consortium, drawing on the most widespread paleoclimate database that has been assembled, published a new reconstruction

of large-scale temperature trends. PAGES 2k Consortium, "Continental-Scale Temperature Variability During the Past Two Millennia," *Nature Geoscience* 6 (2013): 339–346, https://doi.org/10.1038/NGEO1797. They concluded that global temperatures have reached their greatest levels in at least 1,300 years. A direct comparison by a German paleoclimatologist reveals their temperature reconstruction to be virtually identical to the original hockey-stick reconstruction. Stefan Rahmstorf, "Most Comprehensive Paleoclimate Reconstruction Confirms Hockey Stick," *Think Progress*, July 8, 2013, https://archive.thinkprogress.org/most-comprehensive-paleoclimate-reconstruction-confirms-hockey-stick-e7ce8c3a2384. In 2019, a group of scientists extended the conclusion, with similar levels of confidence, back to the past two thousand years. See, for example, George Dvorsky, "Climate Shifts of the Past 2,000 Years Were Nothing Like What's Happening Today," *Gizmodo*, July 24, 2019, https://earther.gizmodo.com/climate-shifts-of-the-past-2-000-years-were-nothing-lik-1836662680. A more tentative pair of peer-reviewed studies has extended this conclusion back to at least the past twenty thousand years. See, for example, "Real Skepticism About the New Marcott 'Hockey Stick,'" *Skeptical Science*, April 10, 2013, https://skepticalscience.com/marcott-hockey-stick-real-skepticism.html.

26. Consider a recent article in the right-wing *Telegraph* of London: Sarah Knapton, "Climate Change: Fake News or Global Threat? This Is the Science," *Telegraph*, October 15, 2019, www.telegraph.co.uk/science/2019/10/15/climate-change-fake-news-global-threat-science. This article promoted a number of discredited criticisms advanced by climate-change deniers against the hockey stick. A panel of experts evaluated the article for the independent climate media watchdog group Climate Feedback and gave the article a negative rating for accuracy, noting the inaccurate claims made about the hockey stick. "Telegraph Article on Climate Change Mixes Accurate and Unsupported, Inaccurate Claims, Misleads with False Balance," Climate Feedback, October 18, 2019, https://climatefeedback.org/evaluation/telegraph-article-misleads-with-false-balance-mixing-in-unsupported-and-inaccurate-claims-sarah-knapton.

27. Carl Sagan, *The Demon-Haunted World* (New York: Random House, 1996).

28. Michael Mann and Tom Toles, *The Madhouse Effect: How Climate Change Denial Is Threatening Our Planet, Destroying Our Politics, and Driving Us Crazy* (New York: Columbia University Press, 2016).

29. These events are described in detail in Mann, *The Hockey Stick and the Climate Wars*, chap. 14.

30. Mann, *The Hockey Stick and the Climate Wars*.

31. Kenneth Li, "Alwaleed Backs James Murdoch," *Financial Times*, January 22, 2010, www.ft.com/content/c33aad22-06c8-11df-b058-00144feabdc0; Sissi Cao, "Longtime Murdoch Ally, Saudi Prince Dumps $1.5B Worth of Fox Shares," *Observer*, November 9, 2017, https://observer.com/2017/11/longtime-murdoch-ally-saudi-prince-dumps-1-5b-worth-of-fox-shares.

32. See Iggy Ostanin, "Exclusive: 'Climategate' Email Hacking Was Carried Out from Russia, in Effort to Undermine Action on Global Warming," *Medium*, June 30, 2019, https://medium.com/@iggyostanin/exclusive-climategate

-email-hacking-was-carried-out-from-russia-in-effort-to-undermine-action
-78b19bc3ca5a.

33. The evidence is laid out in Mann and Toles, *Madhouse Effect*, 164–166.

34. "Putin Says Climate Change Is Not Man-Made and We Should Adapt to It, Not Try to Stop It," Agence France-Presse, March 31, 2017, www.scmp .com/news/world/russia-central-asia/article/2083650/trump-vladimir-putin -says-climate-change-not-man-made.

35. Jonathan Watts and Ben Doherty, "US and Russia Ally with Saudi Arabia to Water Down Climate Pledge," *The Guardian*, December 9, 2018, www.theguardian.com/environment/2018/dec/09/us-russia-ally-saudi-arabia -water-down-climate-pledges-un.

36. "Special Report: Global Warming of 1.5°C," Intergovernmental Panel on Climate Change (IPCC), October 8, 2018, www.ipcc.ch/sr15.

37. Daniel Dale, "Lies, Lies, Lies: How Trump's Fiction Gets More Dramatic over Time," CNN, October 27, 2019, www.cnn.com/2019/10/27/politics /fact-check-lies-trump-fiction-obama-kim-jong-un/index.html.

38. Abel Gustafson, Anthony Leiserowitz, and Edward Maibach, "Americans Are Increasingly 'Alarmed' About Global Warming," Yale Program on Climate Change Communication, February 12, 2019, https://climate communication.yale.edu/publications/americans-are-increasingly-alarmed -about-global-warming.

39. Consider, for example, the 2009 "NIPCC" climate-change-denying report put out by the Heartland Institute, funded by the Koch brothers and the fossil fuel industry: S. Fred Singer and Craig Idso, "Climate Change Reconsidered: 2009 NIPPC Report," Nongovernmental International Panel on Climate Change, 2009, http://climatechangereconsidered.org/climate-change -reconsidered-2009-nipcc-report. The report was formatted to mimic the IPCC report itself. Or the "debate" staged by the Heartland Institute in New York on September 23, 2019, in an attempt to blunt the growing momentum for climate action arising from the United Nations Climate Change Summit of September 2019 and surrounding public awareness campaigns: "Videos: Climate Debate in NYC on Sept. 23—Moderator, John Stossel," Heartland Institute, www.heartland.org/multimedia/videos /climate-debate-in-nyc-on-sept-23---moderator-john-stossel.

40. See, for example, this summary of a study of Australian public opinion on climate change: Naomi Schalit, "Climate Change Deniers Are Rarer Than We Think," *The Conversation*, November 11, 2012, https://theconversation .com/climate-change-deniers-are-rarer-than-we-think-10670.

41. Natacha Larnaud, "'This Will Only Get Worse in the Future': Experts See Direct Line Between California Wildfires and Climate Change," CBS News, October 30, 2019, www.cbsnews.com/news/this-will-only-get-worse -in-the-future-experts-find-direct-line-between-california-wildfires-and -climate-change.

42. Mary Tyler March, "Trump Blames 'Gross Mismanagement' for Deadly California Wildfires," *The Hill*, November 10, 2018, https://thehill.com /homenews/administration/416046-trump-blames-gross-mismanagement

-for-deadly-california-wildfires; Patrick Shanley and Katherine Schaff-stall, "Late-Night Hosts Mock Trump's 'Weird' Trip to California Following Wildfires," *Hollywood Reporter*, November 20, 2018, www.hollywoodreporter.com/live-feed/late-night-hosts-mock-trump-fire-comments-california-trip-1162958.

43. Emily Holden and Jimmy Tobias, "New Emails Reveal That the Trump Administration Manipulated Wildfire Science to Promote Logging: The Director of the US Geological Survey Asked Scientists to 'Gin Up' Emissions Figures for Him," *Mother Jones*, January 26, 2020, www.motherjones.com/environment/2020/01/new-emails-reveal-that-the-trump-administration-manipulated-wildfire-science-to-promote-logging.

44. Michael E. Mann, "Australia, Your Country Is Burning—Dangerous Climate Change Is Here with You Now," *The Guardian*, January 2, 2020, www.theguardian.com/commentisfree/2020/jan/02/australia-your-country-is-burning-dangerous-climate-change-is-here-with-you-now.

45. Indeed, I gave a number of interviews at the time making that connection for viewers and listeners, including ABC Australia. See "'A Tipping Point Is Playing Out Right Now' Says Climate Scientist Michael Mann," YouTube, posted January 2, 2020, ABC Australia, www.youtube.com/watch?v=OYtAGTe9MjY&feature=youtu.be; "Australian Fires: Who Is to Blame?," BBC, January 6, 2020, www.bbc.co.uk/sounds/play/p07zns45.

46. See "The Australian," SourceWatch, www.sourcewatch.org/index.php/The_Australian, accessed January 8, 2020.

47. Lachlan Cartwright, "James Murdoch Slams Fox News and News Corp over Climate-Change Denial," *Daily Beast*, January 14, 2020, www.thedailybeast.com/james-murdoch-slams-fox-news-and-news-corp-over-climate-change-denial.

48. Fiona Harvey, "Climate Change Is Already Damaging Global Economy, Report Finds," *The Guardian*, September 25, 2012, www.theguardian.com/environment/2012/sep/26/climate-change-damaging-global-economy.

49. Nafeez Ahmed, "U.S. Military Could Collapse Within 20 Years Due to Climate Change, Report Commissioned by Pentagon Says," *Vice*, October 14, 2019, www.vice.com/en_us/article/mbmkz8/us-military-could-collapse-within-20-years-due-to-climate-change-report-commissioned-by-pentagon-says.

50. Geoff Dembicki, "DC's Trumpiest Congressman Says the GOP Needs to Get Real on Climate Change," *Vice*, March 25, 2019, www.vice.com/en_us/article/zma97w/matt-gaetz-congress-loves-donald-trump-climate-change.

51. Philip Bump, "Anti-Tax Activist Grover Norquist Thinks a Carbon Tax Might Make Sense—with Some Caveats," *Grist*, November 13, 2012, https://grist.org/politics/anti-tax-activist-grover-norquist-thinks-a-carbon-tax-might-make-sense-with-some-caveats.

52. "Charles Koch—CEO of Koch Industries (#381)," *Tim Ferris Show* (podcast), August 11, 2019, https://tim.blog/2019/08/11/charles-koch.

See also "Charles Koch Talks Environment, Politics, Business and More," Koch Newsroom, August 12, 2019, https://news.kochind.com/news/2019 /charles-koch-talks-environment,-politics,-busi-1.

53. Andrea Dutton and Michael E. Mann, "A Dangerous New Form of Climate Denialism Is Making the Rounds," *Newsweek*, August 22, 2019, www.newsweek.com/dangerous-new-form-climate-denialism-making-rounds -opinion-1455736.

CHAPTER 3: THE "CRYING INDIAN" AND THE BIRTH OF THE DEFLECTION CAMPAIGN

1. James Downie, "The NRA Is Winning the Spin Battle," *Washington Post*, February 20, 2018, www.washingtonpost.com/blogs/post-partisan/wp/2018 /02/20/the-nra-is-winning-the-spin-battle.

2. Dennis A. Henigan, *"Guns Don't Kill People, People Kill People": And Other Myths About Guns and Gun Control* (Boston: Beacon Press, 2016).

3. Joseph Dolman, "Mayor's Promise on Guns Is Noble," *Newsday*, February 15, 2006. In 2017, the most recent year for which we have complete data, 39,773 people died from gun-related injuries in the United States, according to the Centers for Disease Control and Prevention. See John Gramlich, "What the Data Says About Gun Deaths in the U.S.," Pew Research Center, August 16, 2019, www.pewresearch.org/fact-tank/2019/08/16 /what-the-data-says-about-gun-deaths-in-the-u-s.

4. See Patricia Callahan and Sam Roe, "Big Tobacco Wins Fire Marshals as Allies in Flame Retardant Push," *Chicago Tribune*, May 8, 2012, www .chicagotribune.com/lifestyles/health/ct-met-flames-tobacco-20120508-story .html; Patricia Callahan and Sam Roe, "Fear Fans Flames for Chemical Makers," *Chicago Tribune*, May 6, 2012, www.chicagotribune.com/investigations /ct-met-flame-retardants-20120506-story.html.

5. Callahan and Roe, "Big Tobacco Wins Fire Marshals."

6. See, for example, Juliet Eilperin and David A. Fahrenthold, "Va. Climatologist Drawing Heat from His Critics," *Washington Post*, September 17, 2006, www.washingtonpost.com/archive/local/2006/09/17/va-climatologist -drawing-heat-from-his-critics/1bd66873-9fcc-40af-b9af-6e5808649af3.

7. James Pitkin, "Defying a Chemical Lobby, Oregon House Passes Fire-Retardant Ban," *Willamette Week*, June 18, 2009.

8. "Americans for Prosperity," SourceWatch, www.sourcewatch.org/index .php/Americans_for_Prosperity.

9. Callahan and Roe, "Fear Fans Flames."

10. Callahan and Roe, "Fear Fans Flames."

11. Callahan and Roe, "Fear Fans Flames."

12. Deborah Blum, "Flame Retardants Are Everywhere," *New York Times*, July 1, 2014, https://well.blogs.nytimes.com/2014/07/01/flame-retardants-are -everywhere.

13. "Pollution: Keep America Beautiful—Iron Eyes Cody," Ad Council, www.adcouncil.org/Our-Campaigns/The-Classics/Pollution-Keep-America -Beautiful-Iron-Eyes-Cody.

14. Finis Dunaway, "The 'Crying Indian' Ad That Fooled the Environmental Movement," *Chicago Tribune*, November 21, 2017, www.chicagotribune .com/opinion/commentary/ct-perspec-indian-crying-environment-ads -pollution-1123-20171113-story.html. See also Finis Dunaway, *Seeing Green: The Use and Abuse of American Environmental Images* (Chicago: University of Chicago Press, 2015).

15. See Dunaway, *Seeing Green*, 86.

16. Matt Simon, "Plastic Rain Is the New Acid Rain," *Wired*, June 11, 2020, www.wired.com/story/plastic-rain-is-the-new-acid-rain.

17. Dunaway, "'Crying Indian' Ad."

18. "These Things Are Disappearing Because Millennials Refuse to Pay for Them," *Buzznet*, June 27, 2019, www.buzznet.com/2019/06/millennials -refuse-to-buy.

19. David Hagmann, Emily H. Ho, and George Loewenstein, "Nudging Out Support for a Carbon Tax," *Nature Climate Change* 9 (2019): 484–489.

20. Michael E. Mann and Jonathan Brockopp, "You Can't Save the Climate by Going Vegan: Corporate Polluters Must Be Held Accountable," *USA Today*, June 3, 2019, www.usatoday.com/story/opinion/2019/06/03/climate-change -requires-collective-action-more-than-single-acts-column/1275965001.

CHAPTER 4: IT'S YOUR FAULT

1. See Sami Grover, "In Defense of Eco-Hypocrisy," *Medium*, March 21, 2019, https://blog.usejournal.com/in-defense-of-eco-hypocrisy-b71fb86f2b2f.

2. "BP Boss Plans to 'Reinvent' Oil Giant for Green Era," BBC, February 13, 2020, www.bbc.com/news/business-51475379.

3. See Grover, "In Defense of Eco-Hypocrisy."

4. Malcolm Harris, "Shell Is Looking Forward: The Fossil-Fuel Companies Expect to Profit from Climate Change. I Went to a Private Planning Meeting and Took Notes," *New York Magazine*, March 3, 2020, https://nymag.com /intelligencer/2020/03/shell-climate-change.html.

5. Charles Kennedy, "Is Eating Meat Worse Than Burning Oil?," OilPrice .com, October 22, 2019, https://oilprice.com/The-Environment/Global -Warming/Is-Eating-Meat-Worse-Than-Burning-Oil.html.

6. Nathaniel Rich, "Losing Earth: The Decade We Almost Stopped Climate Change," *New York Times Magazine*, August 1, 2018, www.nytimes.com /interactive/2018/08/01/magazine/climate-change-losing-earth.html.

7. Robinson Meyer, "The Problem with *The New York Times*' Big Story on Climate Change," *The Atlantic*, August 1, 2018, www.theatlantic.com/science /archive/2018/08/nyt-mag-nathaniel-rich-climate-change/566525.

8. Hannah Fairfield, "The Facts About Food and Climate Change," *New York Times*, May 1, 2019, www.nytimes.com/2019/05/01/climate/nyt-climate

-newsletter-food.html; Tik Root and John Schwartz, "One Thing We Can Do: Drive Less," *New York Times*, August 28, 2019, www.nytimes.com/2019/08/28 /climate/one-thing-we-can-do-drive-less.html; Andy Newman, "If Seeing the World Helps Ruin It, Should We Stay Home?," *New York Times*, June 3, 2019, www.nytimes.com/2019/06/03/travel/traveling-climate-change.html; Andy Newman, "I Am Part of the Climate-Change Problem. That's Why I Wrote About It," *New York Times*, June 18, 2019, www.nytimes.com/2019/06/18 /reader-center/travel-climate-change.html.

9. Jonathan Safran Foer, "The End of Meat Is Here," *New York Times*, May 21, 2020, www.nytimes.com/2020/05/21/opinion/coronavirus-meat -vegetarianism.html.

10. Editorial Board, "The Democrats' Best Choices for President," *New York Times*, January 19, 2020, www.nytimes.com/interactive/2020/01/19/opinion /amy-klobuchar-elizabeth-warren-nytimes-endorsement.html.

11. S. Lewandowsky, N. Oreskes, J. S. Risbey, B. R. Newell, and M. Smith-son, "Seepage: Climate Change Denial and Its Effect on the Scientific Com-munity," *Global Environmental Change* 33 (2015): 1–13, https://doi.org/10 .1016/j.gloenvcha.2015.02.013.

12. Grover, "In Defense of Eco-Hypocrisy."

13. Jennifer Stock and Geremy Schulick, "Yale's Endowment Won't Divest from Fossil Fuels. Here's Why That's Wrong," *American Prospect*, March 8, 2019, https://prospect.org/education/yale-s-endowment-divest-fossil-fuels.-wrong.

14. Steven D. Hales, "The Futility of Guilt-Based Advocacy," *Quillette*, November 23, 2019, https://quillette.com/2019/11/23/the-futility-of-guilt -based-advocacy.

15. John Schwartz (@jswatz), Twitter, November 14, 2019, 8:38 a.m., https://twitter.com/jswatz/status/1195018223621754880.

16. Clay Evans, "Ditching the Doomsaying for Better Climate Discourse," *Colorado Arts and Sciences Magazine*, December 18, 2019, www.colorado.edu /asmagazine/2019/12/18/ditching-doomsaying-better-climate-discourse.

17. Brad Johnson, "Pete Buttigieg Climate Advisor Is a Fossil-Fuel-Funded Witness for the Trump Administration Against Children's Climate Lawsuit," *Hill Heat*, November 18, 2019, www.hillheat.com/articles/2019/11/18/pete -buttigieg-climate-advisor-is-a-fossil-fuel-funded-witness-for-the-trump -administration-against-childrens-climate-lawsuit; David G. Victor and Charles F. Kennel, "Climate Policy: Ditch the 2°C Warming Goal," *Nature*, October 1, 2014, www.nature.com/news/climate-policy-ditch-the-2-c-warming-goal -1.16018.

18. David G. Victor, "We Have Climate Leaders. Now We Need Followers," *New York Times*, December 13, 2019, www.nytimes.com/2019/12/13/opinion /climate-change-madrid.html.

19. Dr. Genevieve Guenther (@DoctorVive), Twitter, December 13, 2019, 10:38 a.m., https://twitter.com/DoctorVive/status/1205557572528410628.

20. Nathanael Johnson, "Fossil Fuels Are the Problem, Say Fossil Fuel Companies Being Sued," *Grist*, March 21, 2018, https://grist.org/article/fossil -fuels-are-the-problem-say-fossil-fuel-companies-being-sued.

21. See this Twitter thread initiated by journalist Emily Atkin (@emorwee), November 20, 2019, 6:01 a.m., https://twitter.com/emorwee/status/1197152947282612225.

22. See the Discovery Institute's self-described five-year "wedge strategy," archived at AntiEvolution.org, www.antievolution.org/features/wedge.html.

23. Aja Romano, "Twitter Released 9 Million Tweets from One Russian Troll Farm. Here's What We Learned," *Vox*, October 19, 2018, www.vox.com/2018/10/19/17990946/Twitter-russian-trolls-bots-election-tampering.

24. Lucy Tiven, "Where the Presidential Candidates Stand on Climate Change," Attn.com, September 6, 2016, https://archive.attn.com/stories/11189/presidential-candidates-stance-on-climate-change; Brad Plumer, "On Climate Change, the Difference Between Trump and Clinton Is Really Quite Simple," *Vox*, November 4, 2016, www.vox.com/science-and-health/2016/10/10/13227682/trump-clinton-climate-energy-difference.

25. Rebecca Leber, "Many Young Voters Don't See a Difference Between Clinton and Trump on Climate," *Grist*, July 31, 2016, https://grist.org/election-2016/many-young-voters-dont-see-a-difference-between-clinton-and-trump-on-climate.

26. Craig Timberg and Tony Romm, "Russian Trolls Sought to Inflame Debate over Climate Change, Fracking, Dakota Pipeline," *Chicago Tribune*, March 1, 2018, www.chicagotribune.com/nation-world/ct-russian-trolls-climate-change-20180301-story.html.

27. See Nancy LeTourneau, "The Gaslighting Effect of Both-Siderism," *Washington Monthly*, October 8, 2018, https://washingtonmonthly.com/2018/10/08/the-gaslighting-effect-of-both-siderism.

28. Paul Krugman, "The Party That Ruined the Planet: Republican Climate Denial Is Even Scarier Than Trumpism," *New York Times*, December 12, 2019, www.nytimes.com/2019/12/12/opinion/climate-change-republicans.html.

29. The entire conversation can be found at Michael E. Mann (@MichaelEMann), Twitter, December 12, 2019, 5:04 p.m., https://twitter.com/MichaelEMann/status/1205292356208979968.

30. Sandra Laville and David Pegg, "Fossil Fuel Firms' Social Media Fightback Against Climate Action," *The Guardian*, October 11, 2019, www.theguardian.com/environment/2019/oct/10/fossil-fuel-firms-social-media-fightback-against-climate-action; Craig Timberg and Tony Romm, "Russian Trolls Sought to Inflame Debate over Climate Change, Fracking, Dakota Pipeline," *Chicago Tribune*, March 1, 2018, www.chicagotribune.com/nation-world/ct-russian-trolls-climate-change-20180301-story.html.

31. Marianne Lavelle, "'Trollbots' Swarm Twitter with Attacks on Climate Science Ahead of UN Summit," *Inside Climate News*, September 16, 2019, https://insideclimatenews.org/news/16092019/trollbot-Twitter-climate-change-attacks-disinformation-campaign-mann-mckenna-greta-targeted.

32. Oliver Milman, "Revealed: Quarter of All Tweets About Climate Crisis Produced by Bots," *The Guardian*, February 21, 2020, www.theguardian.com/technology/2020/feb/21/climate-tweets-Twitter-bots-analysis.

33. Elisha R .Frederiks, Karen Stenner, and Elizabeth V. Hobman, "Household Energy Use: Applying Behavioural Economics to Understand Consumer Decision-Making and Behaviour," *Renewable and Sustainable Energy Reviews* 41 (2015): 1385–1394, www.sciencedirect.com/science/article/pii/S1364032114007990.

34. Nicole Perlroth, "A Former Fox News Executive Divides Americans Using Russian Tactics," *New York Times*, November 21, 2019, www.nytimes.com/2019/11/21/technology/LaCorte-edition-news.html.

35. See Michael E. Mann and Jonathan Brockopp, "You Can't Save the Climate by Going Vegan. Corporate Polluters Must Be Held Accountable," *USA Today*, June 3, 2019, www.usatoday.com/story/opinion/2019/06/03/climate-change-requires-collective-action-more-than-single-acts-column/1275965001; Michael E. Mann, "Lifestyle Changes Aren't Enough to Save the Planet. Here's What Could," *Time*, September 12, 2019, https://time.com/5669071/lifestyle-changes-climate-change.

36. See, for example, my commentary "Greta Thunberg, Not Donald Trump, Is the True Leader of the Free World," *Newsweek*, September 24, 2019, www.newsweek.com/greta-thunberg-donald-trump-true-leadership-climate-change-free-world-1461147.

37. Nives Dolsak and Aseem Prakash, "Does Greta Thunberg's Lifestyle Equal Climate Denial? One Climate Scientist Seems to Suggest So," *Forbes*, November 14, 2019, www.forbes.com/sites/prakashdolsak/2019/11/14/does-greta-thunbergs-lifestyle-equal-climate-denial-one-climate-scientist-seems-to-suggest-so (subsequently edited by *Forbes*). See also thread at Jerome Foster II (@JeromeFosterII), Twitter, November 15, 2019, 6:35 p.m., https://twitter.com/JeromeFosterII/status/1195530789334798338.

38. Robin McKie, "Climate Change Deniers' New Battle Front Attacked," *The Guardian*, November 9, 2019, www.theguardian.com/science/2019/nov/09/doomism-new-tactic-fossil-fuel-lobby.

39. Dr. Lucky Tran (@luckytran), Twitter, November 15, 2019, 9:28 a.m., https://twitter.com/luckytran/status/1195393067764699137; John Upton (@johnupton), Twitter, November 15, 2019, 8:20 a.m., https://twitter.com/johnupton/status/1195375914365919232.

40. See "Anthony Watts," SourceWatch, www.sourcewatch.org/index.php/Anthony_Watts; "Heartland Institute," SourceWatch, www.sourcewatch.org/index.php/Heartland_Institute; Eric Worrall, "Katharine Hayhoe Attacks Greta Thunberg's Climate 'Shaming' Crusade," *Watts Up with That*, August 19, 2019, https://wattsupwiththat.com/2019/08/19/katharine-hayhoe-attacks-greta-thunbergs-climate-shaming-crusade.

41. The tweet in question is Prof. Katharine Hayhoe (@KHayhoe), Twitter, August 20, 2019, 9:34 a.m., https://twitter.com/KHayhoe/status/1163851874921005057.

42. See, for example, Dictionary.com, www.dictionary.com/e/slang/ok-boomer: "OK *boomer* is a viral internet slang phrase used, often in a humorous or ironic manner, to call out or dismiss out-of-touch or close-minded opinions associated with the baby boomer generation and older people more generally."

43. See, for example, David Roberts, "California Gov. Jerry Brown Casually Unveils History's Most Ambitious Climate Target: Full Carbon Neutrality Is Now on the Table for the World's Fifth Largest Economy," *Vox*, September 12, 2018, www.vox.com/energy-and-environment/2018/9/11/17844896 /california-jerry-brown-carbon-neutral-2045-climate-change.

44. Brown made the announcement at the December 2016 annual meeting of the American Geophysical Union in San Francisco—a conference I attended (I spoke at the meeting).

45. Emily Guerin, "Jerry Brown Is Getting Heckled at His Own Climate Conference," LAist, September 12, 2018, https://laist.com/2018/09/12/la _environmentalists_heckling_jerry_brown_at_climate_conference.php.

46. Mark Hertsgaard, "Jerry Brown vs. the Climate Wreckers: Is He Doing Enough?," *The Nation*, August 29, 2018, www.thenation.com/article/jerry -brown-vs-the-climate-wreckers-is-he-doing-enough.

47. See Arn Menconi (@ArnMenconi), Twitter, November 17, 2019, 6:43 a.m., https://twitter.com/ArnMenconi/status/1196076331269681152.

48. Kate Connolly and Matthew Taylor, "Extinction Rebellion Founder's Holocaust Remarks Spark Fury," *The Guardian*, November 20, 2019, www .theguardian.com/environment/2019/nov/20/extinction-rebellion-founders -holocaust-remarks-spark-fury.

49. Jason Mark, "Yes, Actually, Individual Responsibility Is Essential to Solving the Climate Crisis," Sierra Club, November 26, 2019, www.sierraclub .org/sierra/yes-actually-individual-responsibility-essential-solving-climate -crisis; Mann and Brockopp, "You Can't Save the Climate by Going Vegan."

50. Mann, "Lifestyle Changes Aren't Enough."

51. According to the World Resources Institute, total annual emissions from animal agriculture (production emissions plus land-use change) are about 14.5 percent of all human emissions, of which beef contributes about 41 percent. Richard Waite, Tim Searchinger, and Janet Ranganathan, "6 Pressing Questions About Beef and Climate Change, Answered," World Resources Institute, April 8, 2019, www.wri.org/blog/2019/04/6 -pressing-questions-about-beef-and-climate-change-answered.

52. See Jonathan Kaplan, "There's No Conspiracy in Cowspiracy," Natural Resources Defense Council, April 29, 2016, www.nrdc.org/experts/jonathan -kaplan/theres-no-conspiracy-cowspiracy; Doug Boucher, "Movie Review: There's a Vast Cowspiracy about Climate Change," Union of Concerned Scientists, June 10, 2016, https://blog.ucsusa.org/doug-boucher/cowspiracy -movie-review.

According to the latter, "the 51% figure is key to the film's conspiracy theory. . . . Ironically, in light of *Cowspiracy*'s thesis that environmental NGOs are hiding the science, this study proposing this figure on which they rely so heavily was not published in a scientific journal, but in a report by an environmental organization, the Worldwatch Institute. The report's authors, Jeff Anhang and the late Robert Goodland, were not named in the movie but were described simply as 'two advisers from the World Bank.'"

53. In response to a tweet of mine stressing the urgency of reducing carbon emissions, one vegan or vegetarian activist posed the presumably rhetorical question, "Is @MichaelEMann vegan or vegetarian? Does he drive an EV and have solar at home?" I replied "I don't eat meat, I drive a hybrid, and have a wind-only energy plan. I also think that trying to shame people over lifestyle is presumptuous and unproductive." See Trees (@SolutionsOK), Twitter, November 14, 2018, 10:43 a.m., https://twitter.com/SolutionsOK/status/1062778092345679872.

54. Seth Borenstein, "Climate Scientists Try to Cut Their Own Carbon Footprints," Associated Press, December 8, 2019, https://apnews.com/dde2bf108411ecd973de60bfda5250aa. See also Michael E. Mann (@MichaelEMann), Twitter, November 20, 2019, 8:39 a.m., https://twitter.com/MichaelEMann/status/1197192725445185536.

55. Catherine Brahic, "Train Can Be Worse for Climate Than Plane," New Scientist, June 8, 2009, www.newscientist.com/article/dn17260-train-can-be-worse-for-climate-than-plane.

56. Kaya Chatterjee, The Zero-Footprint Baby: How to Save the Planet While Raising a Healthy Baby (New York: Ig Publishing, 2013).

57. Maxine Joselow, "Quitting Burgers and Planes Won't Stop Warming, Experts Say," Climatewire, E&E News, December 6, 2019, www.eenews.net/stories/1061734031.

58. Seth Borenstein, "Climate Scientists Try to Cut Their Own Carbon Footprints," Associated Press, December 8, 2019, https://apnews.com/dde2bf108411ecd973de60bfda5250aa.

59. George Monbiot, "We Are All Killers," February 28, 2006, www.monbiot.com/2006/02/28/we-are-all-killers.

60. David Freedlander, "The Meteorologist's Meltdown: Eric Holthaus on Deciding to Quit Flying," Daily Beast, October 1, 2013, www.thedailybeast.com/the-meteorologists-meltdown-eric-holthaus-on-deciding-to-quit-flying.

61. See, for example, Eric Berger, "Who Is Eric Holthaus, and Why Did He Give Up Flying Today?," Houston Chronicle, September 27, 2013, https://blog.chron.com/sciguy/2013/09/who-is-eric-holthaus-and-why-did-he-give-up-flying-today; Jason Samenow, "Meteorologist Eric Holthaus' Vow to Never to [sic] Fly Again Draws Praise, Criticism," Washington Post, October 1, 2013, www.washingtonpost.com/news/capital-weather-gang/wp/2013/10/01/meteorologist-eric-holthaus-vow-to-never-to-fly-again-draws-praise-criticism; Will Oremus, "Meteorologist Weeps over Climate Change, Fox News Calls Him a 'Sniveling Beta-Male,'" Slate, October 3, 2013, https://slate.com/technology/2013/10/sniveling-beta-male-fox-news-greg-gutfeld-slams-eric-holthaus-for-giving-up-flying.html.

62. See, for example, Berger, "Who Is Eric Holthaus?"; Eric Holthaus (@EricHolthaus), Twitter, October 1, 2013, 7:22 a.m., https://twitter.com/EricHolthaus/status/385047176851644417, quoted in Samenow, "Meteorologist Eric Holthaus' Vow"; Oremus, "Meteorologist Weeps."

63. "U. Minnesota Scholar Criticizes Air Travel," *Conservative Edition News*, n.d., https://conservativeeditionnews.com/u-minnesota-scholar-criticizes -air-travel; "Doug P.," "Meteorologist Eric Holthaus SHAMES Aviation Buff over Unnecessary Commercial Flight Because It's 'Just as Deadly as a Gun' and 'Should Be Outlawed,'" *Twitchy*, July 25, 2019, https://twitchy.com/dougp -3137/2019/07/25/meteorologist-eric-holthaus-shames-aviation-buff-over -unnecessary-commercial-flight-because-its-just-as-deadly-as-a-gun-and -should-be-outlawed.

64. David Roberts, "Rich Climate Activist Leonardo DiCaprio Lives a Car- bon-Intensive Lifestyle, and That's (Mostly) Fine," *Vox*, March 2, 2016, www .vox.com/2016/3/2/11143310/leo-dicaprios-carbon-lifestyle.

65. See "Beacon Center of Tennessee," SourceWatch, www.sourcewatch .org/index.php/Beacon_Center_of_Tennessee.

66. See David Mikkelson and Dan Evon, "Al Gore's Home Energy Use: Does Al Gore's Home Consume Twenty Times as Much Energy as the Average American House?," Snopes, February 28, 2007, www.snopes.com/fact-check /al-gores-energy-use.

67. Jake Tapper, "Al Gore's 'Inconvenient Truth'?—A $30,000 Utility Bill," ABC News, February 27, 2007, https://abcnews.go.com/Politics/Global Warming/story?id=2906888.

68. See Mikkelson and Evon, "Al Gore's Home Energy Use."

69. Rita Panahi, "Hollywood Hypocrite's Global Warming Sermon," *Herald Sun*, October 7, 2016, www.heraldsun.com.au/blogs/rita-panahi/hollywood -hypocrites-global-warming-sermon/news-story/b4cc2e4b6034c032998 fb3c13e6df4a6.

70. Andrea Peyser, "Leo DiCaprio Isn't the Only Climate Change Hyp- ocrite," *New York Post*, May 26, 2016, https://nypost.com/2016/05/26/leo-di -caprio-isnt-the-only-climate-change-hypocrite.

71. Alison Boshoff and Sue Connolly, "Eco-Warrior or Hypocrite? Leon- ardo DiCaprio Jets Around the World Partying . . . While Preaching to Us All on Global Warming," *Daily Mail*, May 24, 2016.

72. Roberts, "Rich Climate Activist Leonardo DiCaprio."

73. Roberts, "Rich Climate Activist Leonardo DiCaprio."

74. Eric Worrall, "Doh! Climate Messiah Greta Thunberg's Plastic Boat Trip Will Require Four Transatlantic Flights," *Watts Up with That*, August 18, 2019, https://wattsupwiththat.com/2019/08/18/doh-climate-messiah-greta -thunbergs-plastic-boat-trip-will-result-in-two-airline-flights.

75. Naaman Zhou, "Climate Strikes: Hoax Photo Accusing Australian Pro- testers of Leaving Rubbish Behind Goes Viral," *The Guardian*, September 21, 2019, www.theguardian.com/environment/2019/sep/21/climate-strikes -hoax-photo-accusing-australian-protesters-of-leaving-rubbish-behind-goes -viral.

76. Andrew Bolt, "Look, in the Sky! A Hypocrite Called McKibben," *Herald Sun*, April 8, 2013, www.heraldsun.com.au/blogs/andrew-bolt /look-in-the-sky-a-hypocrite-called-mckibben/news-story/164435dd3

d4447ba60bcea92c349edda; Anthony Watts, "Bill McKibben's Excellent Eco-Hypocrisy," *Watts Up with That*, October 5, 2013, https://wattsupwiththat.com/2013/10/05/bill-mckibbens-excellent-eco-hypocrisy.

77. Bill McKibben, "Embarrassing Photos of Me, Thanks to My Right-Wing Stalkers," *New York Times*, August 5, 2016, www.nytimes.com/2016/08/07/opinion/sunday/embarrassing-photos-of-me-thanks-to-my-right-wing-stalkers.html.

78. Isabel Vincent and Melissa Klein, "Gas-Guzzling Car Rides Expose AOC's Hypocrisy Amid Green New Deal Pledge," *New York Post*, March 2, 2019, https://nypost.com/2019/03/02/gas-guzzling-car-rides-expose-aocs-hypocrisy-amid-green-new-deal-pledge.

79. Clover Moore (@CloverMoore), Twitter, December 4, 2019, 10:47 p.m., https://twitter.com/CloverMoore/status/1202479544172630016.

80. Katie Pavlich, "The Frauds of the Climate Change Movement," *The Hill*, October 1, 2019, https://thehill.com/opinion/katie-pavlich/463930-katie-pavlich-the-frauds-of-the-climate-change-movement. The Young America's Foundation has received a substantial amount of money from the Koch brothers and the Koch brothers–funded Donors Trust. See "Young America's Foundation," SourceWatch, www.sourcewatch.org/index.php/Young_America%27s_Foundation.

81. Michael E. Mann (@MichaelEMann), Twitter, November 13, 2018, 1:30 p.m., https://twitter.com/MichaelEMann/status/1062457643313315840.

82. Trees (@SolutionsOK), Twitter, November 14, 2018, 10:43 a.m., https://twitter.com/SolutionsOK/status/1062778092345679872; Michael E. Mann (@MichaelEMann), Twitter, November 14, 2018, 10:47 a.m., https://twitter.com/MichaelEMann/status/1062779071770300418.

83. Ben Penfold (@BenPenfold7), Twitter, January 2, 2020, 8:24 p.m., https://twitter.com/BenPenfold7/status/1212952842656370688.

84. "It's a Fact, Scientists Are the Most Trusted People in World," Ipsos, September 17, 2019, www.ipsos.com/en/its-fact-scientists-are-most-trusted-people-world; Boshoff and Connolly, "Eco-Warrior or Hypocrite?"

85. DiCaprio produced the documentary *Cowspiracy* discussed earlier in this chapter.

86. Shahzeen Z. Attari, David H. Krantz, and Elke U. Weber, "Climate Change Communicators' Carbon Footprints Affect Their Audience's Policy Support," *Climatic Change* 154, no. 3–4 (2019): 529–545; Borenstein, "Climate Scientists Try to Cut Their Own Carbon Footprints."

87. Joselow, "Quitting Burgers."

88. Jeff McMahon, "Greta Is Right: Study Shows Individual Lifestyle Change Boosts Systemic Climate Action," *Forbes*, November 19, 2019, www.forbes.com/sites/jeffmcmahon/2019/11/19/greta-is-right-study-shows-individual-climate-action-boosts-systemic-change/#67fc82e64a54.

89. Cormac O'Rafferty (@CormacORafferty), Twitter, November 21, 2019, 7:08 a.m., https://twitter.com/CormacORafferty/status/1197532349078163456.

90. Mann, "Lifestyle Changes Aren't Enough."

91. Moore has been funded by a number of corporate interests to attack science that proves disadvantageous to them. See "Patrick Moore," SourceWatch, www.sourcewatch.org/index.php/Patrick_Moore, accessed January 5, 2020.

92. Helen Regan, "Watch a GMO Advocate Claim a Weed Killer Is Safe to Drink but Then Refuse to Drink It," *Time*, March 27, 2015, https://time.com/3761053/monsanto-weed-killer-drink-patrick-moore.

93. Paul Wogden (@WogdenPaul), Twitter, September 11, 2019, 10:58 p.m., https://twitter.com/WogdenPaul/status/1172026602374533120.

94. Lavelle, "'Trollbots' Swarm Twitter."

95. Abel Gustafson, Anthony Leiserowitz, and Edward Maibach, "Americans Are Increasingly 'Alarmed' About Global Warming," Yale Program in Climate Change Communication, February 12, 2019, https://climatecommunication.yale.edu/publications/americans-are-increasingly-alarmed-about-global-warming; "Climate Change Opinions Rebound Among Republican Voters: Bipartisan Support for Climate and Clean Energy Policies Remains Strong," Yale Program in Climate Change Communication, May 8, 2018, https://climatecommunication.yale.edu/news-events/climate-change-opinions-rebound-among-republican-voters-bipartisan-support-for-climate-clean-energy-policies-remains-strong.

96. Ramez Naam (@ramez), Twitter, September 30, 2019, 12:53 p.m., https://twitter.com/ramez/status/1178759746075054080.

97. "U. Minnesota Scholar Criticizes Air Travel"; "Doug P.," "Meteorologist Eric Holthaus SHAMES Aviation Buff."

98. Mark, "Yes, Actually, Individual Responsibility Is Essential."

99. Borenstein, "Climate Scientists Try to Cut Their Own Carbon Footprints."

100. Borenstein, "Climate Scientists Try to Cut Their Own Carbon Footprints."

101. See, for example, Jeremy Lovell, "Climate Report Calls for Green 'New Deal,'" Reuters, July 21, 2008, www.reuters.com/article/us-climate-deal/climate-report-calls-for-green-new-deal-idUSL2046100020080721.

102. Salvador Rizzo, "What's Actually in the 'Green New Deal' from Democrats?," *Washington Post*, February 11, 2019, www.washingtonpost.com/politics/2019/02/11/whats-actually-green-new-deal-democrats.

103. Michael E. Mann, "Radical Reform and the Green New Deal," *Nature*, September 18, 2019, www.nature.com/articles/d41586-019-02738.

104. Jeffrey Frankel, "The Best Way to Help the Climate Is to Increase the Price of CO_2 Emissions," *The Guardian*, January 20, 2020, www.theguardian.com/business/2020/jan/20/climate-crisis-carbon-emissions-tax.

105. Mann, "Radical Reform and the Green New Deal."

106. Such was indeed the premise of the right-wing shock jock Glenn Beck's "novel" *Agenda 21*, reviewed by yours truly for *Popular Science*: Michael E. Mann, "What Does a Climate Scientist Think of Glenn Beck's Environmental-Conspiracy Novel?," *Popular Science*, December 12, 2012, www.popsci.com/environment/article/2012-12/what-does-climate-scientist-think-glenn-becks-environmental-conspiracy-novel.

107. Gary Anderson, "Negative Rates, Climate Science and a Fed Warning," Talk Markets, November 21, 2019, https://talkmarkets.com/content

/economics—politics-education/negative-rates-climate-science-and-a-fed
-warning?post=241500#.

108. Dominique Jackson, "The Daily Show Brutally Ridicules Fox's Sean
Hannity for Whining AOC's Green New Deal Will Deprive Him of Hamburgers," *Raw Story*, February 14, 2019, www.rawstory.com/2019/02/watch
-the-daily-shows-trevor-noah-brutally-mocks-sean-hannity-over-thinking-aoc
-wants-outlaw-hamburgers-with-green-new-deal.

109. Antonia Noori Farzan, "The Latest Right-Wing Attack on Democrats: 'They Want to Take Away Your Hamburgers,'" *Washington Post*, March 1,
2019, www.washingtonpost.com/nation/2019/03/01/latest-right-wing-attack
-democrats-they-want-take-away-your-hamburgers.

110. Sam Dorman, "AOC Accused of Soviet-Style Propaganda with Green
New Deal 'Art Series,'" Fox News, August 30, 2019, www.foxnews.com/media
/aoc-green-new-deal-art-series-propaganda; Daniel Turner, "Stealth AOC
'Green New Deal' Now the Law in New Mexico, Voters Be Damned," Fox
News, May 27, 2019, www.foxnews.com/opinion/daniel-turner-stealth-version
-of-aoc-green-new-deal-now-the-law-in-new-mexico-voters-be-damned.

111. Tom Jacobs, "Did Fox News Quash Republican Support for the Green
New Deal?," *Pacific Standard*, May 13, 2019, https://psmag.com/economics/did
-fox-news-quash-republican-support-for-the-green-new-deal.

112. Brian Kahn, "Big Oil Is Scared Shitless," *Gizmodo*, June 18, 2020,
https://earther.gizmodo.com/big-oil-is-scared-shitless-1844084649.

CHAPTER 5: PUT A PRICE ON IT. OR NOT.

1. "Bill McKibben: Actions Speak Louder Than Words," *Bulletin of the
Atomic Scientists* 68, no. 2 (2012): 1–8.

2. Justin Gerdes, "Cap and Trade Curbed Acid Rain: 7 Reasons Why It Can
Do the Same for Climate Change," *Forbes*, February 13, 2012, www.forbes
.com/sites/justingerdes/2012/02/13/cap-and-trade-curbed-acid-rain-7
-reasons-why-it-can-do-the-same-for-climate-change/#6e920874943a.

3. John M. Broder, "'Cap and Trade' Loses Its Standing as Energy Policy
of Choice," *New York Times*, March 25, 2010, www.nytimes.com/2010/03/26
/science/earth/26climate.html.

4. For example, environmental organizations like Greenpeace and climate
scientist and advocate James Hansen. See Paul Krugman, "The Perfect, the
Good, the Planet," *New York Times*, May 17, 2009, www.nytimes.com/2009/05
/18/opinion/18krugman.html.

5. Krugman, "The Perfect, the Good, the Planet."

6. Eric Zimmermann, "Republicans Propose . . . a Carbon Tax?," *The Hill*,
May 14, 2009, https://thehill.com/blogs/blog-briefing-room/news/35719
-republicans-proposea-carbon-tax.

7. Christopher Leonard, "David Koch Was the Ultimate Climate Change
Denier," *New York Times*, August 23, 2019, www.nytimes.com/2019/08/23
/opinion/sunday/david-koch-climate-change.html.

8. Broder, "'Cap and Trade' Loses Its Standing."

9. "C. Boyden Gray," SourceWatch, www.sourcewatch.org/index.php/C ._Boyden_Gray, accessed January 15, 2020.

10. "Americans for Prosperity Foundation (AFP)," Greenpeace, www .greenpeace.org/usa/global-warming/climate-deniers/front-groups/americans -for-prosperity-foundation-afp.

11. Terry Gross, "'Kochland': How The Koch Brothers Changed U.S. Corporate and Political Power," National Public Radio, *Fresh Air*, August 13, 2019, www.wuwm.com/post/kochland-how-koch-brothers-changed-us-corporate -and-political-power#stream/0.

12. Christopher Leonard, "David Koch Was the Ultimate Climate Change Denier," *New York Times*, August 23, 2019, www.nytimes.com/2019/08/23 /opinion/sunday/david-koch-climate-change.html.

13. Broder, "'Cap and Trade' Loses Its Standing."

14. Haroon Siddique, "US Senate Drops Bill to Cap Carbon Emissions," July 23, 2010, www.theguardian.com/environment/2010/jul/23/us-senate-climate -change-bill.

15. "Bob Inglis," John F. Kennedy Presidential Library and Museum, 2015, www.jfklibrary.org/events-and-awards/profile-in-courage-award/award -recipients/bob-inglis-2015.

16. For an excellent review, see Marc Hudson, "In Australia, Climate Policy Battles Are Endlessly Reheated," *The Conversation*, April 9, 2019, https://theconversation.com/in-australia-climate-policy-battles-are-endlessly -reheated-114971, and for a discussion of the comparative politics of climate denial in the United States, Australia, and other Western nations, see Christopher Wright and Daniel Nyberg, "Corporate Political Activity and Climate Coalitions," in *Climate Change, Capitalism, and Corporations: Processes of Creative Self-Destruction* (Cambridge: Cambridge University Press, 2015).

17. Mark Butler, "How Australia Bungled Climate Policy to Create a Decade of Disappointment," *The Guardian*, July 5, 2017, www.theguardian .com/australia-news/2017/jul/05/how-australia-bungled-climate-policy-to -create-a-decade-of-disappointment.

18. See Graham Readfearn, "Australia's Place in the Global Web of Climate Denial," Australian Broadcasting Corporation, June 28, 2011, www.abc.net .au/news/2011-06-29/readfearn—australia27s-place-in-the-global-web-of -climate-de/2775298; Graham Readfearn, "Who Are the Australian Backers of Heartland's Climate Denial?," *DeSmog* (blog), May 21, 2012, www.desmogblog .com/who-are-australian-backers-heartland-s-climate-denial.

19. Julia Baird, "A Carbon Tax's Ignoble End," *New York Times*, July 24, 2014, www.nytimes.com/2014/07/25/opinion/julia-baird-why-tony-abbott-axed -australias-carbon-tax.html; Butler, "How Australia Bungled Climate Policy."

20. Baird, "A Carbon Tax's Ignoble End."

21. See, for example, Turnbull's commentary "Australia's Bushfires Show the Wicked, Self-Destructive Idiocy of Climate Denialism Must Stop," *Time*, January 16, 2020, https://time.com/5765603/australia-bushfires-prime-minister -essay. I had a chance to get to know Turnbull during my sabbatical in Sydney

in early 2020. I found him to be thoughtful, earnest, and honorable—in general, and in his efforts at climate policy. I'm sure he's frustrated that he was unable to convince fellow Liberals to support meaningful action on climate while he was in office. Like Inglis, he seems committed to doing what he can to further the cause of climate action in Australia.

22. Butler, "How Australia Bungled Climate Policy."

23. Jonathan Watts and Ben Doherty, "US and Russia Ally with Saudi Arabia to Water Down Climate Pledge," *The Guardian*, December 10, 2018, www.theguardian.com/environment/2018/dec/09/us-russia-ally-saudi-arabia-water-down-climate-pledges-un.

24. Manuel Roig-Franzia, Rosalind S. Helderman, William Booth, and Tom Hamburger, "How the 'Bad Boys of Brexit' Forged Ties with Russia and the Trump Campaign—and Came Under Investigators' Scrutiny," *Washington Post*, June 28, 2018, www.washingtonpost.com/politics/how-the-bad-boys-of-brexit-forged-ties-with-russia-and-the-trump-campaign--and-came-under-investigators-scrutiny/2018/06/28/6e3a5e9c-7656-11e8-b4b7-308400242c2e_story.html; Richard Collett-White, Chloe Farand, and Mat Hope, "Meet the Brexit Party's Climate Science Deniers," *DeSmog* (blog), May 1 2019, www.desmog.co.uk/2019/05/01/brexit-party-climate-science-deniers.

25. Roman Goncharenko, "France's 'Yellow Vests' and the Russian Trolls That Encourage Them," *Deutsche Welle*, December 15, 2018, www.dw.com/en/frances-yellow-vests-and-the-russian-trolls-that-encourage-them/a-46753388.

26. Emily Atkin, "France's Yellow Vest Protesters Want to Fight Climate Change: Trump Says the Violence Is Proof That People Oppose Environmental Protection. He Couldn't Be More Wrong," *New Republic*, December 10, 2018, https://newrepublic.com/article/152585/frances-yellow-vest-protesters-want-fight-climate-change.

27. Alexander Panetta, "Notorious Russian Troll Farm Targeted Trudeau, Canadian Oil in Online Campaigns," *The Star*, March 18, 2018, www.thestar.com/news/canada/2018/03/18/notorious-russian-troll-farm-targeted-trudeau-canadian-oil-in-online-campaigns.html.

28. Fatima Syed, "The Abuse Catherine McKenna Receives on Twitter Exploded the Day the Carbon Tax Started," *National Observer*, October 25, 2019, www.nationalobserver.com/2019/10/25/news/abuse-catherine-mckenna-receives-Twitter-exploded-day-carbon-tax-started.

29. Ahmed Al-Rawi and Yasmin Jiwani, "Russian Twitter Trolls Stoke Anti-Immigrant Lies Ahead of Canadian Election," *The Conversation*, July 23, 2019, https://theconversation.com/russian-twitter-trolls-stoke-anti-immigrant-lies-ahead-of-canadian-election-119144.

30. Nathalie Graham, "Looks Like Out-of-State Money DID Sway Your Vote, Washington," *The Stranger*, November 7, 2018, www.thestranger.com/slog/2018/11/07/35182424/a-big-night-for-corporate-money-and-a-dismal-display-for-the-carbon-fee; "Sierra Club Position on Carbon Washington Ballot Initiative 732," Sierra Club, September 2016, www.sierraclub.org/washington/sierra-club-position-carbon-washington-ballot-initiative-732; Kate Aronoff, "Why the Left Doesn't Want a Carbon Tax (Or at Least Not This One): The

Battle over a Washington State Ballot Initiative Previews the Future of the Climate Debate," *In These Times*, November 3, 2016, http://inthesetimes.com /article/19592/why-the-left-doesnt-want-carbon-tax-washington-i-732-climate -change-ballot.

31. A good discussion of the relative merits and complementary nature of the two approaches is provided in Fergus Green and Richard Denniss, "Cutting with Both Arms of the Scissors: The Economic and Political Case for Restrictive Supply-Side Climate Policies," *Climatic Change* 150 (2018): 73–87.

32. Lorraine Chow, "These Celebrities Take a Stand Against Dakota Access Pipeline," EcoWatch, September 9, 2016, www.ecowatch.com/justice-league -dakota-access-pipeline-2000093607.html; "Hansen and Hannah Arrested in West Virginia Mining Protest," *The Guardian*, June 24 2009, www.theguardian .com/environment/2009/jun/24/james-hansen-daryl-hannah-mining-protest.

33. Emily Holden, "Harvard and Yale Students Disrupt Football Game for Fossil Fuel Protest," *The Guardian*, November 24, 2019, www.theguardian .com/us-news/2019/nov/23/harvard-yale-football-game-protest-fossil-fuels.

34. See, for example, Rachel M. Cohen, "Will Bernie Sanders Stick with a Carbon Tax in His Push for a Green New Deal?," *The Intercept*, July 3, 2019, https:// theintercept.com/2019/07/03/bernie-sanders-climate-change-policy-carbon-tax.

35. Atkin, "France's Yellow Vest Protesters."

36. "Most Canadian Households to Get More in Rebates Than Paid in Carbon Tax: PBO," *Global News*, February 4, 2020, https://globalnews.ca/news /6504187/canada-carbon-tax-rebate-pbo.

37. Robert W. McElroy, "Pope Francis Brings a New Lens to Poverty, Peace and the Planet," *America: The Jesuit Review*, April 23, 2018, www .americamagazine.org/faith/2018/04/23/pope-francis-brings-new-lens-poverty -peace-and-planet; "Pope Francis Backs Carbon Pricing and 'Radical Energy Transition' to Act Against Global Warming," Australian Broadcasting Corporation, June 14, 2019, www.abc.net.au/news/2019-06-15/pope-backs-carbon -pricing-to-stem-global-warming/11212900.

38. Geoff Dembick, "Meet the Lawyer Trying to Make Big Oil Pay for Climate Change," *Vice*, December 22, 2017, www.vice.com/en_us/article/43qw3j /meet-the-lawyer-trying-to-make-big-oil-pay-for-climate-change.

39. David Hasemyer, "Fossil Fuels on Trial: Where the Major Climate Change Lawsuits Stand Today," *Inside Climate News*, January 17, 2020, https:// insideclimatenews.org/news/04042018/climate-change-fossil-fuel-company -lawsuits-timeline-exxon-children-california-cities-attorney-general.

40. Umair Irfan, "21 Kids Sued the Government over Climate Change. A Federal Court Dismissed the Case," *Vox*, January 17, 2020, www.vox.com /2020/1/17/21070810/climate-change-lawsuit-juliana-vs-us-our-childrens-trust -9th-circuit.

41. For a representative account, see Daryl Roberts, "Nature Conservancy Endorses Fossil Fuel Funded Trojan Horse," Alt Energy Stocks, May 27, 2019, www.altenergystocks.com/archives/2019/05/nature-conservancy-endorses -trojan-horse-tort-liability-waiver.

42. Dana Drugmand, "New Carbon Bills Won't Let Oil Companies Off the Hook for Climate Costs," *Climate Liability News*, July 31, 2019, www .climateliabilitynews.org/2019/07/31/carbon-bills-climate-liability-waiver.

43. I made these arguments in a review of Naomi Klein's book *On Fire* that I wrote for *Nature* in 2019: Michael E. Mann, "Radical Reform and the Green New Deal: Michael E. Mann Examines Naomi Klein's Collection on the Proposed US Policy Aiming to Curb Climate Change," *Nature*, September 19, 2019, www.nature.com/articles/d41586-019-02738-7.

44. See Brad Plumer, "Australia Repealed Its Carbon Tax—and Emissions Are Now Soaring," *Vox*, November 6, 2014, www.vox.com/2014/11/6/7157713 /australia-carbon-tax-repeal-emissions-rise; Brad Plumer, "Australia Is Repealing Its Controversial Carbon Tax," *Vox*, July 17, 2014, www.vox.com/2014 /7/17/5912143/australia-repeals-carbon-tax-global-warming.

45. Brian Kahn, "More Than 600 Environmental Groups Just Backed Ocasio-Cortez's Green New Deal," *Gizmodo*, January 10, 2019, https:// earther.gizmodo.com/more-than-600-environmental-groups-just-backed -ocasio-c-1831640541; "Green New Deal Letter to Congress," January 10, 2019, Scribd, www.scribd.com/document/397201459/Green-New-Deal-Letter-to -Congress.

46. "Sierra Club Position on Carbon Washington Ballot Initiative 732," Sierra Club, September 2016, www.sierraclub.org/washington/sierra-club-position -carbon-washington-ballot-initiative-732.

47. Butler, "How Australia Bungled Climate Policy."

48. Will Steffen, Johan Rockström, Katherine Richardson, Timothy M. Lenton, Carl Folke, Diana Liverman, Colin P. Summerhayes, et al., "Trajectories of the Earth System in the Anthropocene," *Proceedings of the National Academy of Sciences* 115, no. 33 (2018): 8252–8259, https://doi.org/10.1073 /pnas.1810141115.

49. Kate Aronoff, "'Hothouse Earth' Co-Author: The Problem Is Neoliberal Economics," *The Intercept*, August 14, 2018, https://theintercept.com/2018 /08/14/hothouse-earth-climate-change-neoliberal-economics.

50. At this point, you might reasonably be asking what authority I speak with when it comes to such matters. I would simply point out that, while I'm not an economist myself, I have coauthored peer-reviewed research with environmental economists on the topic of carbon pricing and I have attained some degree of familiarity with the discipline. See S. Lewandowsky, M. C. Freeman, and M. E. Mann, "Harnessing the Uncertainty Monster: Putting Quantitative Constraints on the Intergenerational Social Discount Rate," *Global and Planetary Change* 156 (2017): 155–166, https://doi.org/10.1016/j .gloplacha.2017.03.007.

51. Adam Tooze, "How Climate Change Has Supercharged the Left: Global Warming Could Launch Socialists to Unprecedented Power—and Expose Their Movement's Deepest Contradictions," *Foreign Policy*, January 15, 2020, https://foreignpolicy.com/2020/01/15/climate-socialism-supercharged -left-green-new-deal.

52. Mann, "Radical Reform and the Green New Deal."

53. I hasten to mention that I know a thing or two about being "on the front lines." See, for example, my book *The Hockey Stick and the Climate Wars: Dispatches from the Front Lines* (New York: Columbia University Press, 2013).

54. Eric Holthaus (@EricHolthaus), Twitter, November 7, 2019, 9:03 a.m., https://twitter.com/EricHolthaus/status/1192487740279066629.

55. The full thread can be found at Nathalie Molina Niño (@NathalieMolina), Twitter, November 6, 2019, 8:50 p.m., https://twitter.com/NathalieMolina /status/1192303192756936704. For example, one post, from "Patricia," said, "@MichaelEMann has more scientific knowledge in his cerebellum than you have in a hundred Trump heads. Deniers are done. Thanks to people of integrity, like Michael, we are moving on with solutions, including dismantling the fossil fuel industry." "Tenny" posted, "Michael Mann has been on the front lines doing all he can to get the public to understand what's going on, and taking threats of jail from congress critters and the VA AG, been sent white powder in an envelope, etc." "Ursula" tweeted, "As if we don't have enough on our plates. More than not helpful, backdoor approach to aiding and abetting deniers. You are the Front line and have battle scars to prove it. Wonder what her real issue is."

56. Tim Cronin, "Where 2020 Democrats Stand on Carbon Pricing," *Climate X-Change*, November 15, 2019, https://climate-xchange.org/2019/11/15 /where-2020-democrats-stand-on-carbon-pricing.

57. Gillian Tett, "The World Needs a Libor for Carbon Pricing," *Financial Times*, January 24, 2020, www.ft.com/content/20dd6b82-3dd1-11ea-a01a -bae547046735; "Exxon and Friends Still Funding Climate Denial and Obstruction Through IPAA, FTI, Energy in Depth," Climate Investigations Center, December 20, 2019, https://climateinvestigations.org/exxon-and-friends-still -funding-climate-denial-and-obstruction-through-ipaa-fti-energy-in-depth.

58. Tett, "The World Needs a Libor."

59. Thomas Kaplan, "Citing Health Risks, Cuomo Bans Fracking in New York State," *New York Times*, December 17, 2014, www.nytimes.com/2014/12/18 /nyregion/cuomo-to-ban-fracking-in-new-york-state-citing-health-risks.html.

60. Peter Behr, "Grid Chief: Enact Carbon Price to Reach 100% Clean Energy," *Energy and Environment News*, January 23, 2020, www.eenews.net /energywire/2020/01/23/stories/1062152903.

61. "What We Do," New York Independent System Operator, www.nyiso .com/what-we-do.

62. Behr, "Grid Chief."

63. "About the IMF," International Monetary Fund, www.imf.org/en /About.

64. Tett, "The World Needs a Libor"; "Special Report: Global Warming of 1.5°C," Intergovernmental Panel on Climate Change (IPCC), October 8, 2018, www.ipcc.ch/sr15.

65. Gillian Tett, Chris Giles, and James Politi, "US Threatens Retaliation Against EU over Proposed Carbon Tax," *Irish Times*, January 26, 2020, www.irishtimes.com/business/economy/us-threatens-retaliation-against -eu-over-proposed-carbon-tax-1.4151974.

66. Michael E. Mann and Jonathan Brockopp, "You Can't Save the Climate by Going Vegan. Corporate Polluters Must Be Held Accountable," *USA Today*, June 3, 2019, www.usatoday.com/story/opinion/2019/06/03/climate-change -requires-collective-action-more-than-single-acts-column/1275965001.

67. David Mastio (@DavidMastio), Twitter, November 6, 2019, 6:57 a.m., https://twitter.com/DavidMastio/status/1192093712400171011.

68. The full tweet is "11,000 scientists have declared we are in a climate emergency. Among other things, we need to move away from capitalism and instead prioritize 'sustaining ecosystems and improving human well-being by prioritizing basic needs and reducing inequality.'" Nora Biette-Timmons (@biettetimmons), Twitter, November 5, 2019, 7:21 a.m., https://twitter.com /biettetimmons/status/1191737368132366339.

69. George P. Shultz and Ted Halstead, "The Winning Conservative Climate Solution," *Washington Post*, January 16, 2020, www.washingtonpost.com /opinions/the-winning-republican-climate-solution-carbon-pricing/2020/01 /16/d6921dc0-387b-11ea-bf30-ad313e4ec754_story.html.

70. David Roberts (@drvox), Twitter, https://twitter.com/drvox/status /1218316952956997636.

71. Shultz and Halstead, "The Winning Conservative Climate Solution."

72. "Believe It or Not, a Republican Once Led the California Charge on Climate Change," KQED, September 13, 2018, www.kqed.org/science/1931206 /californias-a-team-on-climate-moonbeam-and-the-governator; Edward Helmore, "Angry Schwarzenegger Condemns Trump for Wrecking Clean-Air Standards," *The Guardian*, September 9, 2019, www.theguardian.com/us -news/2019/sep/09/schwarzenegger-trump-california-clean-air-emissions -climate-crisis; J. Edward Moreno, "Schwarzenegger Says Green New Deal Is 'Well Intentioned' but 'Bogus,'" *The Hill*, January 17, 2020, https://thehill .com/homenews/news/478847-schwarzenegger-says-green-new-deal-is-well -intentioned-but-bogus.

73. Myles Wearring and Emily Ackew, "Don't Leave Climate Change Action to the Left, David Cameron Urges Conservatives," Australian Broadcasting Corporation, January 30, 2019, www.abc.net.au/news/2020-01-29 /david-cameron-says-dont-leave-climate-change-action-to-the-left/11907804.

74. Shultz and Halstead, "The Winning Conservative Climate Solution."

75. Tooze, "How Climate Change Has Supercharged the Left."

76. Sheldon Whitehouse and James Slevin, "Carbon Pricing Represents the Best Answer to Our Climate Danger," *Washington Post*, March 10, 2020, www .washingtonpost.com/opinions/carbon-pricing-represents-the-best-answer-to -our-climate-danger/2020/03/10/379693ae-62fb-11ea-acca-80c22bbee96f _story.html.

77. See Kevin Anderson (@KevinClimate), Twitter, January 28, 2020, 12:53 a.m., https://twitter.com/KevinClimate/status/1222080140383080448.

78. This is one of the leading climate denial myths documented at Skeptical Science.com. See "Climate Scientists Would Make More Money in Other Careers," Skeptical Science, https://skepticalscience.com/climate-scientists-in -it-for-the-money.htm.

79. Brian A. Boyle, "Tulsi Gabbard May Not Be a Russian Asset. But She Sure Talks Like One," *Los Angeles Times*, October 25, 2019, www.latimes.com /opinion/story/2019-10-25/tulsi-gabbard-russian-asset-republican.

80. Tim Cronin, "Where 2020 Democrats Stand on Carbon Pricing," *Climate X-Change*, November 15, 2019, https://climate-xchange.org/2019/11/15 /where-2020-democrats-stand-on-carbon-pricing.

81. "Brendan O'Neill, Why Extinction Rebellion Seems So Nuts," *Spiked*, November 11, 2019, www.spiked-online.com/2019/11/07/why-extinction -rebellion-seems-so-nuts; Ben Pile, "Apocalypse Delayed: The IPCC Report Does Not Justify Climate Scaremongering," *Spiked*, October 18, 2019, www .spiked-online.com/2018/10/18/apocalypse-delayed.

82. O'Neill, "Why Extinction Rebellion Seems So Nuts."

83. "Special Report: Global Warming of 1.5°C."

84. George Monbiot, "How US Billionaires Are Fuelling the Hard-Right Cause in Britain," *The Guardian*, December 7, 2018, www.theguardian.com /commentisfree/2018/dec/07/us-billionaires-hard-right-britain-spiked -magazine-charles-david-koch-foundation.

85. James Hansen, "Game over for the Climate," *New York Times*, May 9, 2012, www.nytimes.com/2012/05/10/opinion/game-over-for-the-climate.html.

86. Quoted in Coral Davenport, "Citing Climate Change, Obama Rejects Construction of Keystone XL Oil Pipeline," *New York Times*, November 6, 2015, www.nytimes.com/2015/11/07/us/obama-expected-to-reject-construction -of-keystone-xl-oil-pipeline.html.

87. I made that argument in an interview. See "Dr. Michael Mann on Paris and the Clean Power Plan: 'We're Seeing Real Movement,'" *Climate Reality*, October 27, 2017, www.climaterealityproject.org/blog/dr-michael-mann-paris -and-clean-power-plan-were-seeing-real-movement.

88. Fred Hiatt, "How Donald Trump and Bernie Sanders Both Reject the Reality of Climate Change," *Washington Post*, February 23, 2020, www .washingtonpost.com/opinions/how-donald-trump-and-bernie-sanders-both -reject-the-reality-of-climate-change/2020/02/23/cc657dcc-54de-11ea-9e47 -59804be1dcfb_story.html.

CHAPTER 6: SINKING THE COMPETITION

1. Jocelyn Timperley, "The Challenge of Defining Fossil Fuel Subsidies," *Carbon Brief*, June 12, 2017, www.carbonbrief.org/explainer-the -challenge-of-defining-fossil-fuel-subsidies.

2. Dana Nuccitelli, "America Spends over $20bn per Year on Fossil Fuel Subsidies. Abolish Them," July 30, 2018, www.theguardian.com/environment /climate-consensus-97-per-cent/2018/jul/30/america-spends-over -20bn-per-year-on-fossil-fuel-subsidies-abolish-them.

3. "American Legislative Exchange Council," SourceWatch, www.source watch.org/index.php/American_Legislative_Exchange_Council.

4. See, for example, Kert Davies, "ALEC Lost Membership Worth over $7 Trillion in Market Cap," Climate Investigations Center, November 28, 2018, https://climateinvestigations.org/alec-lost-membership-worth-7-trillion.

5. "ALEC's Latest Scheme to Attack Renewables," *Renewable Energy World*, December 16, 2013, www.renewableenergyworld.com/2013/12/16/alecs-latest-scheme-to-attack-renewables.

6. Camille Erickson, "Bill to Penalize Utilities for Renewable Energy Returns to Wyoming Legislature, Quickly Fails," *Caspar Star-Tribune*, February 12, 2020, https://trib.com/business/energy/bill-to-penalize-utilities-for-renewable-energy-returns-to-wyoming/article_aafdd7cd-5012-5b8d-bf94-28341cea657f.html.

7. Suzanne Goldenberg and Ed Pilkington, "ALEC's Campaign Against Renewable Energy," *Mother Jones*, December 6, 2013, www.motherjones.com/environment/2013/12/alec-calls-penalties-freerider-homeowners-assault-clean-energy; Editorial Board, "The Koch Attack on Solar Energy," *New York Times*, April 26, 2014, www.nytimes.com/2014/04/27/opinion/sunday/the-koch-attack-on-solar-energy.html.

8. Suzanne Goldenberg, "Leak Exposes How Heartland Institute Works to Undermine Climate Science," *The Guardian*, February 15, 2012, www.theguardian.com/environment/2012/feb/15/leak-exposes-heartland-institute-climate.

9. "Heartland Institute," Energy and Policy Institute, www.energyandpolicy.org/attacks-on-renewable-energy-policy-by-fossil-fuel-interests-2013-2014/heartland-institute.

10. Carolyn Fortuna, "The Koch Brothers Have a Mandate to Destroy the EV Revolution—Are You Buying In?," *Clean Technica*, August 22, 2019, https://cleantechnica.com/2019/08/22/the-koch-brothers-have-a-mandate-to-destroy-the-ev-revolution-are-you-buying-in.

11. Ben Jervey, "Senator John Barrasso Parrots Koch Talking Points to Kill Electric Car Tax Credit," *DeSmog* (blog), February 5, 2019, www.desmogblog.com/2019/02/05/senator-john-barrasso-koch-talking-points-electric-car-tax-credit.

12. Fortuna, "The Koch Brothers Have a Mandate."

13. Will Oremus, "North Carolina May Ban Tesla Sales to Prevent 'Unfair Competition,'" *Slate*, May 13, 2013, https://slate.com/technology/2013/05/north-carolina-tesla-ban-bill-would-prevent-unfair-competition-with-car-dealerships.html.

14. Bruce Brown, "Confusing! North Carolina Bans Tesla Sales in Charlotte, Allows Them in Raleigh," *Digital Trends*, May 26, 2016, www.digitaltrends.com/cars/tesla-north-carolina-sales-charlotte-raleigh.

15. Will Oremus, "Free-Market Cheerleader Chris Christie Blocks Tesla Sales in New Jersey," *Slate*, March 12, 2014, https://slate.com/technology/2014/03/new-jersey-tesla-ban-chris-christie-loves-free-market-blocks-direct-car-sales.html.

16. The one "blue state" exception is Connecticut. See Union of Concerned Scientists, "Why You Can't Buy a Tesla in These 6 States," EcoWatch, February 26, 2017, www.ecowatch.com/states-cant-buy-tesla-2278638949.html.

17. Melissa C. Lott, "Solyndra—Illuminating Energy Funding Flaws?," *Scientific American*, September 27, 2011, https://blogs.scientificamerican.com /plugged-in/solyndra-illuminating-energy-funding-flaws.

18. Denise Robbins, "Study: How Mainstream Media Misled on the Success of the Clean Energy Loan Program," Media Matters, April 10, 2014, www.mediamatters.org/new-york-times/study-how-mainstream-media -misled-success-clean-energy-loan-program; Henry C. Jackson, "Program That Funded Solyndra Failure Producing Success Stories," *Washington Post*, December 30, 2014, www.washingtonpost.com/politics/program-that-funded-solyndra -failure-producing-success-stories/2014/12/30/3e896b46-9074-11e4-a900 -9960214d4cd7_story.html.

19. Amy Harder, "Obama Budget Would Pour Funds into Climate, Renewable Energy," *Wall Street Journal*, February 3, 2015, www.wsj.com/articles /obama-budget-would-pour-billions-into-climate-renewable-energy -1422903421.

20. See "The Daily Caller," SourceWatch, www.sourceWatch.org/index.php /The_Daily_Caller.

21. Robbins, "Study: How Mainstream Media Misled."

22. Elliott Negin, "The Wind Energy Threat to Birds Is Overblown," *Live Science*, December 3, 2013, www.livescience.com/41644-wind-energy-threat -to-birds-overblown.html.

23. Wendy Koch, "Wind Turbines Kill Fewer Birds Than Do Cats, Cell Towers," *USA Today*, September 15, 2014, www.usatoday.com/story /money/business/2014/09/15/wind-turbines-kill-fewer-birds-than-cell-towers -cats/15683843.

24. See, for example, my review of Sagan's *The Demon-Haunted World* in "Summer Books," *Nature*, August 3, 2017, www.nature.com/articles /548028a.

25. Simon Chapman, "How to Catch 'Wind Turbine Syndrome': By Hearing About It and Then Worrying," *The Guardian*, November 29, 2017, www .theguardian.com/commentisfree/2017/nov/29/how-to-catch-wind-turbine -syndrome-by-hearing-about-it-and-then-worrying.

26. Sharon Zhang, "Fossil Fuel Knocks the Wind out of Renewable Energy Movement in Ohio," *Salon*, January 5, 2020, www.salon.com/2020/01/05/fossil -fuel-knocks-the-wind-out-of-renewable-energy-movement-in-ohio_partner.

27. Fox Business, *Follow the Money*, November 12, 2010, quoted in Jill Fitzsimmons, "Myths and Facts About Wind Power: Debunking Fox's Abysmal Wind Coverage," *Think Progress*, May 31, 2012, https://archive.thinkprogress .org/myths-and-facts-about-wind-power-debunking-foxs-abysmal-wind-cov erage-e314b70c4059.

28. Philip Bump, "Trump Claims That Wind Farms Cause Cancer for Very Trumpian Reasons," *Washington Post*, April 3, 2019, www.washingtonpost .com/politics/2019/04/03/trump-claims-that-wind-farms-cause-cancer-very -trumpian-reasons.

29. Bump, "Trump Claims That Wind Farms Cause Cancer."

30. John Rodgers, "The Effect of Wind Turbines on Property Values: A New Study in Massachusetts Provides Some Answers," Union of Concerned Scientists, January 22, 2014, https://blog.ucsusa.org/john-rogers/effect-of-wind-turbines-on-property-values-384.

31. "Environmental Impacts of Solar Power," Union of Concerned Scientists, March 5, 2013, www.ucsusa.org/resources/environmental-impacts-solar-power.

32. Noel Wauchope, "A Radioactive Wolf in Green Clothing: Dissecting the Latest Pro-Nuclear Spin," *Independent Australia*, September 20, 2017, https://independentaustralia.net/environment/environment-display/a-radioactive-wolf-in-green-clothing-dissecting-the-latest-pro-nuclear-spin,10735. Among the Breakthrough Institute's original primary funders were the Cynthia and George Mitchell Foundation, which is tied to George Mitchell's fortune derived from natural gas extraction and fracking. "Who Funds Us," Breakthrough Institute, http://thebreakthrough.org/about/funders, accessed July 9, 2015. The foundation advocates for the continued extraction of natural gas. "Shale Sustainability," Cynthia and George Mitchell Foundation, www.cgmf.org/p/shale-sustainability-program.html, accessed July 29, 2015.

33. Clive Hamilton, "Climate Change and the Soothing Message of Luke-Warmism," *The Conversation*, July 25, 2012, https://theconversation.com/climate-change-and-the-soothing-message-of-luke-warmism-8445.

34. Thomas Gerke, "The Breakthrough Institute—Why the Hot Air?," *Clean Technica*, June 17, 2013, https://cleantechnica.com/2013/06/17/the-breakthrough-institute-why-the-hot-air.

35. Michael Shellenberger, "If Solar Panels Are So Clean, Why Do They Produce So Much Toxic Waste?," *Forbes*, May 23, 2018, www.forbes.com/sites/michaelshellenberger/2018/05/23/if-solar-panels-are-so-clean-why-do-they-produce-so-much-toxic-waste.

36. "Environmental Impacts of Solar Power."

37. Michael Shellenberger, "The Real Reason They Hate Nuclear Is Because It Means We Don't Need Renewables," *Forbes*, February 14, 2019, www.forbes.com/sites/michaelshellenberger/2019/02/14/the-real-reason-they-hate-nuclear-is-because-it-means-we-dont-need-renewables.

38. "Solar Energy Plants in Tortoises' Desert Habitat Pit Green Against Green," Fox News, February 20, 2014, www.foxnews.com/us/solar-energy-plants-in-tortoises-desert-habitat-pit-green-against-green.

39. Associated Press, "Environmental Concerns Threaten Solar Power Expansion in California Desert," Fox News, April 18, 2009, www.foxnews.com/story/environmental-concerns-threaten-solar-power-expansion-in-california-desert; Alex Pappas, "Massive East Coast Solar Project Generates Fury from Neighbors," Fox News, February 15, 2019, www.foxnews.com/politics/massive-east-coast-solar-project-generates-fury-from-neighbors-in-virginia; "World's Largest Solar Plant Scorching Birds in Nevada Desert," Fox News, February 15, 2014, www.foxnews.com/us/worlds-largest-solar-plant-scorching-birds-in-nevada-desert.

40. Lee Moran, "Fox News' Jesse Watters Gets Schooled over Nonsensical Winter Solar Panels Claim," *Huffington Post*, February 1, 2019, www.huffpost .com/entry/fox-news-jesse-watters-solar-panels_n_5c540aa7e4b043e25b1b2168.

41. Amy Remeikis, "'Shorten Wants to End the Weekend': Morrison Attacks Labor's Electric Vehicle Policy," *The Guardian*, April 7, 2019, www.theguardian .com/australia-news/2019/apr/07/shorten-wants-to-end-the-weekend-morrison -attacks-labors-electric-vehicle-policy.

42. As noted at "Breakthrough Institute," Wikipedia, https://en.wikipedia .org/wiki/Breakthrough_Institute, accessed February 24, 2020.

43. See, for example, Bill Gates, "Two Videos That Illuminate Energy Poverty," GatesNotes, June 25, 2014, www.gatesnotes.com/Energy/Two-Videos -Illuminate-Energy-Poverty-Bjorn-Lomborg. Tillerson is quoted in Michael Babad, "Exxon Mobil CEO: 'What Good Is It to Save the Planet If Humanity Suffers?,'" *Globe and Mail*, May 30, 2013, www.theglobeandmail .com/report-on-business/top-business-stories/exxon-mobil-ceo-what-good -is-it-to-save-the-planet-if-humanity-suffers/article12258350.

44. Graham Readfearn, "The Millions Behind Bjorn Lomborg's Copenhagen Consensus Center US Think Tank," *DeSmog* (blog), June 24, 2014, www .desmogblog.com/2014/06/25/millions-behind-bjorn-lomborg-copenhagen -consensus-center; "Independent Women's Forum," SourceWatch, www.source watch.org/index.php/Independent_Women%27s_Forum; "Donors Trust: Building a Legacy of Liberty," *DeSmog* (blog), www.desmogblog.com/who-donors -trust; "Claude R. Lambe Charitable Foundation," Conservative Transparency, http://conservativetransparency.org/donor/claude-r-lambe-charitable -foundation; Pete Altman, "House Committee to Vote on Fred Upton's Asthma Aggravation Act of 2011," Natural Resources Defense Council, March 15, 2011, www.nrdc.org/experts/pete-altman/house-committee-vote-fred-uptons -asthma-aggravation-act-2011.

45. "UWA Cancels Contract for Consensus Centre Involving Controversial Academic Bjorn Lomborg," Australian Broadcasting Corporation, May 8, 2015, www.abc.net.au/news/2015-05-08/bjorn-lomborg-uwa-consensus-centre -contract-cancelled/6456708.

46. Graham Readfearn, "Is Bjorn Lomborg Right to Say Fossil Fuels Are What Poor Countries Need?" *The Guardian*, December 6, 2013, www .theguardian.com/environment/planet-oz/2013/dec/06/bjorn-lomborg -climate-change-poor-countries-need-fossil-fuels.

47. Bjorn Lomborg, "Who's Afraid of Climate Change?," Project Syndicate, August 11, 2010, www.project-syndicate.org/commentary/who-s-afraid -of-climate-change.

48. Jonathan Chait, "GOP Senator Upbeat Coronavirus May Kill 'No More Than 3.4 Percent of Our Population,'" *New York Magazine*, March 18, 2020, https://nymag.com/intelligencer/2020/03/gop-senator-no-more-than-3-4-of -our-population-may-die.html.

49. Pope Francis, "Address of His Holiness Pope Francis to the Members of the Diplomatic Corps Accredited to the Holy See," The Holy See, January

13, 2014, http://w2.vatican.va/content/francesco/en/speeches/2014/january
/documents/papa-francesco_20140113_corpodiplomatico.html.

50. R. Jai Krishna, "Renewable Energy Powers Up Rural India," *Wall Street Journal*, July 29, 2015, www.wsj.com/articles/renewable-energy-powers-up
-rural-india-1438193488.

51. Pope Francis, "Address of His Holiness Pope Francis"; "DoD Releases Report on Security Implications of Climate Change," US Department of Defense, DOD News, July 29, 2015, www.defense.gov/news/newsarticle
.aspx?id=129366. The Defense Department report notes that "global climate change will aggravate problems such as poverty, social tensions, environmental degradation, ineffectual leadership and weak political institutions that threaten stability in a number of countries."

52. "World Bank Says Climate Change Could Thrust 100 Million into Deep Poverty by 2030," Fox News, November 8, 2015, www.foxnews.com
/world/2015/11/08/world-bank-says-climate-change-could-thrust-100-million
-into-deep-poverty-by.

53. "Koch Alum's Dark Money Group, 'Power the Future,' Denies Its Own Lobbying Status," *DeSmog* (blog), April 16, 2018, www.desmogblog.com/2018
/04/16/koch-alum-s-dark-money-group-power-future-denies-its-own-lobbying
-status.

54. Nadja Popovich, "Today's Energy Jobs Are in Solar, Not Coal," *New York Times*, April 25, 2017, www.nytimes.com/interactive/2017/04/25/climate
/todays-energy-jobs-are-in-solar-not-coal.html.

55. Sheldon Whitehouse and James Slevin, "Carbon Pricing Represents the Best Answer to Our Climate Danger," *Washington Post*, March 10, 2020, www
.washingtonpost.com/opinions/carbon-pricing-represents-the-best-answer
-to-our-climate-danger/2020/03/10/379693ae-62fb-11ea-acca-80c22
bbee96f_story.html.

56. Sam Haysom, "Michael Moore Talks to Stephen Colbert About His New Climate Change Documentary," *Mashable*, April 22, 2020, https://
mashable.com/video/michael-moore-planet-of-the-humans.

57. See, for example, Lindsey Bahr, "New Michael Moore-Backed Doc Tackles Alternative Energy," Associated Press, August 8, 2019, https://abcnews
.go.com/Entertainment/wireStory/michael-moore-backed-doc-tackles
-alternative-energy-64844048; "Editorial: Michael Moore-Backed Film Criti-cizes Renewable Energy," *Las Vegas Review-Journal*, August 18, 2019, www
.reviewjournal.com/opinion/editorials/editorial-michael-moore-backed-film
-criticizes-renewable-energy-1829377.

58. "Michael Moore Presents: Planet of the Humans," Full Documentary, Directed by Jeff Gibbs, YouTube, posted April 21, 2020, www.youtube.com
/watch?v=Zk11vI-7czE.

59. See, for example, Ketan Joshi, "Planet of the Humans: A Reheated Mess of Lazy, Old Myths," April 24, 2020, https://ketanjoshi.co/2020/04/24/planet
-of-the-humans-a-reheated-mess-of-lazy-old-myths; Leah H. Stokes, "Michael Moore Produced a Film About Climate Change That's a Gift to Big Oil," *Vox*,

April 28, 2020, www.vox.com/2020/4/28/21238597/michael-moore-planet-of
-the-humans-climate-change. A compendium of critical responses is available
at "Moore's Boorish Planet of the Humans: An Annotated Collection," Get Energy Smart Now!, April 25, 2020, http://getenergysmartnow.com/2020/04/25
/moores-boorish-planet-of-the-humans-an-annotated-collection.

60. See Michelle Froese, "Renewables Exceed 20.3% of U.S. Electricity
and Outpace Nuclear Power," *Windpower Engineering*, July 29, 2019, www
.windpowerengineering.com/renewables-exceed-20-3-of-u-s-electricity-and
-outpace-nuclear-power.

61. Mark Z. Jacobson, Mark A. Delucchi, Zack A.F. Bauer, Savannah C.
Goodman, William E. Chapman, Mary A. Cameron, Cedric Bozannat, et al.,
"100% Clean and Renewable Wind, Water, and Sunlight All-Sector Energy
Roadmaps for 139 Countries of the World," *Joule* 1, no. 1 (2017): 108–121,
https://doi.org/10.1016/j.joule.2017.07.005.

62. Michael Moore (@MMFlint), Twitter, April 24, 2020, 6:26 p.m., https://
twitter.com/MMFlint/status/1253857924750999552.

63. See "Frequently Asked Questions: How Much Carbon Dioxide Is Produced per Kilowatthour of U.S. Electricity Generation?," US Energy Information Administration, www.eia.gov/tools/faqs/faq.php?id=74&t=11.

64. Doug Boucher, "Movie Review: There's a Vast Cowspiracy About Climate Change," Union of Concerned Scientists, June 10, 2016, https://blog
.ucsusa.org/doug-boucher/cowspiracy-movie-review.

65. See Biofuelwatch (@biofuelwatch), Twitter, April 27, 2020, 3:42 a.m.,
https://twitter.com/biofuelwatch/status/1254722596530241537.

66. "Editorial: Michael Moore-Backed Film Criticizes Renewable Energy."

67. See Bill McKibben, "Response: Planet of the Humans Documentary,"
350.org, April 22, 2020, https://350.org/response-planet-of-the-humans
-documentary.

68. "Editorial: Michael Moore-Backed Film Criticizes Renewable Energy."

69. "Michael Moore Net Worth," Celebrity Net Worth, www.celebrity
networth.com/richest-celebrities/directors/michael-moore-net-worth.

70. Peter Bradshaw, "Planet of the Humans Review—Contrarian Eco-
Doc from the Michael Moore Stable," *The Guardian*, April 22, 2020, www
.theguardian.com/film/2020/apr/22/planet-of-the-humans-review-environment
-michael-moore-jeff-gibbs.

71. Neal Livingston, "Forget About Planet of the Humans," *Films for Action*, April 24, 2020, www.filmsforaction.org/articles/film-review-forget-about
-planet-of-the-humans.

72. Joshi, "Planet of the Humans: A Reheated Mess."

73. Brian Kahn, "Planet of the Humans Comes This Close to Actually Getting the Real Problem, Then Goes Full Ecofascism," *Gizmodo*, April 20, 2020,
https://earther.gizmodo.com/planet-of-the-humans-comes-this-close-to
-actually-getti-1843024329.

74. AFP, "World's Richest 10% Produce Half of Global Carbon Emissions, Says Oxfam," *The Guardian*, December 2, 2015, www.theguardian

.com/environment/2015/dec/02/worlds-richest-10-produce-half-of-global
-carbon-emissions-says-oxfam.

75. Grant Samms (@grantsamms), Twitter, April 23, 2020, 9:05 a.m., https://twitter.com/grantsamms/status/1253354390943076352.

76. James Delingpole, "Michael Moore Is Now the Green New Deal's Worst Enemy," Breitbart, April 23, 2020, www.breitbart.com/entertainment/2020/04/23/delingpole-michael-moore-is-now-the-green-new-deals-worst-enemy.

77. "Competitive Enterprise Institute," SourceWatch, www.sourcewatch.org/index.php/Competitive_Enterprise_Institute, accessed April 29, 2020; "Heartland Institute," SourceWatch, www.sourcewatch.org/index.php/Heartland_Institute, accessed April 29, 2020; "Anthony Watts," SourceWatch, www.sourcewatch.org/index.php/Anthony_Watts, accessed April 29, 2020.

78. Myron Ebell, "Hurry, See 'Planet of the Humans,' Before It's Banned," Competitive Enterprise Institute, April 24, 2020, https://cei.org/blog/hurry-see-planet-humans-it%E2%80%99s-banned; Donny Kendal, Justin Haskins, Isaac Orr, and Jim Lakely, "In the Tank (Episode 240)—Review: Michael Moore's Planet of the Humans," Heartland Institute, April 24, 2020, www.heartland.org/multimedia/podcasts/in-the-tank-ep240--review-michael-moores-planet-of-the-humans.

79. Anthony Watts, "#EarthDay EPIC! Michael Moore's New Film Trashes 'Planet Saving' Renewable Energy—Full Movie Here!," Watts Up with That, April 22, 2020, https://wattsupwiththat.com/2020/04/22/earthday-epic-michael-moores-new-film-trashes-planet-saving-renewable-energy-full-movie-here.

80. "Steven J. Milloy," SourceWatch, www.sourcewatch.org/index.php/Steven_J._Milloy, accessed April 29, 2020; Steve Milloy (@JunkScience), Twitter, April 27, 2020, 6:55 a.m., https://twitter.com/junkscience/status/1254771076870975491.

81. "Marc Morano," SourceWatch, www.sourcewatch.org/index.php/Marc_Morano, accessed April 29, 2020; Marc Morano (@ClimateDepot), Twitter, June 5, 2020, 4:35 p.m., https://twitter.com/ClimateDepot/status/1269050171809120260.

82. Emily Atkin, "A Party for the Planet('s Destruction): A Powerful Anti-Climate Group Spent Thousands to Promote Michael Moore's Climate Documentary on Facebook This Week," Heated, May 19, 2020, https://heated.world/p/a-party-for-the-planets-destruction.

83. "Film-Maker Michael Moore Visits Julian Assange at Embassy," Irish Independent, June 10, 2016, www.independent.ie/style/celebrity/celebrity-news/film-maker-michael-moore-visits-julian-assange-at-embassy-34788794.html.

84. Adam Tooze, "How Climate Change Has Supercharged the Left: Global Warming Could Launch Socialists to Unprecedented Power—and Expose Their Movement's Deepest Contradictions," Foreign Policy, January 15, 2020, https://foreignpolicy.com/2020/01/15/climate-socialism-supercharged-left-green-new-deal.

85. Emily Atkin, "The Wheel of First-Time Climate Dudes. Or, Alternatively: Why I Don't Want to Review Michael Moore's Climate Change Documentary," *Heated*, April 23, 2020, https://heated.world/p/the-wheel-of-first-time-climate-dudes.

86. Laura Geggel, "Bill Gates 'Discovers' 14-Year-Old Formula on Climate Change," *Live Science*, February 26, 2016, www.livescience.com/53861-bill-gates-climate-formula-not-new.html; Michael E. Mann, "FiveThirtyEight: The Number of Things Nate Silver Gets Wrong About Climate Change," *Huffington Post*, November 24, 2012, www.huffpost.com/entry/nate-silver-climate-change_b_1909482.

87. Giles Parkinson, "How the Tesla Big Battery Has Smoothed the Transition to Zero Emissions Grid," Renew Economy, March 1, 2020. See also Randell Suba, "Tesla 'Big Battery' in Australia Is Becoming a Bigger Nightmare for Fossil Fuel Power Generators," Teslarati, February 28, 2020, www.teslarati.com/tesla-big-battery-hornsdale-australia-cost-savings.

88. Jacobson et al., "100% Clean and Renewable Wind, Water, and Sunlight All-Sector Energy Roadmaps."

89. "Bill Gates Q&A on Climate Change: 'We Need a Miracle,'" *Denver Post*, February 13, 2016, www.denverpost.com/2016/02/23/bill-gates-qa-on-climate-change-we-need-a-miracle.

90. Jacobson et al., "100% Clean and Renewable Wind, Water, and Sunlight All-Sector Energy Roadmaps."

91. "David E. Wojick," SourceWatch, www.sourcewatch.org/index.php?title=David_E._Wojick; David Wojick, "Providing 100 Percent Energy from Renewable Sources Is Impossible," Heartland Institute, February 12, 2020, www.heartland.org/news-opinion/news/providing-100-percent-energy-from-renewable-sources-is-impossible.

92. Patrick Quinn, "After Devastating Tornado, Town Is Reborn 'Green,'" *USA Today*, *Green Living* magazine, April 13, 2013, www.usatoday.com/story/news/greenhouse/2013/04/13/greensburg-kansas/2078901.

93. Will Oremus, "Fox News Claims Solar Won't Work in America Because It's Not Sunny Like Germany," *Slate*, February 7, 2013, https://slate.com/technology/2013/02/fox-news-expert-on-solar-energy-germany-gets-a-lot-more-sun-than-we-do-video.html.

94. Max Greenberg, "Fox Cedes Solar Industry to Germany," Media Matters, February 7, 2013, www.mediamatters.org/fox-friends/fox-cedes-solar-industry-germany.

95. Oremus, "Fox News Claims Solar Won't Work in America."

CHAPTER 7: THE NON-SOLUTION SOLUTION

1. Eillie Anzilotti, "Climate Change Is Inevitable. How Bad It Gets Is a Choice," *Fast Company*, March 12, 2019, www.fastcompany.com/90318242/climate-change-is-inevitable-how-bad-it-gets-is-a-choice.

2. Paul Muschick, "Pennsylvania Is Heating Up Because of Climate Change. Let's Do Something About It," *Morning Call*, November 15, 2019, www .mcall.com/opinion/mc-opi-climate-change-costs-pennsylvania-muschick -20191115-mxaqyumnzfdctnictlskzb6sc4-story.html.

3. Jane Bardon, "How the Beetaloo Gas Field Could Jeopardise Australia's Emissions Target," Australian Broadcasting Corporation, February 29, 2019, www.abc.net.au/news/2020-02-29/beetaloo-basin-gas-field-could-jeopardise -paris-targets/12002164.

4. "Scott Morrison Announces $2 Billion Energy Deal to Boost Gas Use," SBS News, January 31, 2020, www.sbs.com.au/news/scott-morrison -announces-2-billion-energy-deal-to-boost-gas-use.

5. Luke O'Neil, "US Energy Department Rebrands Fossil Fuels as 'Molecules of Freedom,'" *The Guardian*, May 30, 2019, www.theguardian.com/business /2019/may/29/energy-department-molecules-freedom-fossil-fuel-rebranding.

6. Amanda Amos and Margaretha Haglund, "From Social Taboo to 'Torch of Freedom': The Marketing of Cigarettes to Women," *Tobacco Control* 9, no. 1 (2000): 3–8.

7. "What Is the Emissions Impact of Switching from Coal to Gas?," *Carbon Brief*, October 27, 2014, www.carbonbrief.org/what-is-the-emissions-impact -of-switching-from-coal-to-gas.

8. Gayathri Vaidyanathan, "How Bad of a Greenhouse Gas Is Methane?," *ClimateWire*, December 22, 2015, www.scientificamerican.com/article/how -bad-of-a-greenhouse-gas-is-methane.

9. Emily Holden, "Trump Administration to Roll Back Obama-Era Pollution Regulations," *The Guardian*, August 30, 2019, www.theguardian .com/environment/2019/aug/29/trump-administration-roll-back-methane -regulations.

10. Andrew Nikiforuk, "New Study Finds Far Greater Methane Threat from Fossil Fuel Industry," *The Tyee*, February 21, 2020, https://thetyee .ca/News/2020/02/21/Fossil-Fuel-Industry-Far-Greater-Methane-Threat -Study-Finds.

11. Jonathan Mingle, "Atmospheric Methane Levels Are Going Up— And No One Knows Why," *Wired*, May 5, 2019, www.wired.com/story /atmospheric-methane-levels-are-going-up-and-no-one-knows-why.

12. Brendan O'Neill, "Why Extinction Rebellion Seems So Nuts," *Spiked*, November 11, 2019, www.spiked-online.com/2019/11/07/why-extinction -rebellion-seems-so-nuts.

13. Prachi Patel, "New Projects Show Carbon Capture Is Not Dead," *IEEE Spectrum*, January 16, 2017, https://spectrum.ieee.org/energywise/green-tech /clean-coal/carbon-capture-is-not-dead-but-will-it-blossom.

14. Christa Marshall, "Clean Coal Power Plant Killed, Again," *Climatewire*, *E&E News*, February 4, 2015, reprinted by *Scientific American*, www.scientific american.com/article/clean-coal-power-plant-killed-again.

15. Dipka Bhambhani, "Everyone Wants Carbon Capture And Sequestration—Now How to Make It a Reality?," November 21, 2019, www.forbes

.com/sites/dipkabhambhani/2019/11/21/washington-to-wall-street-hears
-harmony-on-ccs-to-address-climate-change/#6d029ffe35da.

16. See Michael Barnard's answer on September 26, 2019, to "Are Industrial Carbon Capture Plants Carbon Neutral in Operation?," Quora, www.quora.com/Are-industrial-carbon-capture-plants-carbon-neutral-in-operation.

17. Brian Kahn, "More Than 600 Environmental Groups Just Backed Ocasio-Cortez's Green New Deal," *Gizmodo*, January 10, 2019, https://earther.gizmodo.com/more-than-600-environmental-groups-just-backed-ocasio-c-1831640541.

18. Robinson Meyer, "The Green New Deal Hits Its First Major Snag," *The Atlantic*, January 18, 2019, www.theatlantic.com/science/archive/2019/01/first-fight-about-democrats-climate-green-new-deal/580543.

19. Meyer, "The Green New Deal Hits Its First Major Snag."

20. James Temple, "Let's Keep the Green New Deal Grounded in Science," *MIT Technology Review*, January 19, 2019, www.technologyreview.com/s/612780/lets-keep-the-green-new-deal-grounded-in-science.

21. "Global Effects of Mount Pinatubo," NASA, Earth Observatory, https://earthobservatory.nasa.gov/images/1510/global-effects-of-mount-pinatubo.

22. Francisco Toro, "Climate Politics Is a Dead End. So the World Could Turn to This Desperate Final Gambit," *Washington Post*, December 18, 2019, www.washingtonpost.com/opinions/2019/12/18/climate-politics-is-dead-end-so-world-could-turn-this-desperate-final-gambit.

23. For a review of the potential pitfalls of sulfate aerosol geoengineering, see "Scientists to Stop Global Warming with 100,000 Square Mile Sun Shade," *The Telegraph*, February 26, 2009, www.telegraph.co.uk/news/earth/environment/globalwarming/4839985/Scientists-to-stop-global-warming-with-100000-square-mile-sun-shade.html.

24. Eli Kintisch, "Climate Hacking for Profit: A Good Way to Go Broke," *Fortune*, May 21, 2010, http://archive.fortune.com/2010/05/21/news/economy/geoengineering.climos.planktos.fortune/index.htm.

25. Gaia Vince, "Sucking CO2 from the Skies with Artificial Trees," BBC, October 4, 2012, www.bbc.com/future/story/20121004-fake-trees-to-clean-the-skies.

26. Johannes Lehmann and Angela Possinger, "Removal of Atmospheric CO_2 by Rock Weathering Holds Promise for Mitigating Climate Change," *Nature*, July 8, 2020, www.nature.com/articles/d41586-020-01965-7.

27. Daniel Hillel, *The Rivers of Eden: The Struggle for Water and the Quest for Peace in the Middle East* (Oxford: Oxford University Press, 1994).

28. Patrick Galey, "Industry Guidance Touts Untested Tech as Climate Fix," Phys.org, August 23, 2019, https://phys.org/news/2019-08-industry-guidance-touts-untested-tech.html.

29. "Fuel to the Fire: How Geoengineering Threatens to Entrench Fossil Fuels and Accelerate the Climate Crisis," Center for International Environmental Law, February 2019, www.ciel.org/wp-content/uploads/2019/02/CIEL_FUEL-TO-THE-FIRE_How-Geoengineering-Threatens-to-Entrench-Fossil-Fuels-and-Accelerate-the-Climate-Crisis_February-2019.pdf.

30. Kate Connolly, "Geoengineering Is Not a Quick Fix for Climate Change, Experts Warn Trump," *The Guardian*, October 14, 2017, www.theguardian.com/environment/2017/oct/14/geoengineering-is-not-a-quick-fix-for-climate-change-experts-warn-trump.

31. See Bjorn Lomborg, "Geoengineering: A Quick, Clean Fix?," *Time*, November 14, 2010, http://content.time.com/time/magazine/article/0,9171,2030804,00.html; Colin McInnes, "Time to Embrace Geoengineering," Breakthrough Institute, June 27, 2013, http://thebreakthrough.org/index.php/programs/energy-and-climate/time-to-embrace-geoengineering.

32. Marc Gunther, "The Business of Cooling the Planet," *Fortune*, October 7, 2011, https://fortune.com/2011/10/07/the-business-of-cooling-the-planet.

33. Benjamin Franta and Geoffrey Supran, "The Fossil Fuel Industry's Invisible Colonization of Academia," *The Guardian*, March 13, 2017, www.theguardian.com/environment/climate-consensus-97-per-cent/2017/mar/13/the-fossil-fuel-industrys-invisible-colonization-of-academia.

34. James Temple, "The Growing Case for Geoengineering," *Technology Review*, April 18, 2017, www.technologyreview.com/s/604081/the-growing-case-for-geoengineering.

35. "David Keith," Breakthrough Institute, https://thebreakthrough.org/people/david-keith; "An Ecomodernist Manifesto," www.ecomodernism.org; George Monbiot, "Meet the Ecomodernists: Ignorant of History and Paradoxically Old-Fashioned," *The Guardian*, February 24, 2015, www.theguardian.com/environment/georgemonbiot/2015/sep/24/meet-the-ecomodernists-ignorant-of-history-and-paradoxically-old-fashioned.

36. Gunther, "The Business of Cooling the Planet"; James Temple, "This Scientist Is Taking the Next Step in Geoengineering," *Technology Review*, July 26, 2017, www.technologyreview.com/s/608312/this-scientist-is-taking-the-next-step-in-geoengineering.

37. Peter Irvine, Kerry Emanuel, Jie He, Larry W. Horowitz, Gabriel Vecchi, and David Keith, "Halving Warming with Idealized Solar Geoengineering Moderates Key Climate Hazards," *Nature Climate Change* 9 (2019): 295–299, https://doi.org/10.1038/s41558-019-0398-8; Peter Irvine, the first author, is Keith's postdoctoral fellow. Keith, the principal investigator, signed his name at the end of the author list.

38. See this Twitter thread: Chris Colose (@CColose), Twitter, March 11, 2019, 3:03 p.m., https://twitter.com/CColose/status/1105227667689951234.

39. Ken Caldeira (@KenCaldeira), Twitter, March 23, 2020, 7:59 a.m., https://twitter.com/KenCaldeira/status/1242103749989949441; Ken Caldeira (@KenCaldeira), Twitter, August 24, 2019, 11:54 a.m., https://twitter.com/KenCaldeira/status/1165336530983874560.

40. Daniel Swain (@Weather_West), Twitter, August 24, 2019, 12:03 p.m., https://twitter.com/Weather_West/status/1165338821468123136.

41. Dr. Jonathan Foley (@GlobalEcoGuy), Twitter, August 24, 2019, 12:01 p.m., https://twitter.com/GlobalEcoGuy/status/1165338382655819776.

42. Matthew Huber (@climatedynamics), Twitter, August 24, 2019, 4:57 p.m., https://twitter.com/climatedynamics/status/1165412731379507200.

43. Michael E. Mann, "If You See Something, Say Something," *New York Times*, January 17, 2014, www.nytimes.com/2014/01/19/opinion/sunday/if-you-see-something-say-something.html.

44. Michael E. Mann (@MichaelEMann), Twitter, March 12, 2019, 6:23 a.m., https://twitter.com/MichaelEMann/status/1105459225852022785.

45. See Michael E. Mann (@MichaelEMann), Twitter, March 12, 2019, 6:53 a.m., https://twitter.com/MichaelEMann/status/1105466799657807872; Michael E. Mann (@MichaelEMann), Twitter, March 12, 2019, 12:14 p.m., https://twitter.com/MichaelEMann/status/1105547711569424385.

46. Toro, "Climate Politics Is a Dead End."

47. See Michael E. Mann (@MichaelEMann), Twitter, December 18, 2019, 1:44 p.m., https://twitter.com/MichaelEMann/status/1207416244460109824; Michael E. Mann (@MichaelEMann), Twitter, December 18, 2019, 1:50 p.m., https://twitter.com/MichaelEMann/status/1207417769119158272.

48. Temple, "The Growing Case for Geoengineering."

49. "Fuel to the Fire."

50. Umair Irfan, "Tree Planting Is Trump's Politically Safe New Climate Plan," *Vox*, February 4, 2020, www.vox.com/2020/2/4/21123456/sotu-trump-trillion-trees-climate-change.

51. Madeleine Gregory and Sarah Emerson, "Planting 'Billions of Trees' Isn't Going to Stop Climate Change: A Popular Study Claims That Reforestation Could Fix Climate Change, But Is That True?," *Vice*, July 16, 2019, www.vice.com/en_au/article/7xgymg/planting-billions-of-trees-isnt-going-to-stop-climate-change.

52. Mark Maslin and Simon Lewis, "Yes, We Can Reforest on a Massive Scale—but It's No Substitute for Slashing Emissions," *Climate Home News*, May 7, 2019, www.climatechangenews.com/2019/07/05/yes-can-reforest-massive-scale-no-substitute-slashing-emissions.

53. Emma Farge and Stephanie Nebehay, "Greenhouse Emissions Rise to More Than 55 Gigatonnes of CO2 Equivalent," *Business Day*, November 26, 2019, www.businesslive.co.za/bd/world/2019-11-26-greenhouse-emissions-rise-to-more-than-55-gigatonnes-of-co2-equivalent.

54. Gregory and Emerson, "Planting 'Billions of Trees' Isn't Going to Stop Climate Change."

55. Andrew Freedman, "Australia Fires: Yearly Greenhouse Gas Emissions Nearly Double Due to Historic Blazes," *The Independent*, January 25, 2020, www.independent.co.uk/news/world/australasia/australia-fires-greenhouse-gas-emissions-climate-crisis-fossil-fuel-a9301396.html.

56. Laura Millan Lombrana, Hayley Warren, and Akshat Rathi, "Measuring the Carbon-Dioxide Cost of Last Year's Worldwide Wildfires," Bloomberg Green, February 10, 2020, www.bloomberg.com/graphics/2020-fire-emissions.

57. Fiona Harvey, "Tropical Forests Losing Their Ability to Absorb Carbon, Study Finds," *The Guardian*, March 5, 2020, www.theguardian.com/environment/2020/mar/04/tropical-forests-losing-their-ability-to-absorb-carbon-study-finds.

58. Roger Harrabin, "Climate Change: UK Forests 'Could Do More Harm Than Good,'" BBC, April 7, 2020, www.bbc.com/news/science-environment-52200045.

59. Leo Hickman, "The History of BECCS," *Carbon Brief*, April 13, 2016, www.carbonbrief.org/beccs-the-story-of-climate-changes-saviour-technology.

60. See Robert Jay Lifton and Naomi Oreskes, "The False Promise of Nuclear Power," *Boston Globe*, July 29, 2019, www.bostonglobe.com/opinion/2019/07/29/the-false-promise-nuclear-power/kS8rzs8f7MAONgXL1fWOGK/story.html.

61. See Lifton and Oreskes, "The False Promise of Nuclear Power."

62. R. Singh, T. Wagener, R. Crane, M. E. Mann, and L. Ning, "A Stakeholder Driven Approach to Identify Critical Thresholds in Climate and Land Use for Selected Streamflow Indices—Application to a Pennsylvania Watershed," *Water Resources Research* 50 (2014): 3409–3427, https://doi.org/10.1002/2013WR014988.

63. M. V. Ramana and Ali Ahmad, "Wishful Thinking and Real Problems: Small Modular Reactors, Planning Constraints, and Nuclear Power in Jordan," *Energy Policy* 93 (2016): 236–245, https://doi.org/10.1016/j.enpol.2016.03.012.

64. James A. Lake, Ralph G. Bennett, and John F. Kotek, "Next Generation Nuclear Power: New, Safer and More Economical Nuclear Reactors Could Not Only Satisfy Many of Our Future Energy Needs but Could Combat Global Warming as Well," *Scientific American*, January 26, 2009, www.scientificamerican.com/article/next-generation-nuclear.

65. Nathanael Johnson, "Next-Gen Nukes: Scores of Nuclear Startups Are Aiming to Solve the Problems That Plague Nuclear Power," *Grist*, July 18, 2018, https://grist.org/article/next-gen-nuclear-is-coming-if-we-want-it.

66. See Lifton and Oreskes, "The False Promise of Nuclear Power."

67. See, for example, this op-ed by four leading climate science colleagues: Ken Caldeira, Kerry Emanuel, James Hansen, and Tom Wigley, "Top Climate Change Scientists' Letter to Policy Influencers," CNN, November 3, 2013, https://edition.cnn.com/2013/11/03/world/nuclear-energy-climate-change-scientists-letter/index.html.

68. "Bob Inglis—Acceptance Speech," John F. Kennedy Presidential Library and Museum, www.jfklibrary.org/node/4466.

69. See Lifton and Oreskes, "The False Promise of Nuclear Power."

70. David Roberts, "Hey, Look, a Republican Who Cares About Climate Change!," *Grist*, July 10, 2012, https://grist.org/article/hey-look-a-republican-who-cares-about-climate-change.

71. "How Fareed Zakaria Became the Most Conservative Liberal of All Time," *Deadline Detroit*, May 29, 2012, www.deadlinedetroit.com/articles/555/how_fareed_zakaria_became_the_most_conservative_liberal_of_all_time; Fareed Zakaria, "Bernie Sanders's Magical Thinking on Climate Change," *Washington Post*, February 13, 2020, www.washingtonpost.com/opinions/bernie-sanderss-magical-thinking-on-climate-change/2020/02/13/3944e472-4ea5-11ea-9b5c-eac5b16dafaa_story.html.

72. See "Nuclear Economics: Critical Responses to Breakthrough Institute Propaganda," World Information Service on Energy (WISE), Nuclear Monitor #840, no. 4630, March 21, 2017, www.wiseinternational.org/nuclear-monitor/840/nuclear-economics-critical-responses-breakthrough-institute-propaganda; "An Ecomodernist Manifesto."

73. See "It's Worse Than You Think—Lower Emissions, Higher Ground," Yang 2020, August 28, 2019, www.yang2020.com/blog/climate-change; Ryan Broderick, "Andrew Yang Wants the Support of the Pro-Trump Internet. Now It Is Threatening to Devour Him," BuzzFeed, March 14, 2019, www.buzzfeednews.com/article/ryanhatesthis/4chan-vs-the-yang-gang.

74. Maya Earls, "Benefits of Adaptation Measures Outweigh the Costs, Report Says," Climatewire, E&E News, September 10, 2019, reprinted at Scientific American, www.scientificamerican.com/article/benefits-of-adaptation-measures-outweigh-the-costs-report-says.

75. Marco Rubio, "We Should Choose Adaptive Solutions," USA Today, August 19, 2019, www.usatoday.com/story/opinion/2019/08/19/rubio-on-climate-change-we-should-choose-adaptive-solutions-column/2019310001.

76. Andrea Dutton and Michael Mann, "A Dangerous New Form of Climate Denialism Is Making the Rounds," Newsweek, August 22, 2019, www.newsweek.com/dangerous-new-form-climate-denialism-making-rounds-opinion-1455736.

77. Francie Diep, "The House Science Committee Just Held a Helpful Hearing on Climate Science for the First Time in Years," Pacific Standard, February 13, 2019, https://psmag.com/news/the-house-science-committee-just-held-its-first-helpful-hearing-on-climate-science-in-years; Tiffany Stecker, "New Climate Panel's Republicans Seek Focus on Adaptation," Bloomberg Energy, March 8, 2019, https://news.bloombergenvironment.com/environment-and-energy/new-climate-panels-republicans-seek-focus-on-adaptation.

78. Steven Mufson, "Are Republicans Coming out of 'the Closet' on Climate Change?," Washington Post, February 4, 2020, www.washingtonpost.com/climate-environment/can-republicans-turn-over-a-new-leaf-on-climate-change/2020/02/03/6a6a6bd8-4155-11ea-aa6a-083d01b3ed18_story.html.

79. Greg Walden, Fred Upton, and John Shimkus, "Republicans Have Better Solutions to Climate Change," Real Clear Policy, February 13, 2019, www.realclearpolicy.com/articles/2019/02/13/republicans_have_better_solutions_to_climate_change_111045.html.

80. Michael Mann, "If There's a Silver Lining in the Clouds of Choking Smoke It's That This May Be a Tipping Point," The Guardian, February 3, 2020, www.theguardian.com/commentisfree/2020/feb/03/if-theres-a-silver-lining-in-the-clouds-of-smoke-its-that-this-could-be-a-tipping-point.

81. "Fire Fight: Tara Brown Finds Out What Australia Can Do to Prevent a Repeat of This Summer's Deadly Bushfires," Nine Network, Australia, February 9, 2020, www.9now.com.au/60-minutes/2020/episode-1.

82. Though, in what could pass for parody, Bolt argued that "it will be good for us." See Van Badham, "Now That Climate Change Is Irrefutable, Denialists Like Andrew Bolt Insist It Will Be Good for Us," *The Guardian*, January 30, 2020, www.theguardian.com/commentisfree/2020/jan/30/now-that-climate -change-is-irrefutable-denialists-like-andrew-bolt-insist-it-will-be-good-for-us.

83. Christopher Wright and Michael E. Mann, "From Denial to 'Resilience': The Slippery Discourse of Obfuscating Climate Action," Sydney Environment Institute of the University of Sydney, February 19, 2020, http://sydney .edu.au/environment-institute/opinion/from-denial-to-resilience.

84. Sarah Martin, "Scott Morrison to Focus on 'Resilience and Adaptation' to Address Climate Change," *The Guardian*, January 14, 2020, www .theguardian.com/environment/2020/jan/14/scott-morrison-to-focus-on -resilience-and-adaption-to-address-climate-change.

85. See "Honest Government Ad: After the Fires," The Juice Media, February 11, 2020, www.thejuicemedia.com/honest-government-ad-the-fires.

86. Graham Readfearn, "Australian PM Scott Morrison Agrees to Permanently Increase Aerial Firefighting Funding," *The Guardian*, January 4, 2020, www.theguardian.com/australia-news/2020/jan/04/australian-pm-scott -morrison-agrees-to-permanently-increase-aerial-firefighting-funding.

87. Sarah Martin, "Coalition Promises $2bn for Bushfire Recovery as It Walks Back from Budget Surplus Pledge," *The Guardian*, January 6, 2020, www.theguardian.com/australia-news/2020/jan/06/coalition-pledges-2bn -for-bushfire-recovery-as-it-walks-back-from-budget-surplus-pledge.

88. "Scott Morrison Announces $2 Billion Energy Deal"; Lucy Barbour and Jane Norman, "Rebel Nationals Wanting New Coal-Fired Power Stations Face Battle with Liberals and Markets," Australian Broadcasting Corporation, February 13, 2020, www.abc.net.au/news/2020-02-13/national-party-rebels -fighting-for-more-coal-power-stations/11959568; David Crowe, "New Resources Minister Calls for More Coal, Gas and Uranium Exports," *Sydney Morning Herald*, February 11, 2020, www.smh.com.au/politics/federal/new-resources -minister-calls-for-more-coal-gas-and-uranium-exports-20200211-p53zu5 .html.

89. M. C. Nisbet, "The Ecomodernists: A New Way of Thinking About Climate Change and Human Progress," *Skeptical Inquirer* 42, no. 6 (2018): 20–24, https://web.northeastern.edu/matthewnisbet/wp-content/uploads/2018/12 /Nisbet2018_TheEcomodernists_SkepticalInquirer_.pdf; "Matthew Nisbet," Breakthrough Institute, https://thebreakthrough.org/people/matthew-nisbet; Matt Nisbet, "Against Climate Change Tribalism: We Gamble with the Future by Dehumanizing Our Opponents," *Skeptical Inquirer* 44, no. 1 (2020), https:// skepticalinquirer.org/2020/01/against-climate-change-tribalism-we-gamble -with-the-future-by-dehumanizing-our-opponents.

90. See, for example, David Roberts, "Why I've Avoided Commenting on Nisbet's 'Climate Shift' Report," *Grist*, April 27, 2011, https://grist.org/climate -change/2011-04-26-why-ive-avoided-commenting-on-nisbets-climate-shift -report.

CHAPTER 8: THE TRUTH IS BAD ENOUGH

1. Justin Gillis, "Climate Model Predicts West Antarctic Ice Sheet Could Melt Rapidly," *New York Times*, March 30, 2016, www.nytimes.com/2016/03/31/science/global-warming-antarctica-ice-sheet-sea-level-rise.html.

2. Michael Mann, "It's Not Rocket Science: Climate Change Was Behind This Summer's Extreme Weather," *Washington Post*, November 2, 2018, www.washingtonpost.com/opinions/its-not-rocket-science-climate-change-was-behind-this-summers-extreme-weather/2018/11/02/b8852584-dea9-11e8-b3f0-62607289efee_story.html.

3. Nicholas Smith and Anthony Leiserowitz, "The Role of Emotion in Global Warming Policy Support and Opposition," *Risk Analysis* 34, no. 5 (2014): 937–948.

4. Clay Evans, "Ditching the Doomsaying for Better Climate Discourse," *University of Colorado Arts and Sciences Magazine*, December 18, 2019, www.colorado.edu/asmagazine/2019/12/18/ditching-doomsaying-better-climate-discourse.

5. Bjorn Lomborg, "Who's Afraid of Climate Change?" Project Syndicate, August 11, 2010, www.project-syndicate.org/commentary/who-s-afraid-of-climate-change.

6. Oliver Milman and Dominic Rushe, "New EPA Head Scott Pruitt's Emails Reveal Close Ties with Fossil Fuel Interests," *The Guardian*, February 23, 2017, www.theguardian.com/environment/2017/feb/22/scott-pruitt-emails-oklahoma-fossil-fuels-koch-brothers; Oliver Milman, "EPA Head Scott Pruitt Says Global Warming May Help 'Humans Flourish,'" *The Guardian*, February 8, 2018, www.theguardian.com/environment/2018/feb/07/epa-head-scott-pruitt-says-global-warming-may-help-humans-flourish.

7. Sam Langford, "'He's Cherry Picking with Intent': Here's What the Climate Scientist Andrew Bolt Keeps Quoting Would Like You to Know," *The Feed*, January 27, 2020, www.sbs.com.au/news/the-feed/he-s-cherry-picking-with-intent-here-s-what-the-climate-scientist-andrew-bolt-keeps-quoting-would-like-you-to-know.

8. See, for example, the op-ed I coauthored on this subject: Michael E. Mann, Susan Joy Hassol, and Tom Toles, "Doomsday Scenarios Are as Harmful as Climate Change Denial," *Washington Post*, July 12, 2017, www.washingtonpost.com/opinions/doomsday-scenarios-are-as-harmful-as-climate-change-denial/2017/07/12/880ed002-6714-11e7-a1d7-9a32c91c6f40_story.html.

9. Ketan Joshi (@KetanJ0), Twitter, January 11, 2020, 2:03 p.m., https://twitter.com/KetanJ0/status/1216118507500457985.

10. Max Hastings, *Winston's War: Churchill, 1940–1945* (New York: Vintage, 2011).

11. JC Cooper (@coopwrJ), Twitter, September 14, 2019, 1:25 p.m., https://twitter.com/CoopwrJ/status/1172969585097621504. Cooper is a materials scientist and zoologist.

12. Jennifer De Pinto, Fred Backus, and Anthony Salvanto, "Most Americans Say Climate Change Should Be Addressed Now—CBS News Poll,"

CBS News, September 15, 2019, www.cbsnews.com/news/cbs-news-poll
-most-americans-say-climate-change-should-be-addressed-now-2019-09
-15. See the cross-tabs for question #8 athttps://drive.google.com/file/d
/0ByVu4fDHYJgVdHFJWFRsbF90TDNSZFV3TklzMkVrRHh0TDNj/view.

13. See "Meet the Team," The Glacier Trust, http://theglaciertrust.org
/people; Joanne Moore, "Family Pay Tribute to Missing Hill Walker," *Gazette
and Herald*, March 8, 2016, www.gazetteandherald.co.uk/news/14328154
.family-pay-tribute-to-missing-hill-walker.

14. Michael E. Mann and Jonathan Brockopp, "You Can't Save the Climate by Going Vegan. Corporate Polluters Must Be Held Accountable,"
USA Today, June 3, 2019, www.usatoday.com/story/opinion/2019/06/03
/climate-change-requires-collective-action-more-than-single-acts
-column/1275965001; Michael E. Mann, June 3, 2019, Facebook, www
.facebook.com/MichaelMannScientist/posts/2327588820630640.

15. Zeke Hausfather and Glen P. Peters, "Emissions—the 'Business as
Usual' Story Is Misleading: Stop Using the Worst-Case Scenario for Climate
Warming as the Most Likely Outcome—More-Realistic Baselines Make for Better Policy," *Nature*, January 29, 2020, www.nature.com/articles/d41586-020
-00177-3.

16. Christopher H. Trisos, Cory Merow, and Alex L. Pigot, "The Projected
Timing of Abrupt Ecological Disruption from Climate Change," *Nature* 580
(2020): 496–501, https://doi.org/10.1038/s41586-020-2189-9.

17. Citizens for Climate Action (@CitFrClimACTION), Twitter, December
13, 2019, https://twitter.com/citfrclimaction/status/1205570515026501633.

18. Darlene "Rethink everything you thought you knew" (@DarleneLily1),
Twitter, https://twitter.com/DarleneLily1/https://twitter.com/citfrclimaction
/status/1205570515026501633. The tweet has since been deleted.

19. Mann and Brockopp, "You Can't Save the Climate by Going Vegan."

20. Raquel Baranow (@666isMONEY), Twitter, September 8, 2019, 12:46
p.m., https://twitter.com/666isMONEY/status/1170785503630614529.

21. Raquel Baranow (@666isMONEY), Twitter, September 13, 2019, 9:14
p.m., https://twitter.com/666isMONEY/status/1172725301576519680.

22. #ForALL CANCEL RENT NOW (@GarrettShorr), Twitter, December
12, 2019, 11:07 p.m., https://twitter.com/GarrettShorr/status/120538380877
9841536.

23. InsideClimate News (@insideclimate), Twitter, April 10, 2020, 1:20
p.m., https://twitter.com/insideclimate/status/1248707322899247104.

24. Michael E. Mann (@MichaelEMann), Twitter, April 10, 2020, 1:32
p.m., https://twitter.com/MichaelEMann/status/1248710415426748422.

25. Bruce Boyes (@BruceBoyes), Twitter, April 11, 2020, 2:36 a.m., https://
twitter.com/BruceBoyes/status/1248907679621251073. Boyes is editor and
lead writer of the award-winning *KM Magazine* in Australia.

26. Wild Talks Ireland (@TalksWild), Twitter, April 10, 2020, 1:36 p.m.,
https://twitter.com/TalksWild/status/1248711522953637896.

27. Jonathan Franzen, "What If We Stopped Pretending? The Climate
Apocalypse Is Coming. To Prepare for It, We Need to Admit That We Can't

Prevent It," *New Yorker*, September 8, 2019, www.newyorker.com/culture
/cultural-comment/what-if-we-stopped-pretending.

28. Ula Chrobak, "Can We Still Prevent an Apocalypse? What Jonathan
Franzen Gets Wrong About Climate Change: What If We Stopped Pretending
the New Yorker's Essay Makes Sense?," *Popular Science*, September 11, 2019,
www.popsci.com/climate-change-new-yorker-franzen-corrections.

29. Jeff Nesbit (@jeffnesbit), Twitter, September 8, 2019, 8:33 a.m., https://
twitter.com/jeffnesbit/status/1170721760678797312.

30. John Upton (@johnupton), Twitter, September 8, 2019, 8:11 a.m.,
https://twitter.com/johnupton/status/1170716277977026560.

31. Dr. Jonathan Foley (@GlobalEcoGuy), Twitter, September 8, 2019, 10:45
a.m., https://twitter.com/GlobalEcoGuy/status/1170755004199657472.

32. Taylor Nicole Rogers, "Scientists Blast Jonathan Franzen's 'Climate
Doomist' Opinion Column as 'the Worst Piece on Climate Change,'" *Business Insider*, September 8, 2019, www.businessinsider.com/scientists-blast
-jonathan-franzens-climate-doomist-new-yorker-op-ed-2019-9.

33. Franzen, "What If We Stopped Pretending?"

34. Alison Flood, "Jonathan Franzen: Online Rage Is Stopping Us Tackling the Climate Crisis," *The Guardian*, October 9, 2019, www.theguardian
.com/books/2019/oct/08/jonathan-franzen-online-rage-is-stopping-us
-tackling-the-climate-crisis.

35. See Dr Tamsin Edwards (@flimsin), Twitter, October 26, 2019, 2:05
p.m., https://twitter.com/flimsin/status/1188199938284539904. For the
lecture by Rupert Read, see "Rupert Read: How I Talk with Children About
Climate Breakdown," YouTube, posted August 13, 2019, www.youtube.com
/watch?v=6Lt0jCDtYSY&feature=youtu.be.

36. Roy Scranton, *We're Doomed. Now What? Essays on War and Climate
Change* (New York: Penguin Random House, 2018).

37. The original tweet was https://twitter.com/royscranton/status/1073
903831870857216.

38. Roy Scranton, "No Happy Ending: On Bill McKibben's 'Falter' and
David Wallace-Wells's 'The Uninhabitable Earth,'" *Los Angeles Review of
Books*, June 3, 2019, https://lareviewofbooks.org/article/no-happy-ending-on
-bill-mckibbens-falter-and-david-wallace-wellss-the-uninhabitable-earth.

39. Alexandria Villaseñor (@AlexandriaV2005), Twitter, January 30, 2019,
3:16 a.m., https://twitter.com/AlexandriaV2005/status/1090569401345236997.

40. Scranton, "No Happy Ending."

41. David Roberts (@drvox), Twitter, https://twitter.com/drvox/status
/1136321633499590656.

42. James Renwick, "Guy McPherson and the End of Humanity (Not),"
Hot Topic (blog), December 11, 2016, http://hot-topic.co.nz/guy-mcpherson
-and-the-end-of-humanity-not.

43. See Twitter exchange at Michael E. Mann (@MichaelEMann), March
21, 2020, 8:47 p.m., https://twitter.com/MichaelEMann/status/124157217
4114304000, which concerns a video that McPherson posted in which he asserted that "by November 1st plus or minus a few months we would be out of

habitat for our species" (at roughly 2:10 into the video). The video is "Edge of Extinction: Coronavirus Update," YouTube, posted February 28, 2020, www.youtube.com/watch?v=vn4PoLOmCME&feature=youtu.be.

44. Scott Johnson, "How Guy McPherson Gets It Wrong," *Fractal Planet* (blog), February 17, 2014, https://fractalplanet.wordpress.com/2014/02/17/how-guy-mcpherson-gets-it-wrong.

45. Catherine Ingram, "Are We Heading Toward Extinction? The Earth's Species—Plants, Animals and Humans, Alike—Are Facing Imminent Demise. How We Got Here, and How to Cope," *Huffington Post*, July 20, 2019, www.huffpost.com/entry/facing-extinction-humans-animals-plants-species_n_5d2ddc04e4b0a873f6420bd3.

46. "Rex Weyler," Greenpeace, www.greenpeace.org/international/author/rex-weyler; Rex Weyler, "Extinction and Rebellion," Greenpeace, May 17, 2019, www.greenpeace.org/international/story/22058/extinction-and-rebellion.

47. Aja Romano, "Twitter Released 9 Million Tweets from One Russian Troll Farm. Here's What We Learned," *Vox*, October 19, 2018, www.vox.com/2018/10/19/17990946/Twitter-russian-trolls-bots-election-tampering.

48. Craig Timberg and Tony Romm, "Russian Trolls Sought to Inflame Debate over Climate Change, Fracking, Dakota Pipeline," *Chicago Tribune*, March 1, 2018, www.chicagotribune.com/nation-world/ct-russian-trolls-climate-change-20180301-story.html.

49. Harry Enten, "Registered Voters Who Stayed Home Probably Cost Clinton the Election," *FiveThirtyEight*, January 5, 2017, https://fivethirtyeight.com/features/registered-voters-who-stayed-home-probably-cost-clinton-the-election.

50. See YouTube archive of McPherson's interviews for American Freedom Radio at www.youtube.com/results?search_query=American+Freedom+Radio+%22Guy+McPherson%22.

51. Guy McPherson, "Why I'm Voting for Donald Trump: McPherson's 6th Stage of Grief (Gallows Humor)," Nature Bats Last, March 11, 2016, https://guymcpherson.com/2016/03/why-im-voting-for-donald-trump-mcphersons-6th-stage-of-grief-gallows-humor.

52. JC Cooper (@Coopwr), Twitter, September 14, 2019, 1:39 p.m., https://twitter.com/CoopwrJ/status/1172973057138315264.

53. Andy Caffrey (@Andy_Caffrey), Twitter, September 5, 2019, 12:50 p.m., https://twitter.com/Andy_Caffrey/status/1169699394574176256.

54. Johnson, "How Guy McPherson Gets It Wrong."

55. Eric Steig (@ericsteig), Twitter, November 11, 2019, 9:49 p.m., https://twitter.com/ericsteig/status/1194130088784093185.

56. Jill (@sooverthis123), Twitter, July 30, 2019, 1:57 p.m., https://twitter.com/sooverthis123/status/1156307730224648193.

57. Dana Nuccitelli, "There Are Genuine Climate Alarmists, but They're Not in the Same League as Deniers," *The Guardian*, July 9, 2018, www.theguardian.com/environment/climate-consensus-97-per-cent/2018/jul/09/there-are-genuine-climate-alarmists-but-theyre-not-in-the-same-league-as-deniers.

58. Scott Johnson, "Once More: McPherson's Methane Catastrophe," *Fractal Planet* (blog), January 8, 2015, https://fractalplanet.wordpress.com/category/science-doing-it-wrong.

59. Ian Johnston, "Earth's Worst-Ever Mass Extinction of Life Holds 'Apocalyptic' Warning About Climate Change, Say Scientists," *The Independent*, March 24, 2017, www.independent.co.uk/environment/earth-permian-mass-extinction-apocalypse-warning-climate-change-frozen-methane-a7648006.html; Howard Lee, "Sudden Ancient Global Warming Event Traced to Magma Flood," *Quanta Magazine*, March 19, 2020, www.quantamagazine.org/sudden-ancient-global-warming-event-traced-to-magma-flood-20200319.

60. Joshua F. Dean, Jack J. Middelburg, Thomas Röckmann, Rien Aerts, Luke G. Blauw, Matthias Egger, S. M. Jetten, et al, "Methane Feedbacks to the Global Climate System in a Warmer World," *Reviews of Geophysics* 56, no. 1 (2018): 207–250; Chris Colose, "Toward Improved Discussions of Methane and Climate," August 1, 2013, www.skepticalscience.com/toward-improved-discussions-methane.html. See also this older but still valid commentary by my colleague David Archer: "Arctic Methane on the Move," Real Climate, March 6, 2010, www.realclimate.org/index.php/archives/2010/03/arctic-methane-on-the-move.

61. See multipart Twitter thread at Michael E. Mann (@MichaelEMann), Twitter, September 14, 2019, 12:04 p.m., https://twitter.com/MichaelEMann/status/1172949203821219841.

62. Ben Heubl, "Arctic Methane Levels Reach New Heights," *Engineering and Technology Magazine*, September 16, 2019, https://eandt.theiet.org/content/articles/2019/09/arctic-methane-levels-reach-new-heights-data-shows.

63. Andrew Nikiforuk, "New Study Finds Far Greater Methane Threat from Fossil Fuel Industry," *The Tyee*, February 21, 2020, https://thetyee.ca/News/2020/02/21/Fossil-Fuel-Industry-Far-Greater-Methane-Threat-Study-Finds.

64. Ed King, "Should Climate Scientists Slash Air Miles to Set an Example?," *Climate Home News*, October 3, 2015, www.climatechangenews.com/2015/03/10/should-climate-scientists-slash-air-miles-to-set-an-example.

65. "About the Committee on Climate Change," Committee on Climate Change, www.theccc.org.uk/about.

66. Kevin Anderson (@KevinClimate), Twitter, January 28, 2020, 12:36 a.m., https://twitter.com/KevinClimate/status/1222076017008836609.

67. Dr Alexandra Jellicow (@alexjellicoe), Twitter, January 28, 2020, 12:42 a.m., https://twitter.com/alexjellicoe/status/1222077373337808897.

68. Kevin Anderson (@KevinClimate), Twitter, January 28, 2020, 12:53 a.m., https://twitter.com/KevinClimate/status/1222080140383080448.

69. Chris Stark (@ChiefExecCCC), Twitter, January 28, 2020, 12:58 a.m., https://twitter.com/ChiefExecCCC/status/1222081581944315904.

70. Dr Tamsin Edwards (@flimsin), Twitter, January 28, 2020, 1:03 a.m., https://twitter.com/flimsin/status/1222082804529364992.

71. Kevin Anderson (@KevinClimate), Twitter, January 28, 2020, 1:54 a.m., https://twitter.com/KevinClimate/status/1222095616140095489.

72. Zing Tsjeng, "The Climate Change Paper So Depressing It's Sending People to Therapy," *Vice*, February 27, 2019, www.vice.com/en_au/article /vbwpdb/the-climate-change-paper-so-depressing-its-sending-people-to-therapy.

73. Jem Bendell, "Deep Adaptation: A Map for Navigating Climate Tragedy," IFLAS Occasional Paper 2, July 27, 2018, available at www.lifeworth .com/deepadaptation.pdf.

74. Tsjeng, "The Climate Change Paper So Depressing It's Sending People to Therapy."

75. Ironically, one of the most thorough debunkings of Bendell's paper was provided by libertarian pundit Ron Bailey (more on that later) in "Good News! No Need to Have a Mental Breakdown over 'Climate Collapse,'" *Reason*, March 3, 2019, https://reason.com/2019/03/29/good-news-no-need-to -have-a-mental-break.

76. Jack Hunter, "The 'Climate Doomers' Preparing for Society to Fall Apart," BBC, March 16, 2020, www.bbc.com/news/stories-51857722.

77. Tsjeng, "The Climate Change Paper So Depressing It's Sending People to Therapy."

78. Hunter, "The 'Climate Doomers' Preparing for Society to Fall Apart."

79. David Wallace-Wells, "Time to Panic," *New York Times*, February 16, 2019, www.nytimes.com/2019/02/16/opinion/sunday/fear-panic-climate-change -warming.html.

80. Sheril Kirshenbaum, "No, Climate Change Will Not End the World in 12 Years: Stoking Panic and Fear Creates a False Narrative That Can Overwhelm Readers, Leading to Inaction and Hopelessness," *Scientific American*, August 13, 2019, https://blogs.scientificamerican.com/observations/no-climate-change -will-not-end-the-world-in-12-years.

81. A transcript of Thunberg's speech appears at *The Guardian*, January 25, 2019, www.theguardian.com/environment/2019/jan/25/our-house-is-on -fire-greta-thunberg16-urges-leaders-to-act-on-climate.

82. Francisco Toro, "Climate Politics Is a Dead End. So the World Could Turn to This Desperate Final Gambit," *Washington Post*, December 18, 2019, www.washingtonpost.com/opinions/2019/12/18/climate-politics-is-dead-end -so-world-could-turn-to-this-desperate-final-gambit.

83. Quoting from their official description at "The Truth," Extinction Rebellion, https://rebellion.earth/the-truth.

84. "Climate Fatalism," Freedom Lab, http://freedomlab.org/climate -fatalism.

85. "About," Freedom Lab, http://freedomlab.org/about-freedomlab.

86. David Roberts (@drvox), Twitter, https://twitter.com/drvox/status /1211713331603611648 (Roberts subsequently deleted the tweet).

87. Jonathan Koomey (@jgkoomey), Twitter, December 30, 2019, 10:48 a.m., https://twitter.com/jgkoomey/status/1211720645928611840.

88. Massimo Sandal (@massimosandal), Twitter, December 30, 2019, 11:47 p.m., https://twitter.com/massimosandal/status/1211916797554937858.

89. Will Steffen, Johan Rockström, Katherine Richardson, Timothy M. Lenton, Carl Folke, Diana Liverman, Colin P. Summerhayes, et al., "Trajectories of the Earth System in the Anthropocene," *Proceedings of the National Academy of Sciences* 115, no. 33 (2018): 8252–8259, https://doi.org/10.1073/pnas.1810141115.

90. Kate Aronoff, "'Hothouse Earth' Co-Author: The Problem Is Neoliberal Economics," *The Intercept*, August 14, 2018, https://theintercept.com/2018/08/14/hothouse-earth-climate-change-neoliberal-economics.

91. Richard Betts, "Hothouse Earth: Here's What the Science Actually Does—and Doesn't—Say," *The Conversation*, August 10, 2018, https://theconversation.com/hothouse-earth-heres-what-the-science-actually-does-and-doesnt-say-101341.

92. Timothy M. Lenton, Johan Rockström, Owen Gaffney, Stefan Rahmstorf, Katherine Richardson, Will Steffen, and Hans Joachim Schellnhuber, "Climate Tipping Points—Too Risky to Bet Against," *Nature*, November 27, 2019, www.nature.com/articles/d41586-019-03595-0.

93. Stephen Leahy, "Climate Change Driving Entire Planet to Dangerous 'Tipping Point,'" *National Geographic*, November 27, 2019, www.nationalgeographic.com/science/2019/11/earth-tipping-point; "Scientists Warn Earth at Dire Risk of Becoming Hellish 'Hothouse,'" *New York Post*, August 7, 2018, https://nypost.com/2018/08/07/scientists-warn-earth-at-dire-risk-of-becoming-hellish-hothouse.

94. David Wallace-Wells, "The Uninhabitable Earth: Famine, Economic Collapse, a Sun That Cooks Us: What Climate Change Could Wreak—Sooner Than You Think," *New York Magazine*, July 2017, https://nymag.com/intelligencer/2017/07/climate-change-earth-too-hot-for-humans.html.

95. See "The 'Doomed Earth' Controversy," Arthur L. Carter Journalism Institute, New York University, November 30, 2017, https://journalism.nyu.edu/about-us/event/2017-fall/the-doomed-earth-controversy.

96. Michael E. Mann, Facebook, July 10, 2017, www.facebook.com/MichaelMannScientist/posts/since-this-new-york-magazine-article-the-uninhabitable-earth-is-getting-so-much-/1470539096335621.

97. Mann et al., "Doomsday Scenarios Are as Harmful as Climate Change Denial."

98. Zeke Hausfather, "Major Correction to Satellite Data Shows 140% Faster Warming Since 1998," *Carbon Brief*, June 30, 2017, www.carbonbrief.org/major-correction-to-satellite-data-shows-140-faster-warming-since-1998.

99. Dana Nuccitelli, "Climate Scientists Just Debunked Deniers' Favorite Argument," *The Guardian*, June 28, 2017, www.theguardian.com/environment/climate-consensus-97-per-cent/2017/jun/28/climate-scientists-just-debunked-deniers-favorite-argume; Benjamin D. Santer, John C. Fyfe, Giuliana Pallotta, Gregory M. Flato, Gerald A. Meehl, Matthew H. England, Ed Hawkins, et al., "Causes of Differences in Model and Satellite Tropospheric Warming Rates," *Nature Geoscience* 10 (2017): 478–485.

100. The workshop was the 2018 Ny-Ålesund Symposium on Navigating Climate Risk. For details, see www.ny-aalesundsymposium.no/2018/Summary

_of_the_Ny-_lesund_symposium_2018.shtml and www.ny-aalesundsymposium .no/artman/uploads/1/Ny-_lesund_Summary_and_Steps_Forward.pdf.

101. "Norwegian Seed Vault Guarantees Crops Won't Become Extinct," *Weekend Edition* with Lulu Garcia-Navarro, National Public Radio, May 21, 2017, www.npr.org/2017/05/21/529364527/norwegian-seed-vault-guarantees -crops-won-t-become-extinct.

102. For the transcript of my full interview with Wallace-Wells, see David Wallace-Wells, "Scientist Michael Mann on 'Low-Probability but Catastrophic' Climate Scenarios," *New York Magazine*, July 11, 2017, https://nymag .com/intelligencer/2017/07/scientist-michael-mann-on-climate-scenarios .html.

103. "Scientists Explain What *New York Magazine* Article on 'The Uninhabitable Earth' Gets Wrong: Analysis of 'The Uninhabitable Earth,' Published in New York Magazine, by David Wallace-Wells on 9 July 2017," Climate Feedback, July 12, 2017, https://climatefeedback.org/evaluation/scientists -explain-what-new-york-magazine-article-on-the-uninhabitable-earth-gets -wrong-david-wallace-wells.

104. "Scientists Explain What *New York Magazine* Article on 'The Uninhabitable Earth' Gets Wrong."

105. See Michael E. Mann (@MichaelEMann), Twitter, July 12, 2017, 10:21 p.m., https://twitter.com/MichaelEMann/status/885368503452262400; the tweet of Roberts to which it is replying is now reported as "not available."

106. "The 'Doomed Earth' Controversy," Arthur L. Carter Journalism Institute, New York University, November 30, 2017, https://journalism.nyu.edu /about-us/event/2017-fall/the-doomed-earth-controversy.

107. David Wallace-Wells, *Uninhabitable Earth: Life After Warming* (New York: Tim Duggan Books / Penguin Random House, 2019), 22.

108. Warren Cornwall, "Even 50-Year-Old Climate Models Correctly Predicted Global Warming," *Science*, December 4, 2019, www.sciencemag.org /news/2019/12/even-50-year-old-climate-models-correctly-predicted-global -warming.

109. "The Uninhabitable Earth," Penguin Random House, www.penguin randomhouse.com/books/586541/the-uninhabitable-earth-by-david-wallace -wells.

110. Yessenia Funes, "HBO Max Is Turning *The Uninhabitable Earth* Into a Fictional Series," *Gizmodo*, January 16, 2020, https://earther.gizmodo.com /hbo-max-is-turning-the-uninhabitable-earth-into-a-ficti-1841048114.

111. "'We Are Entering into an Unprecedented Climate,'" MSNBC, *Morning Joe*, February 20, 2019, www.msnbc.com/morning-joe/watch/-we-are -entering-into-an-unprecedented-climate-1445411907673.

112. Sean Illing, "It Is Absolutely Time to Panic About Climate Change: Author David Wallace-Wells on the Dystopian Hellscape That Awaits Us," *Vox*, February 24, 2019, www.vox.com/energy-and-environment/2019/2/22/18188562 /climate-change-david-wallace-wells-the-uninhabitable-earth.

113. David Wallace-Wells (@dwallacewells), Twitter, September 23, 2019, 9:32 a.m., https://twitter.com/dwallacewells/status/1176172433579159552.

114. Assaad Razzouk (@AssaadRazzouk), Twitter, September 22, 2019, 4:14 p.m., https://twitter.com/AssaadRazzouk/status/1175911365820764161.

115. Richard Betts (@richardabetts), Twitter, September 23, 2019, 12:56 p.m., https://twitter.com/richardabetts/status/1176223928785756161.

116. Eric Steig (@ericsteig), Twitter, September 23, 2019, 6:59 p.m., https://twitter.com/ericsteig/status/1176315272610758656.

117. David Wallace-Wells, "U.N. Climate Talks Collapsed in Madrid. What's the Way Forward?," New York Magazine, December 16, 2019, http://nymag.com /intelligencer/2019/12/cop25-ended-in-failure-whats-the-way-forward.html.

118. "Global CO2 Emissions in 2019," International Energy Agency, February 11, 2020, www.iea.org/articles/global-co2-emissions-in-2019.

119. Adam Vaughan, "China Is on Track to Meet Its Climate Change Goals Nine Years Early," The Guardian, July 26, 2019, www.newscientist.com/article /2211366-china-is-on-track-to-meet-its-climate-change-goals-nine-years-early.

120. Julia Rosen, "Cities, States and Companies Vow to Meet U.S. Climate Goals Without Trump. Can They?," Los Angeles Times, November 4, 2019, www.latimes.com/environment/story/2019-11-04/cities-states-companies -us-climate-goals-trump.

121. See the thread with Kalee Kreider (@kaleekreider), Twitter, December 17, 2019, 6:46 p.m., https://twitter.com/kaleekreider/status/1207129884 339949570.

122. See "UN Climate Pledge Analysis," Climate Interactive, www.climate interactive.org/programs/scoreboard.

123. David Wallace-Wells (@dwallacewells), Twitter, December 17, 2019, 6:08 p.m., https://twitter.com/dwallacewells/status/1207120372585381888.

124. Kalee Kreider (@kaleekreider), Twitter, December 17, 2019, 6:46 p.m., https://twitter.com/kaleekreider/status/1207129884339949570; Kalee Kreider (@kaleekreider), Twitter, December 17, 2019, 6:48 p.m., https://twitter .com/kaleekreider/status/1207130377946587136. See also "U.S.-China Joint Announcement on Climate Change," White House, Office of the Press Secretary, November 11, 2014, https://obamawhitehouse.archives.gov/the-press -office/2014/11/11/us-china-joint-announcement-climate-change.

125. Greta Thunberg (@GretaThunberg), Twitter, December 14, 2019, 12:52 p.m., https://twitter.com/gretathunberg/status/1205953722293145604.

126. Suyin Haynes, "Greta Thunberg Joins Youth Activists on TIME Panel at Davos to Say 'Pretty Much Nothing' Has Been Done on Climate Change," Time, January 21, 2020, https://time.com/5768561/greta-thunberg-davos -panel-time.

127. David Wallace-Wells, "We're Getting a Clearer Picture of the Climate Future—and It's Not as Bad as It Once Looked," New York Magazine, December 20, 2019, https://nymag.com/intelligencer/2019/12/climate-change -worst-case-scenario-now-looks-unrealistic.html.

128. Hausfather and Peters, "Emissions—the 'Business as Usual' Story Is Misleading."

129. Alastair McIntosh (@alastairmci), Twitter, March 16, 2020, 4:10 p.m., https://twitter.com/alastairmci/status/1239690611671916544.

130. Quoted in Christopher J. Bosso, *Pesticides and Politics: The Life Cycle of a Public Issue* (Pittsburgh: University of Pittsburgh Press, 1987), 116; Naomi Oreskes and Eric M. Conway, *Merchants of Doubt: How a Handful of Scientists Obscured the Truth on Issues from Tobacco Smoke to Global Warming* (New York: Bloomsbury Press, 2010).

131. "Dangerous Legacy," Competitive Enterprise Institute, 2016, www .rachelwaswrong.org.

132. Michael Mann, *The Hockey Stick and the Climate Wars: Dispatches from the Front Lines* (New York: Columbia University Press, 2013), 74–77.

133. Vijay Jayaraj, "Opportunistic Doomsayers Compare Climate Change to Coronavirus," CNS News, March 24, 2020, www.cnsnews.com/commentary /vijay-jayaraj/opportunistic-doomsayers-compare-climate-change-coronavi- rus. The Media Research Center received over $400,000 from ExxonMobil between 1998 and 2009. "Factsheet: Media Research Center, MRC," Exxon Secrets.org, www.exxon secrets.org/html/orgfactsheet.php?id=110. Accord- ing to Media Transparency, it received over $3 million from the Sarah Scaife Foundation between 1998 and 2009. Bridge Project, http://mediamattersaction .org/transparency/organization/Media_Research_Center/funders.

134. Mann, *The Hockey Stick and the Climate Wars*, 76.

135. Michael Mann and Lee R. Kump, *Dire Predictions: Understanding Climate Change*, 2nd ed. (New York: DK, 2015), 46–47.

136. Mann, *The Hockey Stick and the Climate Wars*, 160.

137. This is one of the leading climate denial myths documented at Skep- tical Science. See "Climate Scientists Would Make More Money in Other Careers," Skeptical Science, https://skepticalscience.com/climate-scientists-in -it-for-the-money.htm.

138. This is an accusation that was made by Paul Driessen, who has been vari- ously employed by the Center for a Constructive Tomorrow (CFACT), the Cen- ter for the Defense of Free Enterprise, the Frontiers of Freedom, and the Atlas Economic Research Foundation, among others—a virtual cornucopia of industry front groups. See Mann, *The Hockey Stick and the Climate Wars*, 202–203.

139. Alastair McIntosh (@alastairmci), Twitter, March 16, 2020, 4:10 p.m., https://twitter.com/alastairmci/status/1239690611671916544.

140. Ronald Bailey, *Global Warming and Other Eco Myths: How the Envi- ronmental Movement Uses False Science to Scare Us to Death* (Roseville, CA: Prima Lifestyles, 2002); Bailey, "Good News! No Need to Have a Mental Breakdown."

141. Michael E. Mann (@MichaelEMann), Twitter, May 5, 2017, 11:50 a.m., https://twitter.com/MichaelEMann/status/860567443059724288. The comment references Denise Robbins, "New Book Exposes Koch Brothers' Guide to Infiltrating the Media," Media Matters, February 17, 2016, www .mediamatters.org/koch-brothers/new-book-exposes-koch-brothers-guide -infiltrating-media. That discusses the Koch Brothers connection.

142. Michael Bastasch, "Scientists Issue 'Absurd' Doomsday Prediction, Warn of a 'Hothouse Earth,'" *Daily Caller*, August 7, 2018, https://dailycaller .com/2018/08/07/scientists-doomsday-hothouse-earth.

143. Roger A. Pielke Sr (@RogerAPielkeSr), Twitter, August 7, 2018, 10:50 a.m., https://twitter.com/RogerAPielkeSr/status/1026888246305775616.

144. Miranda Devine, "Celebrities, Activists Using Australia Bushfire Crisis to Push Dangerous Climate Change Myth," *New York Post*, January 8, 2020, https://nypost.com/2020/01/08/celebrities-activists-using-australia-bushfire-crisis-to-push-dangerous-climate-change-myth-devine.

145. Kerry Emanuel, "Sober Appraisals of Risk Are Ignored in Critique of Hyperbole," *Boston Globe*, June 5, 2011, http://archive.boston.com/bostonglobe/editorial_opinion/letters/articles/2011/06/05/sober_appraisals_of_risk_are_ignored_in_critique_of_hyperbole.

146. Jeff Jacoby, "I'm Skeptical About Climate Alarmism, but I Take Coronavirus Fears Seriously," *Boston Globe*, March 15, 2020, www.bostonglobe.com/2020/03/14/opinion/im-skeptical-about-climate-alarmism-i-take-coronavirus-fears-seriously.

147. Michael E. Mann, Facebook, July 10, 2017, www.facebook.com/MichaelMannScientist/posts/since-this-new-york-magazine-article-the-uninhabitable-earth-is-getting-so-much-/1470539096335621.

148. Michael E. Mann, "Climatologist Makes Clear: We're Still on Pandemic Path with Global Warming," *Boston Globe*, March 18, 2020, www.bostonglobe.com/2020/03/19/opinion/climatologist-makes-clear-were-still-pandemic-path-with-global-warming.

149. See, for example, Christiana Figueres, Hans Joachim Schellnhuber, Gail Whiteman, Johan Rockström, Anthony Hobley, and Stefan Rahmstorf, "Three Years to Safeguard Our Climate," *Nature*, June 28, 2017, www.nature.com/news/three-years-to-safeguard-our-climate-1.22201.

CHAPTER 9: MEETING THE CHALLENGE

1. See Michael Mann and Tom Toles, *The Madhouse Effect* (New York: Columbia University Press, 2016), 164–166.

2. The provenance of the quote is much-debated, but at least one authoritative source (*Miami Herald*) attributes the quote in this form to Groucho Marx.

3. Brendan Fitzgerald, "Q&A: Michael Mann on Coverage Since 'Climategate,'" *Columbia Journalism Review*, September 19, 2019, www.cjr.org/covering_climate_now/michael-mann-climategate-franzen.php, reprinted at State Impact Pennsylvania, National Public Radio, September 21, 2019, https://stateimpact.npr.org/pennsylvania/2019/09/21/qa-penn-state-climate-scientist-michael-mann-on-news-coverage-since-climategate.

4. Scott Waldman, "Cato Closes Its Climate Shop; Pat Michaels Is Out," *Climatewire, E&E News*, May 29, 2019, www.eenews.net/stories/1060419123.

5. Richard Collett-White, "Climate Science Deniers Planning European Misinformation Campaign, Leaked Documents Reveal," September 6, 2019, www.desmog.co.uk/2019/09/06/climate-science-deniers-planning-coordinated-european-misinformation-campaign-leaked-documents-reveal; Nicholas Kusnetz,

"Heartland Launches Website of Contrarian Climate Science amid Struggles with Funding and Controversy Dogged by Layoffs, a Problematic Spokesperson and an Investigation by European Journalists, the Climate Skeptics' Institute Returns to Its Old Tactics," *Inside Climate News*, March 13, 2020, https://inside climatenews.org/news/12032020/heartland-instutute-climate-change-skeptic.

6. See Connor Gibson, "Heartland's Jay Lehr Calls EPA 'Fraudulent,' Despite Defrauding EPA and Going to Jail," *DeSmog* (blog), September 4, 2014, www.desmogblog.com/2014/09/04/heartland-science-director-jay-lehr-calls -epa-fraudulent-defrauded-epa-himself.

7. Waldman, "Cato Closes Its Climate Shop"; Alexander C. Kaufman, "Pro-Trump Climate Denial Group Lays Off Staff amid Financial Woes, Ex-Employees Say: The Heartland Institute Is the Think Tank Paying the Far-Right German Teen Known as the 'Anti-Greta,'" *Huffington Post*, March 9, 2020, www.huffpost.com/entry/heartland-institute-staff-layoffs-climate-change -denial_n_5e6302a6c5b6670e72f85fa5.

8. "Article by Michael Shellenberger Mixes Accurate and Inaccurate Claims in Support of a Misleading and Overly Simplistic Argumentation About Climate Change," Climate Feedback, https://climatefeedback.org /evaluation/article-by-michael-shellenberger-mixes-accurate-and-inaccurate -claims-in-support-of-a-misleading-and-overly-simplistic-argumentation -about-climate-change.

9. Graham Readfearn, "The Environmentalist's Apology: How Michael Shellenberger Unsettled Some of His Prominent Supporters," *The Guardian*, July 3, 2020, www.theguardian.com/environment/2020/jul/04/the -environmentalists-apology-how-michael-shellenberger-unsettled-some-of -his-prominent-supporters.

10. Kate Yoder, "Frank Luntz, the GOP's Message Master, Calls for Climate Action," *Grist*, July 25, 2019, https://grist.org/article/the-gops-most -famous-messaging-strategist-calls-for-climate-action.

11. Lissa Friedman, "Climate Could Be an Electoral Time Bomb, Republican Strategists Fear," *New York Times*, August 2, 2019, www.nytimes .com/2019/08/02/climate/climate-change-republicans.html.

12. Kusnetz, "Heartland Launches Website of Contrarian Climate Science."

13. Nathanael Johnson, "Fossil Fuels Are the Problem, Say Fossil Fuel Companies Being Sued," *Grist*, March 21, 2018, https://grist.org/article/fossil -fuels-are-the-problem-say-fossil-fuel-companies-being-sued.

14. Matthew Daily, "Dem Climate Plan Would End Greenhouse Gas Emissions by 2050," Associated Press, June 30, 2020, https://apnews.com /f72c5ae628bac72a7732d968a056878d

15. Justin Gillis, "The Republican Climate Closet: When Will Believers in Global Warming Come Out?," *New York Times*, August 12, 2019, www .nytimes.com/2019/08/12/opinion/republicans-environment.html.

16. George P. Shultz and Ted Halstead, "The Winning Conservative Climate Solution," *Washington Post*, January 16, 2020, www.washingtonpost.com /opinions/the-winning-republican-climate-solution-carbon-pricing/2020/01 /16/d6921dc0-387b-11ea-bf30-ad313e4ec754_story.html.

17. Damon Centola, Joshua Becker, Devon Brackbill, and Andrea Baronchelli, "Experimental Evidence for Tipping Points in Social Convention," *Science* 360, no. 6393 (2018): 1116–1119, https://doi.org/10.1126/science .aas8827.

18. "Attitudes on Same-Sex Marriage," Pew Research Center, May 14, 2019, www.pewforum.org/fact-sheet/changing-attitudes-on-gay-marriage.

19. Frank Luntz (@FrankLuntz), Twitter, June 8, 2020, 8:29 a.m., https:// twitter.com/FrankLuntz/status/1270015144337141760.

20. "U.S. Public Views on Climate and Energy," Pew Research Center, November 25, 2019, www.pewresearch.org/science/2019/11/25/u-s-public -views-on-climate-and-energy.

21. Miranda Green, "Poll: Climate Change Is Top Issue for Registered Democrats," *The Hill*, April 30, 2019, https://thehill.com/policy/energy-environment /441344-climate-change-is-the-top-issue-for-registered-democratic-voters.

22. "Why Are Millions of Citizens Not Registered to Vote?," Pew Trusts, June 21, 2017, www.pewtrusts.org/en/research-and-analysis/issue-briefs/2017 /06/why-are-millions-of-citizens-not-registered-to-vote; Aaron Blake, "For the First Time, There Are Fewer Registered Republicans Than Independents," *Washington Post*, February 28, 2020, www.washingtonpost.com/politics/2020/02 /28/first-time-ever-there-are-fewer-registered-republicans-than-independents.

23. Ilona M. Otto, Jonathan F. Donges, Roger Cremades, Avit Bhowmik, Richard J. Hewitt, Wolfgang Lucht, Johan Rockström, et al., "Social Tipping Dynamics for Stabilizing Earth's Climate by 2050," *Proceedings of the National Academy of Sciences* 117, no. 5 (2020): 2354–2365, https://doi.org/10.1073 /pnas.1900577117.

24. Mark Lewis, "Has Saudi Shifted Its Strategy in the Era of Decarbonisation?," *Financial Times*, March 15, 2020, www.ft.com/content/8c17582a-6547 -11ea-a6cd-df28cc3c6a68.

25. Matt Egan, "The Market Has Spoken: Coal Is Dying," CNN Business, September 20, 2019, www.cnn.com/2019/09/20/business/coal-power -dying/index.html; Will Wade, "New York's Last Coal-Fired Power Plant to Retire Tuesday," Bloomberg Green, March 30, 2020, www.bloomberg.com /news/articles/2020-03-30/new-york-s-last-coal-fired-power-plant-will-shut -down-tuesday.

26. Leyland Cecco and agencies, "Canadian Mining Giant Withdraws Plans for C$20bn Tar Sands Project: Teck Resources' Surprise Decision Drew Outrage from Politicians in Oil-Rich Alberta and Cheers from Environmental Groups," *The Guardian*, February 24, 2020, www.theguardian.com/world/2020/feb /24/canadian-mine-giant-teck-resources-withdraws-plans-tar-sands-project.

27. Peter Eavis, "Fracking Once Lifted Pennsylvania. Now It Could Be a Drag," *New York Times*, March 31, 2020, www.nytimes.com/2020/03/31 /business/energy-environment/pennsylvania-shale-gas-fracking.html.

28. Fiona Harvey, "What Is the Carbon Bubble and What Will Happen if It Bursts?," *The Guardian*, June 4, 2018, www.theguardian.com/environment /2018/jun/04/what-is-the-carbon-bubble-and-what-will-happen-if-it-bursts.

29. Coryanne Hicks, "What Is a Fiduciary Financial Advisor? A Fiduciary Is Defined by the Legal and Ethical Requirement to Put Your Best Interest Before Their Own," *US News & World Report*, February 24, 2020, https://money.usnews.com/investing/investing-101/articles/what-is-a-fiduciary-financial-advisor-a-guide-to-the-fiduciary-duty.

30. Sarah Barker, Mark Baker-Jones, Emilie Barton, and Emma Fagan, "Climate Change and the Fiduciary Duties of Pension Fund Trustees—Lessons from the Australian Law," *Journal of Sustainable Finance and Investment* 6, no. 3 (2016): 211–214, https://doi.org/10.1080/20430795.2016.1204687.

31. Scott Murdoch and Paulina Duran, "Australian Pension Funds' $168 Billion 'Wall of Cash' May Lead Overseas," Reuters, September 9, 2019, www.reuters.com/article/us-australia-funds-pensions/australian-pension-funds-168-billion-wall-of-cash-may-lead-overseas-idUSKCN1VV07S.

32. Michael E. Mann (@MichaelEMann), Twitter, March 11, 2020, 12:54 a.m., https://twitter.com/MichaelEMann/status/1237647998575734785.

33. Gillian Tett, "The World Needs a Libor for Carbon Pricing," *Financial Times*, January 23, 2020, www.ft.com/content/20dd6b82-3dd1-11ea-a01a-bae547046735.

34. Huw Jones, "Bank of England Considers Bank Capital Charge on Polluting Assets," Reuters, March 10, 2020, https://uk.reuters.com/article/uk-climatechange-britain-banks/bank-of-england-considers-bank-capital-charge-on-polluting-assets-idUKKBN20X1NU.

35. Christopher Flavelle, "Global Financial Giants Swear Off Funding an Especially Dirty Fuel," *New York Times*, February 12, 2020, www.nytimes.com/2020/02/12/climate/blackrock-oil-sands-alberta-financing.html.

36. Steven Mufson and Rachel Siegel, "BlackRock Makes Climate Change Central to Its Investment Strategy," *Washington Post*, January 14, 2020, www.washingtonpost.com/business/2020/01/14/blackrock-letter-climate-change.

37. Bill McKibben, "Citing Climate Change, BlackRock Will Start Moving Away from Fossil Fuels," *New Yorker*, January 16, 2020, www.newyorker.com/news/daily-comment/citing-climate-change-blackrock-will-start-moving-away-from-fossil-fuels.

38. Juliet Eilperin, Steven Mufson and Brady Dennis, "Major Oil and Gas Pipeline Projects, Backed by Trump, Flounder as Opponents Prevail in Court," *Washington Post*, July 6, 2020, www.washingtonpost.com/climate-environment/2020/07/06/dakota-access-pipeline.

39. Nassim Khadem, "Mark McVeigh Is Taking on REST Super on Climate Change and Has the World Watching," Australian Broadcasting Corporation, January 17, 2020, www.abc.net.au/news/2020-01-18/mark-mcveigh-is-taking-on-rest-super-and-has-the-world-watching/11876360.

40. Richard Knight, "Sanctions, Disinvestment, and U.S. Corporations in South Africa," reprinted and updated from *Sanctioning Apartheid* (Lawrenceville, NJ: Africa World Press, 1990), http://richardknight.homestead.com/files/uscorporations.htm.

41. One Bold Idea, "How Students Helped End Apartheid: The UC Berkeley Protest That Changed the World," University of California, May 2, 2018, www.universityofcalifornia.edu/news/how-students-helped-end-apartheid.

42. See "A New Fossil Free Milestone: $11 Trillion Has Been Committed to Divest from Fossil Fuels," 350.org, https://350.org/11-trillion-divested.

43. Jagdeep Singh Bachher and Richard Sherman, "UC Investments Are Going Fossil Free. But Not Exactly for the Reasons You May Think," *Los Angeles Times*, September 17, 2019, www.latimes.com/opinion/story/2019-09-16/divestment-fossil-fuel-university-of-california-climate-change.

44. The provenance of this quote is murky. See Quote Investigator, https://quoteinvestigator.com/2018/01/07/stone-age.

45. "Global CO2 Emissions in 2019," International Energy Agency, February 11, 2020, www.iea.org/articles/global-co2-emissions-in-2019.

46. "Latest Data Book Shows U.S. Renewable Capacity Surpassed 20% for First Time in 2018. Growth Continues in U.S. Installed Wind and Solar Photovoltaic Capacity, Energy Storage, and Electric Vehicle Sales," National Renewable Energy Laboratory, February 18, 2020, www.nrel.gov/news/program/2020/latest-data-book-shows-us-renewable-capacity-surpassed-20-for-the-first-time-in-2018.html.

47. Seth Feaster and Dennis Wamsted, "Utility-Scale Renewables Top Coal for the First Quarter of 2020," Institute for Energy Economics and Financial Analysis, April 1, 2020, https://ieefa.org/ieefa-u-s-utility-scale-renewables-top-coal-for-the-first-quarter-of-2020.

48. Randell Suba, "Tesla 'Big Battery' in Australia Is Becoming a Bigger Nightmare for Fossil Fuel Power Generators," Teslarati, February 28, 2020, www.teslarati.com/tesla-big-battery-hornsdale-australia-cost-savings.

49. Sophie Vorrath, "South Australia on Track to 100 Pct Renewables, as Regulator Comes to Party," Renew Economy, January 24, 2020, https://reneweconomy.com.au/south-australia-on-track-to-100-pct-renewables-as-regulator-comes-to-party-96366.

50. I coauthored a similarly titled commentary in *Newsweek*: Lawrence Torcello and Michael E. Mann, "Seeing the COVID-19 Crisis Is Like Watching a Time Lapse of Climate Change. Will the Right Lessons Be Learned?," *Newsweek*, April 1, 2020, www.newsweek.com/fake-news-climate-change-coronavirus-time-lapse-1495603.

51. Debora Mackenzie, "We Were Warned—So Why Couldn't We Prevent the Coronavirus Outbreak?," *New Scientist*, March 4, 2020, www.newscientist.com/article/mg24532724-700-we-were-warned-so-why-couldnt-we-prevent-the-coronavirus-outbreak.

52. Harry Stevens, "Why Outbreaks Like Coronavirus Spread Exponentially, and How to 'Flatten the Curve,'" *Washington Post*, March 14, 2020, www.washingtonpost.com/graphics/2020/world/corona-simulator.

53. Ed Yong, "The U.K.'s Coronavirus 'Herd Immunity' Debacle," *The Atlantic*, March 16, 2020, www.theatlantic.com/health/archive/2020/03/coronavirus-pandemic-herd-immunity-uk-boris-johnson/608065.

54. Alex Wickham, "The UK Only Realised 'In the Last Few Days' That Its Coronavirus Strategy Would 'Likely Result in Hundreds of Thousands of Deaths,'" *BuzzFeed*, March 16, 2020, www.buzzfeed.com/alexwickham /coronavirus-uk-strategy-deaths.

55. Michelle Cottle, "Boris Johnson Should Have Taken His Own Medicine," *New York Times*, March 27, 2020, www.nytimes.com/2020/03/27 /opinion/boris-johnson-coronavirus.html.

56. John Burn-Murdoch (@jburnmurdoch), Twitter, April 3, 2020, 2:15 p.m., https://twitter.com/jburnmurdoch/status/1246184639540146178.

57. Patrick Wyman, "How Do You Know If You're Living Through the Death of an Empire? It's the Little Things," *Mother Jones*, March 19, 2020, www.motherjones.com/media/2020/03/how-do-you-know-if-youre-living -through-the-death-of-an-empire.

58. Jonathan Watts, "Delay Is Deadly: What Covid-19 Tells Us About Tackling the Climate Crisis," *The Guardian*, March 24, 2020, www.theguardian .com/commentisfree/2020/mar/24/covid-19-climate-crisis-governments -coronavirus.

59. Saijel Kishan, "Professor Sees Climate Mayhem Lurking Behind Covid-19 Outbreak," Bloomberg Green, March 28, 2020, www.bloomberg.com /news/articles/2020-03-28/professor-sees-climate-mayhem-lurking-behind -covid-19-outbreak.

60. William J. Broad, "Putin's Long War Against American Science," *New York Times*, April 13, 2020, www.nytimes.com/2020/04/13/science/putin -russia-disinformation-health-coronavirus.html.

61. Alex Kotch, "Right-Wing Megadonors Are Financing Media Operations to Promote Their Ideologies," PR Watch, January 27, 2020, www.prwatch .org/news/2020/01/13531/right-wing-megadonors-are-financing-media -operations-promote-their-ideologies; Julie Kelly, "Hockey Sticks, Changing Goal Posts, and Hysteria," American Greatness, March 31, 2020, https:// amgreatness.com/2020/03/31/hockey-sticks-changing-goal-posts-and-hysteria.

62. Benny Peiser and Andrew Montford, "Coronavirus Lessons from the Asteroid That Didn't Hit Earth: Scary Projections Based on Faulty Data Can Put Policy Makers Under Pressure to Adopt Draconian Measures," *Wall Street Journal*, April 1, 2020, www.wsj.com/articles/coronavirus-lessons-from-the -asteroid-that-didnt-hit-earth-11585780465. See "Benny Peiser," Source- Watch, www.sourcewatch.org/index.php/Benny_Peiser; "Andrew Montford," SourceWatch, www.sourcewatch.org/index.php/Andrew_Montford.

63. S. T. Karnick, "Watch Out for Long-Term Effects of Government's Coronavirus Remedies," Heartland Institute, April 2, 2020, www.heartland .org/news-opinion/news/watch-out-for-long-term-effects-of-governments -coronavirus-remedies-1.

64. Nic Lewis, "COVID-19: Updated Data Implies That UK Modelling Hugely Overestimates the Expected Death Rates from Infection," Climate Etc., March 25, 2020, https://judithcurry.com/2020/03/25/covid-19-updated -data-implies-that-uk-modelling-hugely-overestimates-the-expected-death

-rates-from-infection; Monkton, "Are Lockdowns Working?," *Watts Up with That*, April 4, 2020, https://wattsupwiththat.com/2020/04/04/are-lockdowns-working; Marcel Crok (@marcelcrok), Twitter, March 24, 2020, 4:59 a.m., https://twitter.com/marcelcrok/status/1242420742253432832; William M. Briggs, "Coronavirus Update VI: Calm Yourselves," March 24, 2020, https://wmbriggs.com/post/29886.

65. Katelyn Weisbrod, "6 Ways Trump's Denial of Science Has Delayed the Response to COVID-19 (and Climate Change): Misinformation, Blame, Wishful Thinking and Making Up Facts Are Favorite Techniques," *Inside Climate News*, March 19, 2020, https://insideclimatenews.org/news/19032020/denial-climate-change-coronavirus-donald-trump. A video comparison is available. See Michael E. Mann (@MichaelEMann), Twitter, March 25, 2020, 7:05 p.m., https://twitter.com/MichaelEMann/status/1242996165546848256.

66. Scott Waldman, "Obama Blasts Trump over Coronavirus, Climate Change," *Climatewire*, *E&E News*, April 1, 2020, www.eenews.net/stories/1062754487.

67. As pointed out by me on Twitter at Michael E. Mann (@MichaelEMann), Twitter, March 30, 2020, 9:32 a.m., https://twitter.com/MichaelEMann/status/1244663896893513729. Examples for both climate change and coronavirus are provided by these two articles, respectively: Scott Waldman, "Ex-Trump Adviser: 'Brainwashed' Aides Killed Climate Review," *ClimateWire*, *E&E News*, December 4, 2019, www.eenews.net/stories/1061717133; Isaac Chotiner, "The Contrarian Coronavirus Theory That Informed the Trump Administration," *New Yorker*, March 30, 2020, www.newyorker.com/news/q-and-a/the-contrarian-coronavirus-theory-that-informed-the-trump-administration.

68. Weisbrod, "6 Ways Trump's Denial of Science Has Delayed the Response to COVID-19 (and Climate Change)."

69. Jeff Mason, "Do Social Distancing Better, White House Doctor Tells Americans. Trump Objects," Reuters, April 2, 2020, www.reuters.com/article/us-health-coronavirus-trump-birx/do-social-distancing-better-white-house-doctor-tells-americans-trump-objects-idUSKBN21L08A.

70. Laurie McGinley and Carolyn Y. Johnson, "FDA Pulls Emergency Approval for Antimalarial Drugs Touted by Trump as Covid-19 Treatment," *Washington Post*, June 15, 2020, www.washingtonpost.com/health/2020/06/15/hydroxychloroquine-authorization-revoked-coronavirus.

71. Juliet Eilperin, Darryl Fears, and Josh Dawsey, "Trump Is Headlining Fireworks at Mount Rushmore. Experts Worry Two Things Could Spread: Virus and Wildfire," *Washington Post*, June 25, 2020, www.washingtonpost.com/climate-environment/2020/06/24/trump-mount-rushmore-fireworks.

72. The Daily Show (@TheDailyShow), Twitter, April 3, 2020, 11:45 a.m., https://twitter.com/TheDailyShow/status/1246146713523453957.

73. Bobby Lewis and Kayla Gogarty, "Pro-Trump Media Have Ramped Up Attacks Against Dr. Anthony Fauci," Media Matters, March 24, 2020, www.mediamatters.org/coronavirus-covid-19/pro-trump-media-have-ramped-attacks-against-dr-anthony-fauci.

74. Michael Gerson, "The Trump Administration Has Released a Lot of Shameful Documents. This One Might Be the Worst," *Washington Post*, July 13, 2020, www.washingtonpost.com/opinions/fauci-has-been-an-example-of-conscience-and-courage-trump-has-been-nothing-but-weak/2020/07/13/7c9a7578-c52b-11ea-8ffe-372be8d82298_story.html.

75. See video clip and my comment at Michael E. Mann (@MichaelEMann), Twitter, March 23, 2020, 5:49 p.m., https://twitter.com/MichaelEMann/status/1242252283557093377.

76. Matthew Chapman, "Internet Explodes as Fox's Brit Hume Says It's 'Entirely Reasonable' to Let Grandparents Die for the Stock Market," *Rawstory*, March 24, 2020, www.rawstory.com/2020/03/internet-explodes-as-foxs-brit-hume-says-its-entirely-reasonable-to-let-grandparents-die-for-the-stock-market.

77. Bill Mitchell (@mitchellvii), Twitter, April 4, 2020, 7:21 p.m., https://twitter.com/mitchellvii/status/1246623932767141890.

78. The remarkable parallels between the various stages of denial and inactivism with both climate change and coronavirus were explored in an exchange between me and *Sydney Morning Herald* environmental journalist Peter Hannam. See Michael E. Mann (@MichaelEMann), Twitter, April 5, 2020, 2:08 p.m., https://twitter.com/MichaelEMann/status/1246907494221451270; Peter Hannam (@p_hannam), Twitter, April 5, 2020, 2:02 p.m., https://twitter.com/p_hannam/status/1246906143932215297.

79. Mike MacFerrin (@IceSheetMike), Twitter, March 25, 2020, 6:40 a.m., https://twitter.com/IceSheetMike/status/1242808580350177287.

80. Dan Rather (@DanRather), Twitter, March 25, 2020, 9:10 a.m., https://twitter.com/DanRather/status/1242846264963678209.

81. Michael E. Mann (@MichaelEMann), Twitter, March 25, 2020, 9:23 a.m., https://twitter.com/MichaelEMann/status/1242849532402114563.

82. Waldman, "Obama Blasts Trump over Coronavirus, Climate Change."

83. Steve Schmidt (@SteveSchmidtSES), Twitter, April 4, 2020, 3:05 p.m., https://twitter.com/SteveSchmidtSES/status/1246559631067021313.

84. Nicole Acevedo, "Democratic Lawmakers Want Answers to Trump Administration's Coronavirus Response in Puerto Rico," NBC News, April 3, 2020, www.nbcnews.com/news/latino/democratic-lawmakers-want-answers-trump-administration-s-coronavirus-response-puerto-n1175801.

85. Rebecca Hersher, "Climate Change Was the Engine That Powered Hurricane Maria's Devastating Rains," National Public Radio, April 17, 2019, www.npr.org/2019/04/17/714098828/climate-change-was-the-engine-that-powered-hurricane-marias-devastating-rains.

86. Torcello and Mann, "Seeing the COVID-19 Crisis Is Like Watching a Time Lapse of Climate Change."

87. Yasmeen Abutaleb, Josh Dawsey, Ellen Nakashima, and Greg Miller, "The U.S. Was Beset by Denial and Dysfunction as the Coronavirus Raged," *Washington Post*, April 4, 2020, www.washingtonpost.com/national-security/2020/04/04/coronavirus-government-dysfunction.

88. Ellen Knickmeyer and Tom Krisher, "Trump Rollback of Mileage Standards Guts Climate Change Push," Associated Press, March 31, 2020, https://apnews.com/98f311a6d4275334a9e4d3a804cd2e1a; Alexander C. Kaufman, "States Quietly Pass Laws Criminalizing Fossil Fuel Protests amid Coronavirus Chaos," *Huffington Post*, March 27, 2020, www.huffpost.com/entry/pipeline-protest-laws-coronavirus_n_5e7e7570c5b6256a7a2aab41.

89. Mark Kaufman, "Earth Scorched in the First 3 Months of 2020," *Mashable*, April 6, 2020, https://mashable.com/article/climate-change-2020-records.

90. Denise Chow, "Great Barrier Reef Hit by Third Major Bleaching Event in Five Years," NBC News, March 23, 2020, https://nbcnews.com/science/environment/great-barrier-reef-hit-third-major-bleaching-event-five-years-n1166676.

91. Noted environmental historian Naomi Oreskes, for example, declared that "coronavirus has killed neoliberalism." Naomi Oreskes (@NaomiOreskes), Twitter, April 5, 2020, 10:32 a.m., https://twitter.com/NaomiOreskes/status/1246853279507775488.

92. Michael E. Mann (@MichaelEMann), Twitter, April 5, 2020, 10:52 a.m., https://twitter.com/MichaelEMann/status/1246858265524342785.

93. John Vidal, "Destroyed Habitat Creates the Perfect Conditions for Coronavirus to Emerge," *Scientific American*, March 18, 2020, www.scientificamerican.com/article/destroyed-habitat-creates-the-perfect-conditions-for-coronavirus-to-emerge.

94. James E. Lovelock and Lynn Margulis, "Atmospheric Homeostasis by and for the Biosphere: The Gaia Hypothesis," *Tellus* 26, no. 1–2 (1974): 2–10.

95. Madeleine Stone, "Carbon Emissions Are Falling Sharply Due to Coronavirus. But Not for Long," *National Geographic*, April 6, 2020, www.nationalgeographic.co.uk/environment-and-conservation/2020/04/carbon-emissions-are-falling-sharply-due-coronavirus-not-long.

96. Michael E. Mann (@MichaelEMann), Twitter, March 18, 2020, 1:13 a.m., https://twitter.com/MichaelEMann/status/1240189600510799872.

97. Swati Thiyagarajan, "Covid-19: Planet Earth Fights Back," *Daily Maverick*, March 17, 2020, https://conservationaction.co.za/recent-news/covid-19-planet-earth-fights-back.

98. For examples, see Bill Black, "The El Paso Shooter's Manifesto Contains a Dangerous Message About Climate Change," *The Week*, August 6, 2019, https://theweek.com/articles/857100/el-paso-shooters-manifesto-contains-dangerous-message-about-climate-change; Charlotte Cross, "Extinction Rebellion Disowns 'Fake' East Midlands Group over Coronavirus Tweet," ITV News, www.itv.com/news/central/2020-03-25/extinction-rebellion-disowns-east-midlands-group-over-coronavirus-tweet.

99. I published a commentary expressing these thoughts: Michael E. Mann, "Climatologist Makes Clear: We're Still on Pandemic Path with Global Warming," *Boston Globe*, March 18, 2020, www.bostonglobe.com/2020/03/19/opinion/climatologist-makes-clear-were-still-pandemic-path-with-global-warming.

100. Jeremy Miller, "Trump Seizes on Pandemic to Speed Up Opening of Public Lands to Industry," *The Guardian*, April 30, 2020.

101. Nicholas Kusnetz, "BP and Shell Write Off Billions in Assets, Citing Covid-19 and Climate Change," *Inside Climate News*, July 2, 2020, https://insideclimatenews.org/news/01072020/bp-shell-coronavirus-climate-change.

102. David Iaconangelo, "100% Clean Energy Group Launches, with Eyes on Coronavirus," *Climatewire, E&E News*, April 2, 2020, www.eenews.net/energywire/2020/04/02/stories/1062762687.

103. Tina Casey, "And So It Begins: World's 11th-Biggest Economy Pitches Renewable Energy for COVID-19 Recovery," *Clean Technica*, April 5, 2020, https://cleantechnica.com/2020/04/05/and-so-it-begins-worlds-11th-biggest-economy-pitches-renewable-energy-for-covid-19-recovery.

104. Adam Morton, "Australia's Path to Net-Zero Emissions Lies in Rapid, Stimulus-Friendly Steps," *The Guardian*, April 3, 2020, www.theguardian.com/environment/2020/apr/04/australias-path-to-net-zero-emissions-lies-in-small-stimulus-friendly-steps.

105. Alison Rourke, "Greta Thunberg Responds to Asperger's Critics: 'It's a Superpower,'" The Guardian, September 2, 2019, www.theguardian.com/environment/2019/sep/02/greta-thunberg-responds-to-aspergers-critics-its-a-superpower.

106. Desmond Butler and Juliet Eilperin, "The Anti-Greta: A Conservative Think Tank Takes on the Global Phenomenon: How a Group Allied with the Trump Administration Is Paying a German Teen to Question Established Climate Science," *Washington Post*, February 23, 2020, www.washingtonpost.com/climate-environment/2020/02/23/meet-anti-greta-young-youtuber-campaigning-against-climate-alarmism.

107. "UK Parliament Declares Climate Change Emergency," BBC, May 1, 2019, www.bbc.com/news/uk-politics-48126677; "Climate Change: Ireland Declares Climate Emergency," BBC, May 9, 2019, www.bbc.com/news/world-europe-48221080.

108. Matthew Taylor, "Majority of UK Public Back 2030 Zero-Carbon Target—Poll," *The Guardian*, November 7, 2019, www.theguardian.com/environment/2019/nov/07/majority-of-uk-public-back-2030-zero-carbon-target-poll.

109. Christopher Walsh, "Teens Mobilize to 'Save the Planet,'" *East Hampton Star*, August 15, 2019, www.easthamptonstar.com/2019815/teens-mobilize-save-planet.

110. Walsh, "Teens Mobilize to 'Save the Planet'"; Jerome Foster II (@JeromeFosterII), Twitter, August 18, 2019, 12:51 p.m., https://twitter.com/JeromeFosterII/status/1163176641016868865.

111. Gregor Hagedorn, Peter Kalmus, Michael Mann, Sara Vicca, Joke Van den Berge, Jean-Pascal van Ypersele, Dominique Bourg, et al., "Concerns of Young Protesters Are Justified," *Science* 364, no. 6436 (2019): 139–140, https://doi.org/10.1126/science.aax3807; Haley Ott, "Thousands of Scientists Back 'Young Protesters' Demanding Climate Change Action," CBS News, April 12, 2019, www.cbsnews.com/news/youth-climate-strike-protests-backed-by-scientists-letter-science-magazine.

112. Walsh, "Teens Mobilize to 'Save the Planet.'"

113. Luke Henriques-Gomes, "Andrew Bolt's Mocking of Greta Thunberg Leaves Autism Advocates 'Disgusted,'" *The Guardian*, August 2, 2019, www.theguardian.com/media/2019/aug/02/andrew-bolts-mocking-of-greta-thunberg-leaves-autism-advocates-disgusted.

114. See "Christopher Caldwell," Claremont Institute, Leadership and Staff, www.claremont.org/leadership-bio/christopher-caldwell, accessed April 7, 2020; "Claremont Institute for the Study of Statesmanship and Political Philosophy," SourceWatch, www.sourcewatch.org/index.php/Claremont_Institute_for_the_Study_of_Statesmanship_and_Political_Philosophy, accessed April 7, 2020; Christopher Caldwell, "The Problem with Greta Thunberg's Climate Activism: Her Radical Approach Is at Odds with Democracy," *New York Times*, August 2, 2019, www.nytimes.com/2019/08/02/opinion/climate-change-greta-thunberg.html.

115. "CO_2 Coalition," SourceWatch, www.sourcewatch.org/index.php?title=CO2_Coalition; Scott Waldman, "Climate Critics Escalate Personal Attacks on Teen Activist," *Climatewire*, *E&E News*, August 9, 2019, www.eenews.net/stories/1060889513.

116. Jonathan Watts, "'Biggest Compliment Yet': Greta Thunberg Welcomes Oil Chief's 'Greatest Threat' Label," *The Guardian*, July 5, 2019, www.theguardian.com/environment/2019/jul/05/biggest-compliment-yet-greta-thunberg-welcomes-oil-chiefs-greatest-threat-label.

117. Watts, "'Biggest Compliment Yet.'"

118. Max Boykoff, "The Kids Are All Right. Adults Are the Climate Change Problem," Center for Science and Technology Policy Research, reprinted from the *Daily Camera*, September 5, 2019, https://sciencepolicy.colorado.edu/ogmius/archives/issue_54/ogmius_exchange1.html.

119. Malcolm Harris, "Shell Is Looking Forward: The Fossil-Fuel Companies Expect to Profit from Climate Change. I Went to a Private Planning Meeting and Took Notes," *New York Magazine*, March 3, 2020, https://nymag.com/intelligencer/2020/03/shell-climate-change.html.

120. Michael E. Mann (@MichaelEMann), Twitter, April 10, 2020, 1:32 p.m., https://twitter.com/MichaelEMann/status/1248710415426748422.

121. "Wicked Problem," Wikipedia, https://en.wikipedia.org/wiki/Wicked_problem, accessed April 10, 2020.

122. Jonathan Gilligan (@jg_environ), Twitter, August 17, 2019, 11:01 a.m., https://twitter.com/jg_environ/status/1162786503208177664.

123. Paul Price (@swimsure), Twitter, December 24, 2017, 6:55 a.m., https://twitter.com/swimsure/status/944944603060424704.

124. Peter Jacobs (@past_is_future), Twitter, August 28, 2019, 8:54 a.m., https://twitter.com/past_is_future/status/1166740808810344450.

125. Michael E. Mann (@MichaelEMann), Twitter, August 17, 2019, 8:03 p.m., https://twitter.com/MichaelEMann/status/1162923050444165120.

126. Thomas (@djamesalicious), Twitter, April 9, 2020, 10:35 a.m., https://twitter.com/djamesalicious/status/1248303643985674240.

127. Michael E. Mann (@MichaelEMann), Twitter, April 9, 2020, 3:55 p.m., https://twitter.com/MichaelEMann/status/1248384126572355586.

128. Thomas (@djamesalicious), Twitter, April 9, 2020, 10:35 a.m., https://twitter.com/djamesalicious/status/1248303643985674240.

129. John Kruzel, "Was Joe Biden a Climate Change Pioneer in Congress? History Says Yes," Politifact, May 8, 2019, www.politifact.com/factchecks/2019/may/08/joe-biden/was-joe-biden-climate-change-pioneer-congress-hist.

130. Shane Harris, Ellen Nakashima, Michael Scherer, and Sean Sullivan, "Bernie Sanders Briefed by U.S. Officials That Russia Is Trying to Help His Presidential Campaign," *Washington Post*, February 21, 2020, www.washingtonpost.com/national-security/bernie-sanders-briefed-by-us-officials-that-russia-is-trying-to-help-his-presidential-campaign/2020/02/21/5ad396a6-54bd-11ea-929a-64efa7482a77_story.html.

131. See "About This Book" on my website at https://michaelmann.net/content/tantrum-saved-world-carbon-neutral-kids-book. Available for purchase at https://world-saving-books.myshopify.com.

132. Graham Readfearn, "Great Barrier Reef's Third Mass Bleaching in Five Years the Most Widespread Yet," *The Guardian*, April 6, 2020, www.theguardian.com/environment/2020/apr/07/great-barrier-reefs-third-mass-bleaching-in-five-years-the-most-widespread-ever.

133. I said this on Twitter at Michael E. Mann (@MichaelEMann), May 7, 2019, 3:52 p.m., https://twitter.com/MichaelEMann/status/1125896127662960640.

134. Will Wade, "Going 100% Green Will Pay for Itself in Seven Years, Study Finds," Bloomberg News, December 20, 2019, www.bloomberg.com/news/articles/2019-12-20/going-100-green-will-pay-for-itself-in-seven-years-study-finds.

135. Lauri Myllyvirta, "Analysis: Coronavirus Temporarily Reduced China's CO_2 Emissions by a Quarter," *Carbon Brief*, February 19, 2020, www.carbonbrief.org/analysis-coronavirus-has-temporarily-reduced-chinas-co2-emissions-by-a-quarter.

136. Glen Peters (@Peters_Glen), Twitter, April 10, 2020, 12:36 a.m., https://twitter.com/peters_glen/status/1248515055836160000.

137. See Anthony Leiserowitz, Edward Maibach, Seth Rosenthal, John Kotcher, Matthew Ballew, Matthew Goldberg, Abel Gustafson, and Parrish Bergquist, "Politics and Global Warming, April 2019," Yale Program on Climate Change Communication, May 16, 2019, https://climatecommunication.yale.edu/publications/politics-global-warming-april-2019/2.

138. Michael J. Coren, "Americans: 'We Need a Carbon Tax, but Keep the Change,'" *Quartz*, January 22, 2019, https://qz.com/1529997/survey-finds-americans-want-a-carbon-tax; Michael E. Mann (@MichaelEMann), Twitter, April 13, 2020, 10:31 a.m., https://twitter.com/MichaelEMann/status/1249751975824146432.

139. Michael E. Mann (@MichaelEMann), Twitter, April 11, 2020, 10:19 a.m., https://twitter.com/MichaelEMann/status/1249024208484601856; Michael E. Mann (@MichaelEMann), Twitter, April 11, 2020, 10:31 a.m., https://twitter.com/MichaelEMann/status/1249027242858090501.

140. David Roberts (@drvox), Twitter, April 12, 2020, 12:22 p.m., https://twitter.com/drvox/status/1249417607289036801.

141. Fred Hiatt, "How Donald Trump and Bernie Sanders Both Reject the Reality of Climate Change," *Washington Post*, February 23, 2020, www .washingtonpost.com/opinions/how-donald-trump-and-bernie-sanders-both -reject-the-reality-of-climate-change/2020/02/23/cc657dcc-54de-11ea-9e47 -59804be1dcfb_story.html.

142. One user, for example, stated, at Laura Neish (@laurajneish), Twitter, April 12, 2020, 12:37 p.m., https://twitter.com/laurajneish/status /1249421480946892801:

Carbon pricing got hijacked by right wing extremists who want to:
-treat it as the *one* policy in place of any other regulatory action,
-tack on indemnity for ff companies
-make it an incredibly regressive tax one way or another
-use it to replace Corp and/or income tax

143. See "Beliefs and Principles," Unitarian Universalist Association, www .uua.org/beliefs/what-we-believe.

144. Sarah T. Fischell (@estee_nj), Twitter, April 12, 2020, 1:15 p.m., https://twitter.com/estee_nj/status/1249430997801873411.

145. Harris et al., "Bernie Sanders Briefed by U.S. Officials."

146. Jeff Goodell, "The Climate Crisis and the Case for Hope," *Rolling Stone*, September 16, 2019, www.rollingstone.com/politics/politics-news /climate-crisis-the-case-for-hope-884063.

Epilogue

1. Coral Davenport, "Climate Change Legislation Included in Coronavirus Relief Deal," *New York Times*, December 21, 2020, www.nytimes .com/2020/12/21/climate/climate-change-stimulus.html.

2. Andrew Freedman, "Climate Activists Score a Third ExxonMobil Board Seat," Axios, June 2, 2021, www.axios.com/exxonmobil-board-seat-activist -investor-1f019368-3e8a-4da4-97a6-181922b53df5.html; "Shell: Netherlands Court Orders Oil Giant to Cut Emissions," BBC, May 26, 2021, www.bbc.com/news /world-europe-57257982; "Developer Officially Cancels Keystone XL Pipeline Project Blocked by Biden," Reuters, June 10, 2021, www.reuters.com/business /energy/tc-energy-terminates-keystone-xl-pipeline-project-2021-06-09.

3. Michael E. Mann, "Wildfires, Floods and Extreme Heat: It Is Time to Heed Warnings from Climate Scientists," *The Hill*, July 22, 2021, https://the hill.com/opinion/energy-environment/564304-wildfires-floods-and-extreme -heat-it-is-time-to-heed-warnings-from.

4. Julia Musto, "What Is a Heat Dome? Pacific Northwest Boils Under Its Effects," Fox News, June 29, 2021, www.foxnews.com/science /what-is-a-heat-dome-pacific-northwest-boils-under-effects.

5. Brad Plumer and Ivan Penn, "Electric Utilities, Formed Decades Ago, Struggle to Meet Climate Crisis," *New York Times*, July 29, 2021, www.ny times.com/2021/07/29/climate/electric-utilities-climate-change.html; Rebecca

Solnit, "Our Climate Change Turning Point Is Right Here, Right Now," *The Guardian*, July 12, 2021, www.theguardian.com/commentisfree/2021/jul/12/our-climate-change-turning-point-is-right-here-right-now.

6. For a video of the full June 8, 2021, hearing, see "Examining Climate Change: A Threat to the Homeland," Committee on Homeland Security, https://homeland.house.gov/activities/hearings/examining-climate-change_a-threat-to-the-homeland.

7. Michael E. Mann (@MichaelEMann), Twitter, June 8, 2021, 12:23 p.m., https://twitter.com/MichaelEMann/status/1402344713789784068.

8. James Taylor, "Yes, the Climate Is Changing. No, It's Not an Emergency," *Newsweek*, August 3, 2021, www.newsweek.com/yes-climate-changing-no-its-not-emergency-opinion-1615632.

9. Naomi Oreskes, Michael E. Mann, Gernot Wagner, Don Wuebbles, Andrew Dessler, Andrea Dutton, Geoffrey Supran, et al., "That 'Obama Scientist' Climate Skeptic You've Been Hearing About...," *Scientific American*, June 1, 2021, www.scientificamerican.com/article/that-obama-scientist-climate-skeptic-youve-been-hearing-about.

10. Michael E. Mann, "Why Biden's Actions Are Good News from Front Lines of the Climate Change War," *USA Today*, February 2, 2021, www.usatoday.com/story/opinion/2021/02/02/michael-mann-biden-takes-good-first-steps-climate-change-war-column/4336625001.

11. Hyung-Jin Kim, "US, China Agree to Cooperate on Climate Crisis with Urgency," Associated Press, April 18, 2021, https://apnews.com/article/business-science-general-news-china-climate-905125d79b6c31940b8747df86c2a87a.

12. Emma Newburger, "Biden Pledges to Slash Greenhouse Gas Emissions in Half by 2030," CNBC, www.cnbc.com/2021/04/22/biden-pledges-to-slash-greenhouse-gas-emissions-in-half-by-2030.html.

13. See Mann, "Wildfires, Floods"; Michael E. Mann, "The Right Path Forward on Climate Change," *Newsweek*, February 23, 2021, www.newsweek.com/right-path-forward-climate-change-opinion-1571169; Caroline Vakil, "Climate Activists Target Manchin," *The Hill*, October 16, 2021, https://thehill.com/policy/energy-environment/577059-climate-activists-target-manchin.

14. "The Great Reset," World Economic Forum, n.d., www.weforum.org/great-reset.

15. Fiona Harvey, "Trillions of Dollars Spent on Covid Recovery in Ways That Harm Environment," *The Guardian*, July 15, 2021, www.theguardian.com/business/2021/jul/15/trillions-of-dollars-spent-on-covid-recovery-in-ways-that-harm-environment.

16. Joel Clement, "Why Biden's Interior Department Isn't Shutting Down Oil and Gas," *The Hill*, July 23, 2021, https://thehill.com/opinion/energy-environment/564359-why-bidens-interior-department-isnt-shutting-down-oil-and-gas.

17. Katie Bo Williams and Maegan Vazquez, "Biden Addresses Intelligence Community for First Time as President," CNN, July 27, 2021, www.cnn.com/2021/07/27/politics/biden-intelligence-community/index.html.

18. Caleb Ecarma, "Republicans Want to Protect Your Right to Die to Own the Libs: State Lawmakers in Montana and Alabama Have Passed Laws Protecting People Who Choose Not to Get the COVID-19 Vaccine from 'Discrimination,' and Others Are Following Suit," *Vanity Fair*, July 12, 2021, www.vanityfair.com/news/2021/07/republicans-want-unvaccinated-protected-group.

19. Michael E. Mann (@MichaelEMann), Twitter, September 6, 2020, 7:14 a.m., https://twitter.com/MichaelEMann/status/1302611141315756033.

20. Michael Mann, "After Australia, Reading the Lies About West Coast Fires Feels Like Déjà Vu," *Newsweek*, September 14, 2020, www.newsweek.com/lies-climate-change-west-coast-fires-1531786.

21. David Goldman and Matt Egan, "A Third Climate Activist Is Expected to Be Elected to Exxon's Board," CNN, June 2, 2021, www.cnn.com/2021/06/02/investing/exxon-shareholders-climate-activist/index.html.

22. "Shell: Netherlands Court Orders Oil Giant to Cut Emissions," BBC, May 26, 2021, www.bbc.com/news/world-europe-57257982.

23. Jennifer Laidlaw, "Fossil Fuel Financing Falls in 2020 amid COVID-19, Environmental Groups Find," S&P Global, April 8, 2021, www.spglobal.com/marketintelligence/en/news-insights/latest-news-headlines/fossil-fuel-financing-falls-in-2020-amid-covid-19-environmental-groups-find-63302209.

24. "Investors Look for Greener Markets to Reduce Fossil Fuel Exposure," Power Engineering International, July 21, 2021, www.powerengineeringint.com/renewables/global-investment-in-the-energy-transition-hits-500-billion-energy-transition.

25. "Pathway to Critical and Formidable Goal of Net-Zero Emissions by 2050 Is Narrow but Brings Huge Benefits, According to IEA Special Report," International Energy Agency, press release, May 18, 2021, www.iea.org/news/pathway-to-critical-and-formidable-goal-of-net-zero-emissions-by-2050-is-narrow-but-brings-huge-benefits.

26. Elizabeth Piper and Markus Wacket, "In Climate Push, G7 Agrees to Stop International Funding for Coal," Reuters, May 21, 2021, www.reuters.com/business/energy/g7-countries-agree-stop-funding-coal-fired-power-2021-05-21.

27. George Ferns and Marcus Gomes, "G7: Major Economies Not Ready to Break with Fossil Fuel Industry," UPI International, June 10, 2021, www.upi.com/Top_News/Voices/2021/06/10/britain-G7-fossil-fuels-energy-economy/3091623331112.

28. Bill McKibben, "How Does Bill Gates Plan to Solve the Climate Crisis?," *New York Times*, February 15, 2021, www.nytimes.com/2021/02/15/books/review/bill-gates-how-to-avoid-a-climate-disaster.html.

29. See Ariel Cohen, "A Bill Gates Venture Aims to Spray Dust into the Atmosphere to Block the Sun. What Could Go Wrong?," *Forbes*, January 11, 2021, www.forbes.com/sites/arielcohen/2021/01/11/bill-gates-backed-climate-solution-gains-traction-but-concerns-linger/?sh=3c3c9721793b; Catherine Clifford, "How Bill Gates' Company TerraPower Is Building Next-Generation Nuclear Power," CNBC, April 8, 2021, www.cnbc.com/2021/04/08/bill-gates-terrapower-is-building-next-generation-nuclear-power.html.

30. Mann, "Right Path Forward."

31. Philippa Nuttal Jones, "The Rise of the Climate Dude," *New Statesman*, February 17, 2021, www.newstatesman.com/bill-gates-avoid-climate-disaster-michael-mann-new-climate-war-review.

32. See Michael E. Mann (@MichaelEMann), Twitter, April 26, 2021, 8:48 a.m., https://twitter.com/MichaelEMann/status/1386708888725200898.

33. See, for example, this reply from a best-selling author to a Barack Obama tweet: Matthew Todd (@MrMatthewTodd), Twitter, March 5, 2021, 4:13 p.m., https://twitter.com/MrMatthewTodd/status/1367991754557886467. And this comment from a journalist at the *New Republic* with regard to Gates's speaking role at the climate summit: Kate Aronoff (@KateAronoff), Twitter, May 4, 2021, 7:50 a.m., https://twitter.com/KateAronoff/status/1389593242786729984; Barack Obama (@BarackObama), Twitter, March 5, 2021, 9:01 a.m., https://twitter.com/BarackObama/status/1367882853451759616; Jeff Mason, "Bill Gates, Dozens of World Leaders to Attend Biden Climate Summit—Source," Reuters, April 20, 2021, www.reuters.com/business/sustainable-business/bill-gates-dozens-world-leaders-attend-biden-climate-summit-source-2021-04-20.

34. John Carey, "Why Bill Gates and John Kerry Are Wrong About Climate Change," *Bulletin of the Atomic Scientists*, May 25, 2021, https://thebulletin.org/2021/05/why-bill-gates-and-john-kerry-are-wrong-about-climate-change.

35. See, for example, Mark Z. Jacobson, Mark A. Delucchi, Zack A.F. Bauer, Savannah C. Goodman, William E. Chapman, Mary A. Cameron, Cedric Bozonnat, et al., "100% Clean and Renewable Wind, Water, and Sunlight All-Sector Energy Roadmaps for 139 Countries of the World," *Joule* 1 (September 6, 2017): 108–121, https://web.stanford.edu/group/efmh/jacobson/Articles/I/CountriesWWS.pdf; "The US Can Reach 90 Percent Clean Electricity by 2035, Dependably and Without Increasing Consumer Bills," Berkeley Public Policy, Goldman School, press release, June 9, 2020, https://gspp.berkeley.edu/faculty-and-impact/news/recent-news/the-us-can-reach-90-percent-clean-electricity-by-2035-dependably-and-without-increasing-consumer-bills.

36. Michael E. Mann (@MichaelEMann), Twitter, May 20, 2021, 6:53 a.m., https://twitter.com/MichaelEMann/status/1394772004037857281.

37. Michael Mann, "It's Not Too Late for Australia to Forestall a Dystopian Future That Alternates Between Mad Max and Waterworld," *The Guardian*, March 23, 2021, www.theguardian.com/commentisfree/2021/mar/24/catastrophic-fires-and-devastating-floods-are-part-of-australias-harsh-new-climate-reality.

38. National Academies of Sciences, Engineering, and Medicine, *Reflecting Sunlight: Recommendations for Solar Geoengineering Research and Research Governance* (Washington, DC: National Academies Press, 2021).

39. Ray Pierrehumbert and Michael Mann, "Some Say We Can 'Solar-Engineer' Ourselves Out of the Climate Crisis. Don't Buy It," *The Guardian*, April 22, 2021, www.theguardian.com/commentisfree/2021/apr/22/climate-crisis-emergency-earth-day.

40. Michael Mazengarb, "Chevron Concedes CCS Failures at Gorgon, Seeks Deal with WA Regulators," *Renew Economy*, July 19, 2021,

https://reneweconomy.com.au/chevron-concedes-ccs-failures-at-gorgon
-seeks-deal-with-wa-regulators.

41. Molly Taft, "The Only Carbon Capture Coal Plant in the
U.S. Just Closed," *Gizmodo*, February 2, 2021, https://gizmodo.com
/the-only-carbon-capture-plant-in-the-u-s-just-closed-184617777.

42. Michael Mann and Susan Joy Hassol, "Glasgow's Hope at a Critical
Moment in the Climate Battle," *Los Angeles Times*, November 13, 2021, www
.latimes.com/opinion/story/2021-11-13/cop26-glasgow-climate-change.

Index

used against climate change
   denialists, 117
used by progressives, 118–119
See also deflection campaigns
Weisbrod, Katelyn, 242
Weyler, Rex, 192
Whitehouse, Sheldon, 118
"wicked problem" idea, 257
WikiLeaks, 36, 141
wildfires, 42–43, 180
   in Amazon forest, 43, 166–167,
      180
   Australian bushfires, 43, 87, 103,
      121, 166, 175–177, 182, 220,
      260
   carbon emissions and, 166
   conservative acceptance of climate
      change and, 175–176
   reforestation and, 167

wind energy, 128–129, 137. *See also*
   renewable energy
"wind turbine syndrome," 129
Wojick, David, 144
Wolf, Christa, 223–224
World Economic Forum (Davos,
   2019), 252
World War II, 183–184
Wyman, Patrick, 240

Yanukovych, Viktor, 40
Yellow Vest protests, 106, 108, 186
Young America's Foundation, 86–87,
   289n80
youth climate activism, 6, 251–255,
   258–259
   attacks on youth activists, 85–86,
      191, 253–255, 259
   See also Thunberg, Greta

MICHAEL E. MANN is Distinguished Professor of Atmospheric Science at Penn State, with joint appointments in the Department of Geosciences and the Earth and Environmental Systems Institute. He has received many honors and awards, including the National Oceanic and Atmospheric Administration's outstanding publication award in 2002 and selection by *Scientific American* as one of the fifty leading visionaries in science and technology that same year. Additionally, he contributed, with other Intergovernmental Panel on Climate Change authors, to the award of the 2007 Nobel Peace Prize. In 2018, he received the Award for Public Engagement with Science from the American Association for the Advancement of Science as well as the Climate Communication Prize from the American Geophysical Union. He received the Tyler Prize for Environmental Achievement in 2019, and in 2020 he was elected to the National Academy of Sciences. He is the author of numerous books, including *Dire Predictions: Understanding Climate Change*, *The Hockey Stick and the Climate Wars: Dispatches from the Front Lines*, and *The Madhouse Effect: How Climate Change Denial Is Threatening Our Planet, Destroying Our Politics, and Driving Us Crazy*. He lives in State College, Pennsylvania.

PublicAffairs is a publishing house founded in 1997. It is a tribute to the standards, values, and flair of three persons who have served as mentors to countless reporters, writers, editors, and book people of all kinds, including me.

I. F. STONE, proprietor of *I. F. Stone's Weekly*, combined a commitment to the First Amendment with entrepreneurial zeal and reporting skill and became one of the great independent journalists in American history. At the age of eighty, Izzy published *The Trial of Socrates*, which was a national bestseller. He wrote the book after he taught himself ancient Greek.

BENJAMIN C. BRADLEE was for nearly thirty years the charismatic editorial leader of *The Washington Post*. It was Ben who gave the *Post* the range and courage to pursue such historic issues as Watergate. He supported his reporters with a tenacity that made them fearless and it is no accident that so many became authors of influential, best-selling books.

ROBERT L. BERNSTEIN, the chief executive of Random House for more than a quarter century, guided one of the nation's premier publishing houses. Bob was personally responsible for many books of political dissent and argument that challenged tyranny around the globe. He is also the founder and longtime chair of Human Rights Watch, one of the most respected human rights organizations in the world.

. . .

For fifty years, the banner of Public Affairs Press was carried by its owner Morris B. Schnapper, who published Gandhi, Nasser, Toynbee, Truman, and about 1,500 other authors. In 1983, Schnapper was described by *The Washington Post* as "a redoubtable gadfly." His legacy will endure in the books to come.

Peter Osnos, *Founder*